Advanced Series in Physical Chemistry Vol. 13

PROGRESS IN

EXPERIMENTAL AND THEORETICAL STUDIES OF CLUSTERS

Advanced Series in Physical Chemistry

Published

Vol. 1: Physical Chemistry of Solids: Basic Principles of Symmetry and Stability of Crystalline Solids
H. F. Franzen

Vol. 2: Modern Electronic Structure Theory
ed. D. R. Yarkony

Vol. 3: Progress and Problems in Atmospheric Chemistry
ed. J. R. Barker

Vol. 4: Molecular Dynamics and Spectroscopy by Stimulated Emission Pumping
eds. H.-L. Dai and R. W. Field

Vol. 5: Laser Spectroscopy and Photochemistry on Metal Surfaces
eds. H.-L. Dai and W. Ho

Vol. 6: The Chemical Dynamics and Kinetics of Small Radicals
eds. K. Liu and A. Wagner

Vol. 7: Recent Developments in Theoretical Studies of Proteins
ed. R. Elber

Vol. 8: Charge Sensitivity Approach to Electronic Structure and Chemical Reactivity
R. F. Nolewajski and J. Korchowiec

Vol. 9: Vibration-Rotational Spectroscopy and Molecular Dynamics
ed. D. Papoušek

Vol. 10: Photoionization and Photodetachment
ed. C.-Y. Ng

Vol. 11: Chemical Dynamics in Extreme Environments
ed. R. A. Dressler

Vol. 12: Chemical Applications of Synchrotron Radiation
ed. T.-K. Sham

Advanced Series in Physical Chemistry Vol. 13

PROGRESS IN

EXPERIMENTAL AND THEORETICAL STUDIES OF CLUSTERS

Editors

Tamotsu Kondow & Fumitaka Mafuné

Toyota Technological Institute

World Scientific

New Jersey • London • Singapore • Hong Kong

Published by

World Scientific Publishing Co. Pte. Ltd.
5 Toh Tuck Link, Singapore 596224
USA office: Suite 202, 1060 Main Street, River Edge, NJ 07661
UK office: 57 Shelton Street, Covent Garden, London WC2H 9HE

British Library Cataloguing-in-Publication Data
A catalogue record for this book is available from the British Library.

PROGRESS IN EXPERIMENTAL AND THEORETICAL STUDIES OF CLUSTERS

ISBN 981-02-3893-2

Typeset by Stallion Press

Printed in Singapore by World Scientific Printers (S) Pte Ltd

ADVANCED SERIES IN PHYSICAL CHEMISTRY

INTRODUCTION

Many of us who are involved in teaching a special-topic graduate course may have the experience that it is difficult to find suitable references, especially reference materials put together in a suitable text format. Presently, several excellent book series exist and they have served the scientific community well in reviewing new developments in physical chemistry and chemical physics. However, these existing series publish mostly monographs consisting of review chapters of unrelated subjects. The modern development of theoretical and experimental research has become highly specialized. Even in a small subfield, experimental or theoretical, few reviewers are capable of giving an in-depth review with good balance in various new developments. A thorough and more useful review should consist of chapters written by specialists covering all aspects of the field. This book series is established with these needs in mind. That is, the goal of this series is to publish selected graduate texts and stand-alone review monographs with specific themes, focusing on modern topics and new developments in experimental and theoretical physical chemistry. In review chapters, the authors are encouraged to provide a section on future developments and needs. We hope that the texts and review monographs of this series will be more useful to new researchers about to enter the field. In order to serve a wider graduate student body, the publisher is committed to making available the monographs of the series in a paperbound version as well as the normal hardcover copy.

Cheuk-Yiu Ng

PREFACE

Since the discovery of renowned Kubo's size effect, nano-sized materials have been investigated extensively such as a cluster which is an aggregate of less than 10^{3-4} atoms or molecules with a diameter of 1 nm or less. Because of an intermediate nature of the cluster between gas and condensed phases, the cluster is sometimes called a material of the fourth phase after gas, liquid and solid phases. The cluster is attractive not only because of its dramatic size-dependent properties but also providing a model for the theory of a few body system, whose size is too small to be treated by statistical physics and thermodynamics but too large and too complicated to be handled by quantum chemistry.

It is always important that theory and experiment should cooperate to establish a rising new field such as the cluster science, but we should admit that the cooperation is not always easy. The easiest task for the theorists is often the most difficult task for the experimentalists and vice versa. In any way, we need to formulate a clear picture of clusters through joint interactions between experimentalists and theorists. At this opportune time, Professor Cheuk-Yiu Ng of Iowa State University suggested that we publish a book describing both a firm base and recent advances in cluster science, which are comprehensible to graduate students of physics and chemistry and researchers of the other fields who are interested. World Scientific Publishing Co. kindly gave us the opportunity to edit this book entitled "Recent Advances of Experimental and Theoretical Studies of Clusters", and generously waited until our dream came true.

The chapters of this book are categorized into three sections. The first two chapters are dedicated to explaining general features of clusters theoretically and experimentally. The next three chapters discuss several contemporary methodologies. The final two chapters examine topics related to condensed materials. The first chapter of the book describes the general properties and theoretical treatments of the cluster noting that the cluster represents a few body system. It is also shown that the cluster is not simply a new class of material, but provides an important and powerful medium for the study of how chemical reactions proceed. Chapter two deals with experiential studies on the physical and chemical properties of metal clusters. It is shown that transition metal clusters have remarkable size- and temperature-dependent magnetic properties and reactivity, for instance. In Chapter three, ultra sensitive and high-resolution spectroscopies applied to cluster studies are explained: the methods, technologies and limitations. Since femtosecond spectroscopy has proven to be one of the most powerful tools for the studies of dynamics, Chapter four is devoted to describing the application of the femtosecond spectroscopic methods such as NeNePo and STIRAP to the studies of metal-cluster dynamics. In Chapter five, core-level excitation of clusters is reviewed as a unique approach to probe element- and site-specificity in size-dependent properties of clusters in the gas phase. In Chapter six, the levitation of a single aerosol for the study of homogeneous and heterogeneous condensation processes through clusters is discussed. Chapter seven introduces the production of clusters from liquids, which can scarcely be considered in a conventional method, using the liquid beam technique.

Finally, we have to give some important closing remarks. We are grateful to the contributors to this book for their efforts and patience. The completion of this book would not have been possible without the superb editorial assistance provided by Ms. Mari Yamaguchi. She rearranged every manuscript into the style requested by the publishers. We would like to offer our most heartfelt thanks to her at this time.

Tamotsu Kondow and Fumitaka Mafuné
May 9, 2002

CONTENTS

CHAPTER 1

SURVEY OF STRUCTURE, ENERGETICS AND DYNAMICS OF CLUSTERS

R. S. Berry

Department of Chemistry and the James Franck Institute,
The University of Chicago,
5735 South Ellis Aveñue, Chicago, Illinois 60637, USA
Email: berry@uchicago.edu

R. D. Levine

The Fritz Haber Research Center for Molecular Dynamics,
The Hebrew University, Jerusalem 91904, Israel

This chapter provides a survey of properties and behavior of atomic and molecular clusters. Beginning with a review of the means available for studying clusters, the discussion goes on to the kinds of bonding they exhibit and the kinds of structures that result. The next section discusses the thermodynamics of clusters, especially their phase behavior and then the relation of that behavior to bulk phase transitions. The next section describes the electronic properties of clusters and the theoretical tools for analyzing electronic properties. This discussion includes not only the methodology, but also a summary of what is known about the metal-insulator transition in small systems. The kinetics of free clusters is the subject of the next section; this leads into the final section, on chemical reactions of clusters.

Contents

1. Introduction

Clusters are aggregates of atoms or molecules, which may contain any number of component particles from three to tens or hundreds of millions. When such aggregates reach sizes with diameters of nanometers, we generally cease to call them clusters, but refer instead to nanoscale particles. Their properties, as we shall see, lie between those of individual molecules and those of bulk matter; some are much like molecular properties, others are similar to bulk behavior, and some are unique to clusters. Clusters generally may add or lose particles much more readily than do conventional molecules, and are similar to bulk matter in this regard. Furthermore clusters of any given size typically may take on a variety of geometric structures, even any of a very large number of structures. They may be composed of atoms or of molecules or of some of each. They may consist of a single element, or have the composition of a single compound, or may be comprised of a random mixture, or may be anything between. As we shall see, they may be held together by any of a variety of forces. They are a bridge between isolated molecules and bulk matter; we encounter them whenever we see a mist form from a vapor, as the free molecules collect together into pairs, triples and larger clusters, becoming droplets or snowflakes. Because we can use the powerful methods developed to study molecules in order to investigate clusters, and then use that insight to help understand how clusters are related to their macroscopic cousins, the forms of bulk condensed matter, we gain new kinds of insights into the nature of matter itself by studying these tiny bits.

Let us begin by reviewing some of what is known about clusters and summarize literature that has become doctrine in the field. Clusters of atoms and molecules have increasingly become subjects of intense study, from both theoretical and experimental approaches, since the 1970s. One theoretical work by Kubo has sometimes been considered especially seminal.[1] Another laid the groundwork for simulations by molecular dynamics,[2] although the tools to carry out such calculations began with the statistical sampling method known as the "Monte Carlo" approach.[3] The principle and demonstration that such calculations were feasible by solving equations of motion, the "molecular dynamics" approach, was established by Alder and Wainwright.[4–6] One now-classic treatise dealt with thermodynamics of small systems.[7,8] Theoretical, and particularly simulation studies began to flourish then.[9–34] It was some time before experiments began to catch up with the theoretical and computational studies.

Experiments with clusters started to appear slowly in the 1960s, although not nearly as fast as the computations and theory.[35–37] Then, in the next decade, they began at an increasing rate.[38–59] Reviews of charged clusters appeared in this period.[60,61] However, the history of scientific interest in clusters goes much further back, at least to Faraday's experiments with colloidal gold.[62] Thoughts were turning to the structures of rare gas clusters even in the 1930s.[63] Theoretical studies were addressing finite-size effects,[64–67] albeit not specifically directed toward clusters at that time.

Today, we have variety of methods that enable us to make clusters; the choice depends largely on both what information we seek and on what kinds of forces bind the cluster particles together. One aspect is common to virtually all those methods to study clusters in the laboratory: we must carry out these procedures in a protected environment, typically a vacuum chamber. We can bombard solids with energetic ions to drive bunches of atoms out as clusters. We can push a vapor, under pressure of a gas, through a small aperture or nozzle, cooling the vapor enough to form clusters. We can sweep up vapor of a metal or semiconductor from an oven, using an inert gas as the carrier, and drive that through a nozzle to cool the metal to a cluster-forming temperature but leaving the inert gas as free molecules. We can study the clusters in the gas phase, or deposit them onto a surface[68] or even bury them and trap them in a solid. We can examine the clusters by probing their structures with electrons,[69,70] or by studying how they respond to radiation. Using ultraviolet light, we can ionize or fragment the clusters and determine what species are produced, e.g. with a mass spectrometer,[71] or we can determine the energies of any electrons or ions that we produce.[72–79] We can also carry out "conventional" spectroscopy, to determine what wavelengths of radiation the clusters absorb, whether infrared (in vibrations)[80,81] or visible or ultraviolet (in electronic degrees of freedom).[82,83] From another approach, we can model the behavior of clusters by simulating them on computers, as aggregations either of classical particles or of quantum particles.[84] And finally, we can of course apply any and all the tools of theory that we recognize as relevant. All of these methods have proved to be useful to help us understand clusters. To this day, it remains a hallmark of the work in this field that experiments and theory proceed hand in hand. Not everything that we imagine, analyze or compute can be measured and many observations still provide significant challenges for their interpretation. In this chapter, we will survey what is the secure ground from which one can proceed forward.

1.1. *Bonding Types*

We may approach the subject of the modern studies of clusters from several perspectives. One is the nature of the forces that hold them together — their bonding. The classes of clusters viewed this way are much like the classes of chemical compounds, with some interesting variations. There are the very weakly bound clusters held by van der Waals forces, the interactions due to correlated fluctuations in the electron clouds of the atoms comprising the cluster. The simplest, most studied examples of these are the clusters of rare gas atoms. Other very weakly bound clusters are those in which attractive interactions of nonspherical charge distributions contribute much or most of the bonding force. Clusters of nitrogen molecules are bound by a combination of quadrupole–quadrupole attractions, arising from the ellipsoidal, rather than spherical, shape of the molecule and from van der Waals forces; clusters of CO or HCl molecules have significant contributions from dipole–dipole attractions as well as dipole-quadrupole, quadrupole–quadrupole and van der Waals attractions, because, although they are electrically neutral, they have small excess negative charges "at one end" and equal positive charges at the other. More precisely, the center or mean position of the negatively charged electrons in these molecules is not at the same point as the center of the positive charge.

The common characteristic of these weak forces is that they are due to nonoverlapping electronic clouds of their constituents. They differ in the important aspect of their directional character. The atom–atom van der Waals force is directed along the line of centers of the two atoms. The forces between molecules with anisotropic charge distributions induce a preferred orientation of the molecules toward one another. One can mimic the effect of these by distributing discrete charges (and dipoles, quadrupoles and, in principle, higher multipoles) within each molecule[85–88] and calculating the resulting electrostatic force. Unlike the stereochemistry of the chemical bond, where we can rationalize the geometrical structure in terms of orbital interactions, it is still not easy to rationalize trends such as that in the weak van der Waals complexes of hydrogen halides with CO_2 molecules, from collinear to bent.

At an extreme from the weak van der Waals clusters are the strongly-bound clusters such as those of silicon atoms, with covalent bonds between their atoms that give these clusters very high binding energies. The distinction between clusters and molecules begins to be lost with some covalent clusters. Clusters of Si may have many stable shapes, and Si clusters of different sizes have roughly the same binding energy per atom, and are properly called "clusters". However this is not the case for some carbon clusters. Even small carbon clusters, omnipresent in carbon vapor, species such as C_2, C_3 and C_5, tend to have unique structures. The best known of the larger carbon clusters are the fullerenes, the shells of carbon atoms of which the smallest is C_{60}, buckminsterfullerene.[89–94] This now-famous species, with its soccer-ball structure, is so much more stable than any other isomer of 60 carbon atoms or than any other species of about that size that it is much more usefully thought of as a molecule, as are its higher homologues such as C_{70} and C_{76} (which is optically active, with left- and right-handed enantiomeric forms) and C_{84}.[95–97] Another variety of "cluster-molecules" are the metal-carbohedrenes or "met-cars" with the general formula $C_{20}M_{12}$.[98] Whether classed as molecules or clusters, systems held by covalent bonds are tightly held indeed.

Metallic clusters are also strongly bound, not necessarily as strongly as covalent clusters. Clusters of alkali metals or of silver atoms are examples of such systems. However small clusters of alkaline earth or mercury atoms are a different matter; these are formed of atoms with filled outer *s*-subshells, and hence do not have electrons readily available to form bonds or even to go into analogues of conduction bands. In fact small clusters of alkaline earth, mercury and similar atoms are van der Waals clusters, not metallic clusters, and one of the active subjects of current research involves determining how and at what sizes such clusters become metal-like.[99–104] A very complex class of clusters are those held together by hydrogen bonds. By far the most important of these are clusters of water molecules. The multitude of structures available to water clusters, the directionality of the hydrogen bond and the intermediate scale of energies for such bonds, between the covalent or metallic and the very weak, makes the understanding of water clusters a major challenge, despite considerable research on this topic. Even the choice of a suitable force field to represent interactions between water molecules is a subject of much discussion and even controversy.

The last class of clusters in this list consists of ionic species, most clearly illustrated by the alkali halides. That alkali halide vapors contain many clusters was proved in the 1950s.[53,105] (See especially Chap. 1 of Ref. 105.) These tend to have the same structures as the corresponding bulk solids, the rocksalt or NaCl structure in most cases,[55–57,106–109] and to be very strongly bound.[110] Their electronic structures are virtually the same as those of their bulk counterparts, that is, consisting of alkali cations and halide anions packed to give highly stable structures.

1.2. *Size Scales*

Clusters fall rather neatly into three classes of sizes.[111–113] The smallest sizes have properties that are often unique to clusters and depend sensitively on the specific number of particles comprising the cluster. This is true of both structural and electronic properties. Properties (such as ionization energy and electron affinity) of 19-atom, 20-atom and 21-atom clusters of sodium differ considerably from one another. Likewise the properties of rare gas clusters of 12, 13 and 14 atoms differ considerably and not monotonically; for example the binding energy per particle of the 13-atom cluster is significantly larger than that of either of the other two, because the 13-atom cluster has the structure of a complete icosahedron. Those sizes that have closed-shell structures, whether geometric as with Ar_{13}, or electronic, as with Na_{20}, are said to be "magic numbers". Clusters of these sizes are generally unusually stable. Since the origins of this stability may lie either with the way the atoms or molecules fit together, or with the way the electrons "fit together", there may be some competition between the two kinds of stabilization. In most cases, geometric stability dominates the smaller clusters, and electronic shell structure, the larger clusters.

The intermediate range of sizes includes clusters whose properties vary rather monotonically with cluster size, yet may be quite different from those of bulk matter. For example rare gas clusters of hundreds of atoms have, as their most stable forms, polyhedral structures based, in most cases, on icosahedra. By contrast, bulk rare gas solids have face-centered cubic (fcc) close-packed structures. The transition from icosahedral to close-packed fcc structures is a reasonable (but not unique) indication of passage from clusters of intermediate size to large clusters.

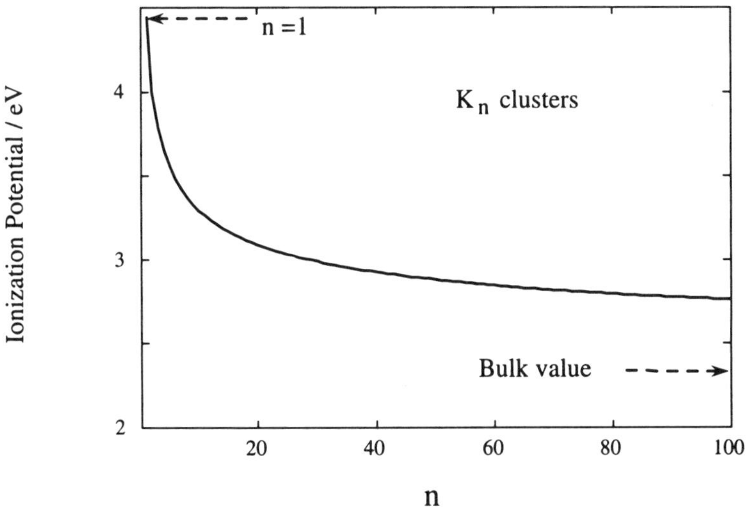

Fig. 1. A smooth representation of the cluster size effect: the ionization potential of K_n clusters versus the size n. At lower values of n there are deviations of the experimental values from this monotonic decline and this is one piece of evidence. for "magic numbers" as further discussed in Sec. 5.

In the intermediate regime one can try to smoothly extrapolate from the limit of the monomer to that of the bulk. An example of such interpolation is shown in Fig. 1 for the ionization potential (IP) of potassium clusters.[114–116]

The isolated potassium atom has an ionization potential (IP) of 4.44 eV while the work function of the bulk metal, the energy required to extract an electron with zero kinetic energy, is 2.3 eV. The experimental measurements of IP as a function of cluster size show evidence for magic numbers so that to discuss this figure realistically, one needs a model at least as sophisticated as the jellium model discussed later in Sec. 5. Here we just want to concentrate on the mean smooth behavior of the ionization curve. The energy necessary to remove an electron from a neutral sphere of radius R to infinity is $e^2/2R$. The radius of a spherical cluster should scale with the number n of atoms as $R \sim n^{1/3}$. It is therefore suggestive to take the ionization potential to scale with size as W (eV) $= 2.3 + a/n^{1/3}$. Here W is the work function or IP of the cluster; this relation, with a constant $a = 2.14$ eV (fitted to the data for the bulk and the single atom) is the smooth line shown in Fig. 1. If one wants to be a shade more ambitious, one can estimate the radius of the cluster from the number of atoms as $n = [V_{\text{cluster}}/V_{\text{atom}}]$ where the atomic volume V_{atom} can be taken (approximately) from the bulk density of the metal. An energy of $e^2/2R$ is then equivalent to 2.75 eV/$n^{1/3}$. Much of the discrepancy with the fitted value of 2.14 eV/$n^{1/3}$ can be explained as due to our neglect of the image potential seen by the electron which leaves behind a metallic sphere.[117–119]

The large clusters are essentially small particles of bulk matter. Their properties vary little with size. They differ from ordinary bulk matter primarily in having large surface areas and enough curvature of those surfaces to make their equilibrium vapor pressures and other related properties a bit different from those of macroscopic samples of the same materials.

Clusters of most sizes may exhibit a variety of stable structures, and most sizes of clusters can be found experimentally. However certain kinds of clusters are so stable, in very specific structural forms, that they may be considered molecules with definite shapes and compositions. The best known of these are the fullerenes. Otherwise however, clusters have alternative, low-energy stable structures. This will have a profound effect to be discussed in Sec. 3.

2. Structures of Clusters

An important characteristic of clusters distinguishing them from molecules is the variety of structures they exhibit, both in the varieties of structures available to many systems containing specific numbers of component atoms or molecules, and in the capacity of most substances to form clusters of any number of components. Determining the lowest-energy structures of clusters is a topic that has naturally attracted considerable attention, to which we shall turn shortly. We shall however be also interested in higher energy isomers. Until one actually counts[17,18,120–122] or otherwise estimates the number of locally stable structures,[123] it is not apparent how rapidly the number of such structures increases with n, the number of particles comprising the cluster. We illustrate this behavior in Fig. 2 To a first approximation, the number of structures increases as $\exp(n^{\beta})$, with the value of β slightly greater than 1.

It is possible to infer these structures, that of lowest energy and others of higher energy as well, from simple models for some kinds of clusters. This is particularly so for atomic clusters whose binding forces are well represented by pairwise interactions such as the Lennard–Jones potentials that approximate van der Waals interactions,[125–127] or the Born–Mayer interactions that describe the forces between the ions in alkali halide clusters.[105,128,129] Many molecular clusters also behave much like their simple models, models

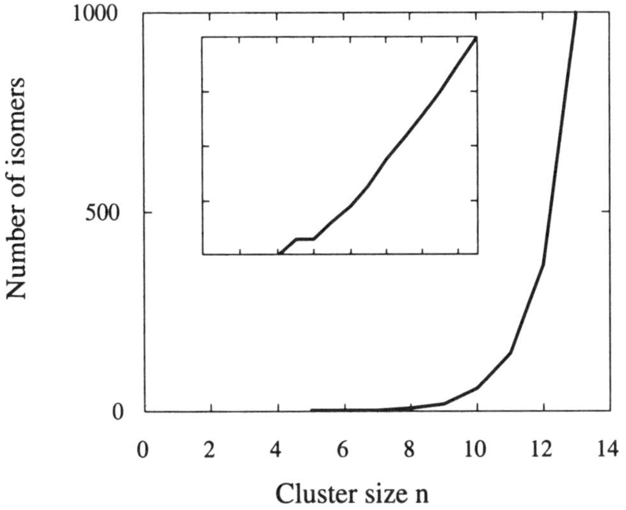

Fig. 2. The large number of geometric isomers for (Lennard–Jones bound) clusters: number of isomers versus n. The insert shows a logarithmic plot. Adapted from Refs. 123 and 124.

that may include Lennard–Jones interactions between all pairs of atoms on different molecules, sometimes Coulomb interactions between small charges introduced to model polar molecules or bonds, and rigid structures for each molecule in the cluster. However, for truly accurate representations of real clusters, a note of caution is needed, and more on this in Sec. 5: potentials are seldom pairwise additive. The presence of one partner nearby influences the bonding between two other partners. This is invariably so for chemical interactions and even van der Waals potentials do have a three-body contribution.[130–138] Nonetheless, the three-body and higher-order interactions contribute relatively little to total binding energies (except in the case of helium, especially ^{3}He) so that their inclusion typically only shifts energy levels a little, and hence only affects the energy sequence of levels if they are quite close together.

Clusters of metal, semiconductor and non-ionic insulator species are more challenging, and require more subtle or complex models. Models for metal clusters span many levels of sophistication. The simplest are spherical representations of the atomic cores as smooth, continuous, positively-charged and rigid background in which the valence electrons dwell — the "spherical jellium" model — and more elaborate versions of the jellium picture in which the core is either ellipsoidal or polyhedral. These are two ways to retain most of the simplicity of a jellium picture yet reduce the symmetry and split the electronic quantum states to make them more realistic. Models for clusters in which "chemical" bonding forces operate must allow for the familiar saturation of the bonding capacity of any atom, and for the directionality of such forces. We will discuss some explicit models in Sec. 5, but at present, obtaining a realistic force model is still a subject of active research.

A long standing, still challenging problem is one of relating the geometric structures of clusters to their counterparts in bulk systems. Some, notably ionic clusters, have the same structures in almost the smallest clusters they form as they do in bulk. Most of the alkali halide clusters, for example, have as their lowest-energy or global minimum structures the same rocksalt forms as bulk alkali halides. Even $(NaCl)_4$ is, in its lowest-energy structure, a tiny bit of rocksalt.

Many other kinds of clusters differ sharply from this simple relationship. Clusters of rare-gas atoms such as argon tend to have structures based on icosahedral geometry. This geometry cannot be the basis for a lattice; it simply does not have the translational symmetry necessary to build a lattice. The spacings between neighbors in shells distant from the central atom differ from those near the core. The cluster sizes for which complete, filled icosahedra can be made, namely 13, 55, 137,..., are called "magic numbers" and the structures are called Mackay icosahedra.[139] Not all the most stable structures of such clusters have icosahedral structures, and the specific structure of the most stable form depends on the forces that bind the cluster together.[140,141] For example, a number of clusters bound by Lennard–Jones[127] or Morse[142] potentials, potentials $V(R)$ that depend only on the distance R between pairs of particles. The first of these has the form

$$V(R) = 4D_e\left[\left(\frac{\sigma}{R}\right)^{12} - \left(\frac{\sigma}{R}\right)^{6}\right]. \qquad (1)$$

As Fig. 3 illustrates, this is really a parameter-free potential in the sense that if $V(R)$ is expressed in units of D_e, and R in units of σ, then there is only one universal Lennard–Jones (12, 6) potential. Such universality appears in the Law of Corresponding States, the relation in which all the characteristic properties of any gas, including its condensation and critical behavior, depend only on its critical temperature T_c, critical pressure P_c and critical density ρ_c (or specific volume ν_c). While this law is only approximate for real gases, it would be strictly true for a gas whose particles interact

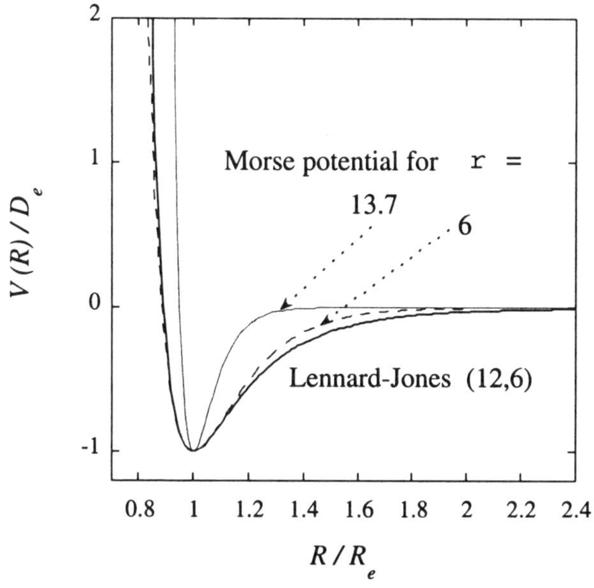

Fig. 3. Reduced plot of the Lennard–Jones (12, 6) potential and Morse potentials for two different values of ρ versus the scaled distance R/R_e. Note that the scaled variables $V(R)/D_e$ have the same meaning: D_e is the dissociation energy of the pair of particles bound by the potential and the function is plotted such that the potential goes to zero at a large distance. However the scaled distances αR and R/σ do not have quite the same meaning; the equilibrium distance, (the potential minimum), for the Morse potential occurs at $R = R_e$, while that of the Lennard–Jones potential occurs at $R_{e'} = 2^{1/6}\sigma$.

through the Lennard–Jones potential. The Morse potential,[142] often used to represent the interaction of covalently-bonded atoms (for example Ref. 143 and later supplements to this work).

$$V(R) = D_e[1-\exp(-\alpha(R-R_e))]^2 = D_e\left[1-\exp\left(-\rho\left(\frac{R}{R_e}-1\right)\right)\right]^2 \quad (2)$$

has one variable parameter, ρ, and can therefore represent many sorts of interactions. By decoupling the length scale α from the equilibrium bond distance R_e, one can fit the rotational spectrum, determined by R_e, without affecting the curvature at the minumum of the well. This is not possible for the Lennard–Jones potential. It is useful to use as the adjustable Morse parameter the scaled distance $\rho = \alpha R_e$. The diatomic molecules have values of ρ between about 3 and 7, with larger values corresponding to shorter ranges of the potential.

Figure 3 shows Morse potentials, all with the same depth and equilibrium interparticle distance, for several values of the range parameter. If ρ is 6, the Morse potential has the same curvature at its minimum as the Lennard–Jones, and generates model systems extremely similar to Lennard–Jones systems. In general, the shorter the range of the potential, the rougher and the more local minima there are on the multidimensional surface of a system consisting of several (or many) particles interacting through pairwise forces and potentials. The roughest known Morse surfaces are those of clusters of C_{60} molecules; for this system, $\rho = 13.7$.[144]

Some clusters described by these potentials have global minimum states with dodecahedral structures, rather than icosahedral.[141] Among those dodecahedral structures are the face-centered cubic structures that lie on the lattice of the crystal structure of bulk rare gases. The challenging problem that these structures present is determining how the transition occurs, as the size of the clusters grows, from a polyhedral geometry for that global minimum, to the lattice structure of that face-centered cubic (fcc) close-packed structure. Solving this problem will require more than just determining the structure of the global minima for clusters of different sizes; it will be a problem of determining the relative thermodynamic stabilities of different structures of clusters of many sizes, at many temperatures. We shall discuss this in the next section.

3. Thermodynamics of Clusters

Sometimes one sees naive statements such as "Thermodynamics cannot describe the behavior of clusters because they are so small". These statements are correct in a very restricted sense, a sense easy to understand in the contexts of Maxwellian and Gibbsian conceptions of statistical thermodynamics. A single cluster, viewed as a thermodynamic system in and of itself, is indeed too small to display the kind of behavior predicted by classical thermodynamics. The fluctuations are too large, and other properties discussed below that are taken as universal truths for large systems simply do not apply to a single cluster. However a perfectly sound thermodynamic description for clusters emerges as soon as we ask about the properties of an *ensemble* of clusters. As with all statistical, ensemble-based descriptions intended for a thermodynamic kind of interpretation, one must use care in making the choice of kind of system to be described and the corresponding choice of ensemble. To describe clusters produced in a jet, after they are isolated and no longer collide with other particles, a microcanonical (constant-energy) ensemble is appropriate. Clusters of a single size in a heat bath correspond to the systems in a canonical or constant-temperature ensemble. Clusters in an ensemble in which evaporation and condensation, as well as thermal equilibration, occur are properly described by a grand canonical ensemble. We simply have to choose whatever ensemble is appropriate for the particular situation to be described; the essential consideration is using an ensemble of clusters to develop the thermodynamic properties. Such descriptions depend on the validity of the assumption that they are ergodic, i.e. that the ensemble average of any equilibrium property of all the systems in the ensemble is the same as the time average of the same property for a single system in the ensemble. With the possible exception of some very small clusters or systems at such low temperatures that their vibrational modes of motion are essentially uncoupled from each other, so energy cannot flow from one kind of vibration to another (at which temperatures a quantum-mechanical description is probably necessary), the evidence indicates that the assumption of ergodicity is valid.

Thermodynamics of clusters lends itself to a traditional statistical-mechanical approach, either through analytic models or computer-based simulations. Often these two have been used to complement each other. Some of the statistical models that mimic simulations quite well are based on surprisingly simple constructs. A simple quantum-statistical model including only very rudimentary vibrational information[145] gave a reasonably quantitative reproduction of the melting behavior of small argon clusters from simulations.[22,24,26–28,32,146,147] An even simpler model, based on only a very few states of the cluster but with arbitrarily variable degeneracies that carry much of the crucial information, displays the coexistence property so characteristic of phase changes of clusters.[148] A very similar model, based only on a single state with adjustable degeneracy for each phase-like form, reproduces the simulations of phase changes of the three-phase Ar_{55} cluster and the two-phase $(KCl)_{32}$ cluster.[149] It seems that the details of the vibrational and rotational contributions to the partition function are not particularly important for the caloric curve or the relative amounts of the species at equilibrium (equivalent to a chemical equilibrium constant, to which we shall return), certainly not in comparison with the configurational degeneracy and mean energy of each phase-like form.

3.1. *"Smooth" Phase Changes*

Now we turn to one of the most widely noted of the physical properties of clusters, the phenomenon often described as the "absence of a sharp melting point". We must think of a phase-like form of a cluster as a form that satisfies all the dynamic and structural criteria of a homogeneous bulk phase that are consistent with the finite size of the system. Thus a liquid cluster can be expected to have a large diffusion coefficient, a relative nearest-neighbor distance that fluctuates more than 10% of its mean value (the Lindemann criterion[150]), and at least a few normal modes of extremely low frequency. These properties are all frequently-used diagnostic indices that can distinguish whether a cluster is behaving as a solid or a liquid.[22,24,26–28,32] Other properties that sometimes serve the same function include the pair distribution function and the three-particle angular distribution function.[147] However a liquid cluster can obviously not have continuous translational symmetry as a bulk liquid would, just because it is a small, finite system, nor can a cluster with a regular, lattice-like structure be said to have the periodic translational symmetry of a macroscopic crystal, for just the same reason.

3.1.1. *"Surface Melting"*

Typically, small individual clusters, e.g. of less than about 45 or 50 particles, may exhibit only homogeneous phases. Larger clusters may have surfaces that exhibit liquid properties, so are sometimes said to be able to exhibit surface melting — and so it was thought, so long as numerical indices and snapshots were the only probes.[34,151] It is straightforward to use simulations to distinguish properties associated specifically with the surface and not with the interior of a cluster. One may carry out simulations and their analyses with the particles on the surface distinguished from the others, and thus evaluate the properties of interest for the two sets of particles separately. If one does this for a cluster of about 45 atoms or fewer, no distinction arises between the two sets, at any temperature. Clusters of more than 45 atoms behave differently within a small range of temperature just below the temperature T_f, at which all the atoms display the same, liquid-like behavior when the cluster is liquid-like. In this small range of just a few degrees K if one scales the system to correspond to argon, during extended periods, the surface atoms show the high diffusion coefficients and relative mean square displacements from their equilibrium interparticle separations, and the large-amplitude motions that characterize liquids, while the interior atoms undergo small-amplitude vibrations about their equilibrium sites, with virtually no diffusion. Moreover snapshots of clusters exhibiting these quantitative characteristics show amorphous outer layers and well-ordered polyhedral cores.

All this information led to the conception that the surfaces of such clusters were behaving like amorphous swarms, i.e. like a classical concept of a liquid. However *animations* showed that "surface melted" argon clusters of more than about 45 atoms have a character of their own: most of the atoms in the surface layer undergo large-amplitude, anharmonic motions which, examined for any very short intervals, such as a fraction of a vibrational period, would be consistent with the notion that the surface atoms form an amorphous swarm around a well-structured core. Animations show that this is not really correct. The atoms in the outermost layer vibrate *collectively*, around a well-defined polyhedral structure. A few atoms in the outer layer, about 1 in 30, are promoted out of that layer and move on the surface of the cluster. There, they are much freer to move about than are the atoms in the outer layer. It is these promoted atoms, or "floaters", that are responsible for the low frequency vibrational modes and for the large-amplitude motions that make the surface seem liquid-like.[152,153] The floaters do exchange with surface atoms, roughly every few thousand vibrations, on average; hence, eventually all atoms of the surface layer spend intervals as floaters, and all the surface particles permute with one another. In this sense, the surface fulfills criteria and exhibits properties we expect of liquids. Neveretheless this behavior is simply not that we normally associate with a true liquid. As a result, it has become most convenient to consider the "surface melted" form of clusters as a separate phase or phase-like form, distinct from either solid or liquid.

3.1.2. *Phase Coexistence*

At very low temperatures or energies, the most stable form of all clusters except those of helium and possibly hydrogen is a solid, typically a well-structured solid, based on some polyhedral or lattice structure. The reason is simply that the energies of such structures are the lowest of all the possible structures of the clusters, and the entropic contribution to the free energy at low temperatures is negligible. Hence the most stable structures are solid-like and have the form of the structure lying at the global minimum of the cluster's effective potential surface, as defined within the context of the Born–Oppenheimer approximation. At higher temperatures, the entropic contributions to the free energy cannot be neglected, and the density of states associated with the multiplicity of available locally-stable structures. That is, the cluster may spend long intervals of time in any of many local minima on the potential surface, provided it has or can get enough energy to move to those minima. If enough minima are accessible, and the durations of the residences in the minima are short, then the system is liquid-like. When the temperature is high enough, these conditions are satisfied so well that no observable fraction of the clusters in an ensemble can be found in a deep, solid-like well for a period long enough for a cluster to develop equilibrium-like, solid-like properties, and the entire ensemble must be said to be liquid. These are the simplest, qualitative considerations that indicate why clusters are solid at low temperatures and liquid at higher temperatures — provided their vapor pressures are low enough that liquid clusters can persist in approximate equilibrium with a fixed composition. Fortunately, there seems to be a temperature range for many clusters for which this is the case; such clusters can remain for nanoseconds without evaporating, even when their temperature is high enough to make them liquid.

These arguments are precisely those for the existence of bulk solids at low temperatures and bulk liquids at high temperatures. To describe bulk behavior, we must go one large step further, in order to explain at a microscopic level why the transition between solid and liquid is sharp, in the sense that the two phases may exist at only a single temperature, at any chosen pressure. More generally, we may ask why the Gibbs phase rule exists, why it is that the number of phases, p, the number of components, c, and the number of degrees of freedom, n, are related by this rule, $n = c - p + 2$. We shall return to this question in the next section; meanwhile, we can address the nature of phase coexistence of clusters, based on the qualitative considerations of the previous paragraphs.

Consider an ensemble of clusters, e.g. of a single species and single size, in equilibrium at a fixed temperature. A realization of this might be a monodisperse, gaseous collection of clusters in a bath of inert atoms that maintains a constant temperature. At very low temperatures, these clusters are all found vibrating around the global minimum of the potential energy surface. At higher temperatures, many clusters will be found in other minima, in some cases for time intervals long enough to develop observable equilibrium properties of those minima, such as full vibrational equipartition. In other cases, as mentioned above, the cluster may visit many minima and wander throughout a wide range of configuration space, partly by low-frequency, large-amplitude motions characteristic of liquid compliance.

If the clusters do spend times long enough to exhibit equilibrium properties characterisic of a phase, then we can justifiably identify that phase as a stable species in the ensemble. If two such phase-like forms appear by satisfying the condition of observability in equilibrium, then we may identify the condition of the ensemble as one of phase equilbrium. However this is as much an equilibrium of chemical isomers as it is of phases, in the sense that each of the persistent forms can be assigned a well-defined free energy. Given that, we can immediately construct a traditional equilibrium constant K_{eq} for the ratio of concentrations or partial pressures of the two species,

$$K_{eq} = e^{-\Delta F/kT} = \frac{[\text{liquid}]}{[\text{solid}]} \tag{3}$$

if the two species happen to be liquid and solid. They may also be two solid forms, or a conventional solid and a soft solid,[154] or any two persistent forms. Moreover since $\Delta F = F_{\text{liquid}} - F_{\text{solid}}$ in the example given above, and both F_{liquid} and F_{solid} are functions of temperature, ΔF is a finite function of temperature. So long as $\Delta F/kT$ is in range of roughly ± 10 or even perhaps ± 20, there may be detectable quantities of *both* phases present in the ensemble. It is not necessary that ΔF be zero to have the two phases coexisting in equilibrium. The two forms may coexist over a range of temperatures, at any chosen pressure at which the two persist for observable time intervals.[7,8] In the following discussion, it will be useful to use a transformation of K_{eq} in order to work with a variable whose range is bounded. We introduce the distribution D_{eq} thus:

$$\begin{aligned} D_{eq} &= [K_{eq} - 1]/[K_{eq} + 1] \\ &= \{[\text{liquid}] - [\text{solid}]\}/\{[\text{liquid}] + [\text{solid}]\} \end{aligned} \tag{4}$$

a quantity that varies between -1, when the ensemble of systems is all solid, to $+1$, when all the systems in the ensemble are liquid.

It is possible in many situations to perform measurements that pick out and distinguish each of the phases present in the ensemble. However it is also often possible to choose a method of measurement slow enough that the measurement sees each system in all its phase-like forms during the course of the measurement. Naturally, in such situations the observations show the average properties of all the phases, weighted by the proportion of the sample in each, or by the time individual clusters spend in each form. This is a vivid illustration of a situation underlying virtually all of thermodynamics, but customarily left tacit because only one time scale is relevant for traditional bulk systems. However one can pose apparent paradoxes by questioning such tacit assumptions; for example, without careful consideration of time scales for different phenomena, the notion of a "reversible, adiabatic process" can be conceived as an oxymoron.[155] Clusters change this situation because the time scales for different classes of phenomena are typically much closer together than they are for bulk systems. This means that an ensemble may look like it consists of a mixture of solid and liquid clusters, if it is observed reasonably rapidly, e.g. in less than about 50 ns, but would look like a slush if observed by a measurement that requires 1 μs or more.

Now we turn to a more precise examination of phase coexistence in clusters. We begin by supposing that the free energy $F(T) = E(T,\gamma) - TS(T,\gamma)$ of the cluster can be written as a function of temperature, pressure and another parameter γ, indicating the degree of nonrigidity, analogous to a Landau order parameter. (We use E here for internal energy and S for entropy.) It will be convenient to think of scaling this parameter so that its range is from 0 to 1, with 0 the rigid limit and 1, the limit of nonrigidity. At any fixed temperature and pressure, $F(\gamma, T, P)$ has at least one minimum. Any minimum corresponds to a locally stable state, provided the dwell time of the cluster in the region of that minimum is long enough for the system to equilibrate vibrationally, i.e. so that the average energy is the same in each mode of vibration. Clusters exhibit multiple forms, corresponding to whatever local minima the free energy may have in which this persistence condition is met, like any chemically distinct species such as isomers of the same chemical composition. Like molecules, if clusters are very floppy or fluxional, then they can be said to have only average structures, but if they remain long enough in one or more minimum to be observed, then each of these minima corresponds to a separate, recognizable species. The relative amounts of the species, as stated previously for solid and liquid clusters, is fixed by the exponentials of the free energies of the each species, in units of kT.

The temperature dependence of the stability and relative concentrations of these species emerges when we examine how the free energy is a consequence of the form of the density of states, more specifically of the density of accessible states.[156–159] Almost all clusters have a unique configuration of lowest energy, which supports at least a few vibrational levels deeper in energy than states associated with any higher-energy configuration. At very low temperatures or energies, the only occupied states are those vibrational states associated with that lowest-energy configuration. Such clusters are very solid-like. They "reside" typically deep in a rather narrow potential energy well, approximately parabolic in its lower reaches. We can define our measure of nonrigidity so that the clusters have a single minimum in their free energy at or near $\gamma = 0$.

The density of states of the solid rises fairly rapidly with energy, as more and more vibrational states, especially combination states, come into accessibility. The density of states of the liquid-like cluster, i.e. the density of states near $\gamma = 1$, is typically zero at very low energies, if for no other reason than that at low energies there are no available states for a liquid. Without an input of energy, it is simply not possible to create the defects, whatever they may be, that give rise to the large-amplitude motions of atoms or molecules that are responsible for the flow and compliance of liquids. The density of states of the liquid, although it first becomes nonzero only at some energy well above the global minimum where the solid has its lowest-energy, rises much faster with energy than does that of the solid. Figure 4 shows this behavior schematically.

This property of densities of solid and liquid states means that the free energy at low temperatures has only a single minimum at or near $\gamma = 0$, but,

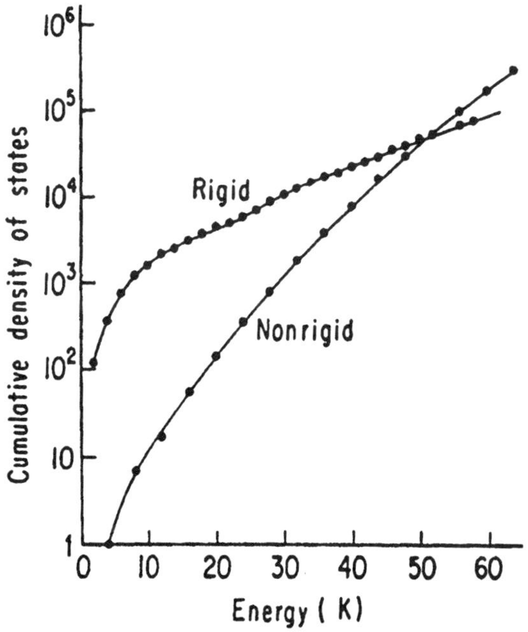

Fig. 4. Schematic representation of the densities of states of solid and liquid forms of a cluster of a given size, typically for at least 6 or 7 particles.

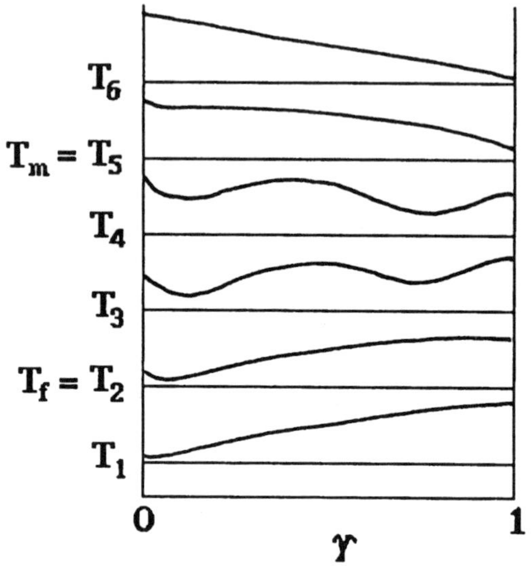

Fig. 5. A schematic representation of the free energy of a cluster as a function of an order parameter for nonrigidity, at a succession of temperatures.

with increasing temperature, the curve of $F(\gamma, T, P)$ flattens as γ increases. This continues until a point of inflection develops at a temperature we may call T_f, the "freezing limit", below which only the solid is stable, and above which the free energy has two local minima, one corresponding to the solid and the other, to the liquid. As the temperature is taken still higher, the curve of $F(\gamma, T, P)$ tilts more and more toward $\gamma = 1$, until the minimum near $\gamma = 0$ turns into a point of inflection, at a temperature we may call T_m, the melting limit. Above T_m, only the liquid form is stable because $F(\gamma, T, P)$ has only a single minimum there. Between T_f and T_m, both phases exist in stable thermodynamic equilibrium.[160–163] Figure 5 is a schematic representation of the free energy of a cluster as a function of an order parameter for nonrigidity, at a succession of temperatures. Neither of these is the equivalent of the classical bulk freezing/melting point, but both do provide sharp limits to the range of thermodynamic stability of a given phase.

Between T_f and T_m, there is a point on the temperature scale, T_{equil}, at which $F_{\text{solid}}\,(\gamma, T, P) = F_{\text{liquid}}\,(\gamma, T, P)$ and the two species occur in equal amounts or with equal probability. It is this point that becomes the classical transition temperature of the bulk, when N, the number of particles in the cluster, becomes very large. We shall examine how this happens in the next section.

The theory just presented indicates that below T_f, only solid clusters are stable, and that above T_m, only liquid clusters are stable, and that between these two, both phases may be found. This means that there should be a discontinuity in $K_{\rm eq}$ and $D_{\rm eq}$ at each of these temperatures, $K_{\rm eq}$ from 0 to some value $\exp[-\Delta F(\gamma, T, P)/kT]$ at $T = T_f$, and from $\exp[-\Delta F(\gamma; T_m, P)/kT_m]$ to ∞ at T_m. Likewise, $D_{\rm eq}$ changes from -1 to some intermediate value of $\tanh[-\Delta F(\gamma, T_f, P)/2kT_f]$ at T_f, and from tanh $[-\Delta F(\gamma, T_m, P)/2kT_m]$ to $+1$ at T_m. At a point where $\Delta F(\gamma, T, P) = 0$, $K_{\rm eq} = 1$ and $D_{\rm eq} = 0$. Between the two limits of T_f and T_m, the equilibrium concentration should shift continuously toward more liquid as the temperature goes up.

In the normal situation, $T_{\rm eq}$ falls between T_f and T_m. We shall see in the next section how this leads to the normal behavior of bulk phase changes. It is possible, however, that $T_{\rm eq}$ falls outside the region between T_f and T_m. In this case, we shall see that the less favored phase, the one with the higher free energy and hence the liquid phase in this example, is simply not stable in the bulk limit. This situation is just what gives rise to phases that are thermodynamically stable for clusters but not for bulk matter. It is a fascinating situation that arises because differences between small numbers and very large numbers of particles that are formally only quantitative actually become qualitative differences with regard to observability of phenomena. Figure 6 is a schematic representation of the free energy for several temperatures in this situation. One apparent example of this behavior occurs with a monoclinic phase of clusters of TeF_6 molecules, a phase found for clusters but not for bulk.

The argument regarding the link between local minima in the free energy and the coexistence of two phases over a band of temperarure and pressure (or of any two state variables) is not restricted to only two phases. Just as more than two chemical isomers may coexist, more than two phases of a cluster may coexist. There is nothing essential that prohibits the appearance of multiple minima in the free energy. This is perhaps clearest if the order parameter is not a simple scalar but a vector characterizing more than one property of the system. While a variety of kinds of clusters have been objects of study for surface melting,[34,151–153,164–183] the best example is found with argon clusters, notably $(Ar)_{55}$, which has been examined in detail. In particular, it is notable for its region of three-phase coexistence — solid, liquid and "surface melted".[175,176] One can use a double index,

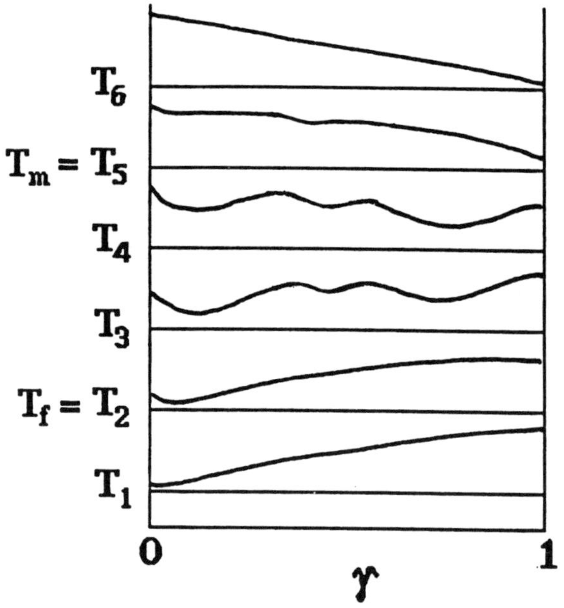

Fig. 6. A schematic representation of the free energy of a cluster as a function of an order parameter for nonrigidity, for a case of a phase that is observable for clusters but not for bulk matter. The reason is that in the range of conditions under which this phase is locally stable and present in observable concentrations for small systems, it is never the most stable phase, and hence, in the bulk limit, is never present in observable concentrations.

such as the mean square displacements of atoms, per unit time, of the surface and core atoms, in order to distinguish the three phase-like forms such clusters may exhibit. A phenomenological representation of the free energy of such systems, in which order parameter indices of surface and core are independent, shows that it is possible, in principle, to have a variety of situations. These range from one in which no more than two phases are locally stable under any given conditions, and only two-phase coexistence is possible, to one in which solid, surface melted and liquid forms all coexist over a wide range of conditions. Argon clusters lie between the two limits: in a small range, solid and surface melted forms may coexist; then there is a band at slightly higher temperatures within which all three forms may coexist like three isomeric forms; at still higher temperatures, only surface melted and liquid forms coexist, and above T_m, of course only the liquid is a stable condensed phase. It should also be possible for clusters to exist

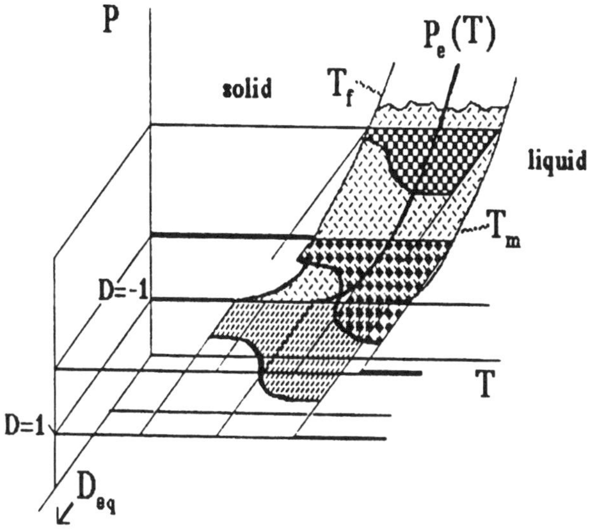

Fig. 7. A schematic representation of a phase diagram for a cluster, with axes corresponding to pressure P, temperature T and distribution $D_{\rm eq}$.

with solid surfaces and liquid cores, if the liquid form is denser than the solid, as in gallium, indium or water.

3.2. *Phase Diagrams*

At this point it is useful to introduce two ways to construct phase diagrams for clusters. One is a direct extension of the way conventional phase diagrams are constructed for bulk matter, in which the conditions for coexistence (in observable concentrations) are satisfied only along a curve, e.g. in the space of pressure and temperature.[184–186] The other is a diagram of the locus of points of extrema in the free energy as a function of temperature and one or more order parameters.[175,176,185,186]

In the first kind of phase diagram, sketched in Fig. 7, we simply augment the space of variables to include the distribution $D_{\rm eq}$ as a third variable, so that the graph becomes one in a three-dimensional space instead of just a two-dimensional plane. The variable in the third dimension, $D_{\rm eq}$, has a range from only -1 to $+1$. Let us use the melting freezing phase change as our example. At low temperatures, below T_f, the system in this 3-space is simply a part of the plane of $D_{\rm eq} = -1$, expressing all the states accessible to the pure solid phase. Likewise, above T_m, the point representing the system lies anywhere in the corresponding part of the plane of $D_{\text{eq}} = +1$, reflecting that only the liquid is stable. Since only P and T are truly independent variables, they determine other quantities such as density, in the usual way, just as they determine the possible values of D_{eq}. Thus at each fixed pressure, there is only a single curve of D_{eq}. At any chosen pressure and at the $T_f(P)$ for that pressure, the curve of stability jumps from $D_{\text{eq}} = -1$ to some other, larger value of D_{eq} which corresponds to the equilibrium partition between solid and liquid just at the appearance of a stable liquid phase. Similarly, at any chosen pressure P (for which phase equilibrium is possible, of course), as the temperature falls from just above $T_m(P)$ to just below it, D_{eq} shows a discontinuity from $D_{\text{eq}} = +1$ to some smaller value corresponding to the equilibrium partition at the highest temperature that can sustain a stable solid phase. Between these two limits, at any fixed pressure, the curve bends its way outward, corresponding to an increasing equilibrium fraction of liquid. The envelope of the curves for all pressures generates a curved surface that lies between $D_{\text{eq}} = -1$ and $D_{\text{eq}} = +1$. As drawn, the surface passes through $D_{\text{eq}} = 0$, as one would expect for any cluster exhibiting normal melting or freezing.

The second kind of phase diagram is a locus in the space of temperature and one or more order parameters, for a chosen pressure. The locus is the set of points for which the free energy is a minimum or maximum as a function of order parameter, for each temperature. These curves have been drawn, in fact, with the inverse temperature as the vertical axis, and in the illustration shown in Fig. 8, an order parameter ρ_s indicating the density of defects in the surface layer, and another, ρ_c, indicating the corresponding density of defects in the core. The assumption that was used in the model was that the free energy of each phase could be expanded as a series in the defect densities, at any given pressure and temperature.[175,176] If one starts following the curve from the lowest temperature, at the top of the figure, the regions of a stable stationary point are the portions of the curve in which one moves downward; regions in which the curve moves upward are unstable, in precise analogy with the region of instability of a van der Waals isotherm in its two-phase region. The parameters to construct this curve to represent Ar_{55} were taken from molecular dynamics simulations.

The curve of Fig. 8 shows a low-temperature region at the top where only a single branch of the curve is present. The curve drops, turns upward and drops again as it moves to higher temperatures and higher values of ρ_s, still remaining at very low values of ρ_c. In words, this means that the surface may have a moderate density of defects or a low density of defects,

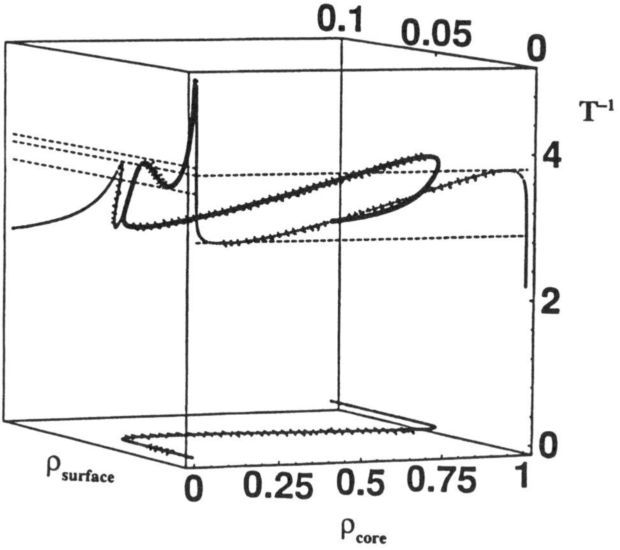

Fig. 8. A schematic representation of a "stability locus" phase diagram with parameters chosen to correspond to a cluster of Ar_{55}.

in either of two stable forms, while the core remains free of defects. The high-surface-defect form is indeed the "surface melted" phase-like form. Then, as we follow the curve of stationary points further, it moves upward again, this time to higher values of ρ_c, and then turns downward once more, revealing a stable truly-liquid phase. At any temperature at which two or more "downward" branches occur, there are stable phases corresponding to each of those branches. Hence the first coexistence region is that of only solid and surface melted phases; then there is a region of three-phase coexistence, followed by another two-phase coexistence region for surface melted and liquid phases only.

Diagrams of the second type have been constructed for model systems with model sets of parameters but the only realistic system for which this approach has been applied is to argon clusters. The model parameters indicate that it should be possible to find cases in which ρ_c increases at low temperatures and ρ_s, only at higher temperatures. This would be the situation of a solid shell stable around a liquid core, in contrast to the usual case one would expect, of a liquid surface around a solid core.

4. Implications of Cluster Thermodynamics for Bulk

4.1. *First-Order and Second-Order Transitions*

The first way we may link the behavior of clusters undergoing phase changes to that of bulk matter is simply to examine how a phase diagram of the type of Fig. 7 becomes a standard *P-T* phase diagram for a first-order transition of bulk matter. This is most straightforward. As the system becomes larger and larger, the (horizontal) curve of $D_{\text{eq}}(T)$ shows progressively smaller and smaller jumps at its points of discontinuity, and the curve remains closer and closer to its limiting values of -1 and $+1$, except in a narrower and narrower zone where it makes its transition from near -1 to near $+1$. In the limit of macroscopic systems, the true discontinuities at T_f and T_m become unobservably small, and the continuous transition through $D_{\text{eq}} = 0$ becomes so sharp that it appears as a discontinuity. This is precisely how a first-order phase transition emerges as systems grow larger and larger, eventually to macroscopic scale.

The second way we may link our knowledge of phase changes in clusters to our concepts of macroscopic phase transitions is by using information about transitions that are presumably second-order or continuous transitions in the bulk limit. Here the information is still limited, but enough is known to permit us to gain some new insights. The systems for which small system counterparts of second-order transitions have been studied are molecular clusters, particularly structural (solid–solid) transitions of octahedral molecules such as TeF_6, in clusters and in bulk.

The experimental evidence from diffraction indicates that solid TeF_6 has two solid phases, a low-temperature phase with an orientationally-ordered, base-centered monoclinic structure and C_{2h} symmetry, and a higher-temperature phase with a body-centered cubic structure and cubic O_h symmetry. The experiments with clusters of TeF_6, done by electron diffraction, show both of these and at temperatures between the regions of stability of these phases, a monoclinic phase with only partial orientational order as well. Simulations of the clusters show that the transition from body-centered cubic to monoclinic involves both rotational and translational motion of the motion, while the lower-temperature transition, between monoclinic and base-centered monoclinic, requires only coupling of the molecular rotational motions.[187–190] Furthermore the transition at higher temperature in clusters shows dynamic coexistence of two phases in equilibrium over a range of at least a few degrees of temperature, which immediately implies that this transition involves two local minima in the free energy. However in the bulk system the transition between the two bulk phases is apparently second-order, meaning traditionally that the free energy has only a single minimum which simply shifts as the temperature shifts. These can be reconciled only by supposing that the two minima of the small system converge as the system grows larger. The dependence on cluster size of the separation of minima in the free energy was examined to the extent that sufficient criteria — that the defects responsible for the non-rigid, high-entropy phase either attract each other or lower the frequencies of vibration of the cluster — are now established as a condition that the minima remain apart as the cluster grows to macroscopic size.[191] Presumably, if the defects repel or shift the frequencies of vibration upward, then the minima in the free energy converge or at least approach as the size of the cluster increases.

The lower-temperature transition of the clusters, in contrast, appears from simulations and symmetry arguments, to involve only a single minimum in the free energy because it exhibits no sign of coexisting phases.[187–190] This has now been interpreted to mean that although in clusters all three phases can be found, in the bulk limit the two transitions occur in concert and the intermediate phase does not appear. Moreover this example demonstrates that there may be two kinds of bulk second-order phase changes: those emerging from small systems with two minima that converge, and those for which even small systems have only a single minimum.

These inferences raise further questions that have yet to be answered. Do the two minima converge at a finite number of constituent particles n? Do they truly converge only as n grows to infinity? Do they actually converge at all, or do they only approach asymptotically, or perhaps only approach some small separation? If they only approach a small separation, is the zero point energy of the bulk system above or below the saddle separating the two minima? Do other second-order transitions, notably magnetic transitions, behave in a fashion similar to structural transitions? We raise these to illustrate the extent to which this subject is still very open, with many challenges.

4.2. *Other Phase Changes*

Much the same situation occurs with surface melting. While the only kind of surface melting found thus far for clusters involves promotion of surface atoms to become "floaters" above but bound to the surface, this is only one of several possible mechanisms that could lead to what has been called "surface melting" of bulk matter.[180] The question is still very open as to which one or ones among these mechanisms may occur. It is quite possible that different substances or even a given substance under different conditions will exhibit different mechanisms of surface melting.

Helium clusters, and possibly clusters of hydrogen molecules, constitute a very special subset of all atomic clusters. The first reason is simply that helium clusters seem to show no solid-like form. While it is conceivable that helium clusters might solidify under very high pressures, there is no evidence for this, and it is difficult to imagine a medium, apart from solid H_2, in which such an experiment might be conducted.

The second reason helium and possibly hydrogen clusters are special is their capability to transform to a *superfluid* state. This is a state in which some or all of the component particles are in their lowest quantum state, a state common to all the particles in that state. This is only possible for particles with integral values of their total spin, particles known as *bosons*. The common isotope ^{4}He is one such particle. At a sufficiently low temperature, bulk ^{4}He goes into its superfluid state, in which a finite fraction of the atoms are in their lowest quantum state. This gives the very cold liquid helium special and sometimes dramatic properties, such as the capability to climb walls of a container, and to spout up through a capillary to form a fountain.

It was conjectured for some time that clusters of ^{4}He could also show superfluidity. The hallmarks of the condition would be the finite amount of energy required to excite an atom from the ground state of the superfluid up to a "single-particle excitation" state, and the consequent decoupling of the motions of the atoms of the cluster from the motions of any foreign molecule contained in the cluster.[192–195] Soon thereafter, experiments showed both the isolation of the foreign molecules, and of the energy gap expected of a boson condensate.[196–198] Whether clusters of hydrogen molecules also show this behavior is not yet known.[199]

5. Electronic Properties

There are many reasons why we would like to compute the electronic properties of clusters. The equilibrium geometry of the cold cluster and also that of any higher energy isomers can be determined in this way. Such a computation will also determine the energetics of these structures and predict any "magic numbers" for which the cluster is particularly stable. We can predict or verify the energy levels and spectrum of the cluster, both electronic and vibrational; it is this set of energy levels that determines the cluster's thermodynamic properties. When one can implement such computations for clusters of various sizes, we learn how cluster properties evolve from the monomer to the bulk. In principle, almost any property of the cluster can be computed from knowledge of its electronic structure. The reason we say "almost" rather than "all" is a practical one. Approaches to computing the electronic properties very often invoke the Born–Oppenheimer approximation, which we discuss below. Indeed, the simple idea of an electronic structure relies on this approximation. Yet, for certain types of clusters, this approximation may fail and one then has to go beyond the Born–Oppenheimer approximation.[200,201]

Let us begin from a quantum chemical[202–207] point of view. A cluster is a collection of atoms. Whether the cluster is an assembly of atoms or of more complex molecular monomers from which the cluster itself is made does not immediately matter. One first fixes the position of each atom's nucleus and solves the Schrödinger equation for the energy of the electrons for that configuration of the nuclei. Next, the positions of the atoms are shifted to a new configuration and the process is repeated. After a sufficient number of repetitions we have determined the electronic energy as a function of the positions of the atoms. This is what we mean by finding a potential energy function or potential energy surface. It is almost always the case that the electrons move much faster than the nuclei. We then argue that as the atomic nuclei move, the electrons respond fast enough to rearrange themselves into their equilibrium distribution appropriate for each instantaneous position of the nuclei. The total energy of the electrons, plus the potential energy of interaction of the nuclei (their Coulomb repulsion), is then the potential energy $\mathrm{V}(\mathrm{R}_1, \ldots, \mathrm{R}_\mathrm{n})$ under which the nuclei will move. Here R_j is the set of coordinates of the jth nucleus. This is the Born–Oppenheimer approximation; it assumes that there is a unique set of electronic states associated with each configuration of the nuclei. Often we want to know about the ground state of the cluster; in that case, we are interested in the lowest electronic state for each nuclear configuration. (Strictly, it may happen that one electronic state is lowest over some range of configurations, but in a different range, the system could have a different electronic state as its lowest. This seems to be rare for clusters but one must be aware that it may occur.)

Why is the quantum chemical approach particularly challenging for clusters? The basic answer is that, by its very definition, a cluster consists of many atoms. This has two immediate implications. One is that solving the electronic Schrödinger equation may well be a very large task: the number of electrons in a cluster is typically high, and the effort of a quantum chemical computation scales as the number of electrons to some power, typically higher than unity. Quantum chemists are hard at work to reduce this scaling, but we are not quite there yet. Of course, one quite reasonable approximation to reduce the effective number of electrons is to "freeze" the cores of the atoms. A clear illustration is a sodium cluster, Na_n. In our simplification, we take each Na atom to have only one valence electron outside a frozen rare-gas (Ne) core, and treat only the n valence electrons, one from each atom. This provides a drastic simplification (n versus $11n$ electrons) at a modest price: we need to know the potential seen by the valence electrons. This is not quite the Coulomb potential of the Na^+ core because of two effects: the valence electron can penetrate the core and the effect of the Pauli exclusion principle is to exclude the valence electron from the neighborhood of the core electrons. The *de facto* repulsion due to the Pauli principle is the more significant one and so the effective potential seen by the valence electron will typically have a repulsive inner core.[205,208]

The other inherent difficulty created by the need to deal with many atoms is that the electronic energy is a function of the position of all these many atoms. In order to sample a range of configurations it is therefore necessary to repeat the computation of the electronic energy many, many times over. Of course, the results at one configuration can be used as a good starting point for an adjacent configuration. Moreover, even a single computation, when done in a suitable manner[209–213] can yield not only the electronic energy but also its derivatives with respect to the nuclear displacements. Since the electronic energy is the potential for the motion of the nuclei, the first derivative is the force on the nuclei in that direction and the second derivative gives the force constant k in the harmonic contribution to the potential energy, $kx^2/2$.

The limit of a large cluster is a solid. It therefore comes to mind to overcome the difficulties of scaling with size by using the methods of solid state physics. That is indeed so, and monographs dealing with a synthesis of the quantum chemical and solid state points of views are available.[207] However solid state physics has its own limitations. Firstly, the methodology takes advantage of the fact that it deals with an infinitely extended system, so surface effects are of no direct concern. But an essential point about clusters is that a high fraction of all atoms are at the surface, in an environment where their electronic structure is not simply that determined by a full shell of near neighbors. Moreover, in a zeroth approximation, the methods of solid state science suppose that the solid is periodic. Large-amplitude motions of the atoms are of little concern in solids (unless one considers either diffusion in solids or systems near their melting points and indeed solid state theory of melting is not an easy problem). For clusters, we must allow large displacements; otherwise they cannot rearrange, evaporate, etc.

The very fascination of clusters, that they present us with an intermediate state of matter, conforming to neither the bulk nor the small molecule model, means that determining the electronic structure is not going to be quite simple. On the other hand, the very fact that one can speak of the evolution of cluster properties with size means that there are regularities. Some of those regularities do not depend on the specific size and must therefore appear also in more approximate theories. We will therefore proceed from the simpler points of view to the more elaborate ones. It is however, to be expected that the simpler models must be tailored to the bonding type in the cluster and that what will work for clusters held together by van der Waals forces will not do for metallic clusters, and so forth.

5.1. *Clusters as Quantum Dots*

An equally good title for this section is the electronic shell structure of clusters. This is sometimes also known as the jellium model[115,214–216] or the electron gas model and has its origins in atomic nuclear and solid state physics. Our title reflects the recent support for the model we are about to discuss which has been provided by the spectroscopy of so-called nanodots.[217–219] These are clusters that are prepared by wet chemical methods and are studied in solution or as two- or three-dimensional arrays. They are prevented from collapsing into an extended solid by coating their surface with an organic functional group so that individual dots can be studied. The older experimental evidence for the model that we will discuss is the idea of magic numbers, that certain cluster sizes have an enhanced stability. Unlike the atoms (2, 10, 18, etc.) or nuclei, the sets of magic numbers for clusters are not identical for all bonding types; we would like the theory to apply, whenever possible, for more than one kind of cluster and to tell us why and when this is possible and when it is not.

The "quantum" in our title reflects the essential idea of the model: the energies of the (valence) electrons in the cluster are determined by their confinement to the volume of the cluster. The positive cores determine the detailed shape of the confining potential and the valence electrons move as a gas of non-interacting particles in this potential, hence "electron gas". In the "jellium" model, we replace that set of positive cores with a smooth background, a continuous distribution of positive charge, which acts like a jelly through which the electrons move. The system behaves like a metal if it is large enough that the electronic energy levels are very closely spaced — hence "jellium", like a name for a metal. The density of this background is often taken to be the density of the bulk material. The electrons are thus confined to a sort of box. The lowest approximation supposes that box to be a sphere; other approximations allow the sphere to distort into an ellipsoid. The most realistic distortions are those that account for the fact that the real cluster has a polyhedral or lattice geometry.[216]

The simplest confining potential is a square well and this model has been extensively employed in the spectroscopy of quantum dots. The essential feature is that energy levels of a well potential are inversely proportional to the square of its radius. This is most easily seen by a semiclassical argument. An electron in a stationary state with a quantum number n moving in a well extending from $-L/2$ to $L/2$, must satisfy $n(\lambda/2) = L$ where λ is the deBroglie wavelength $\lambda = \hbar/p_n$. Here p_n is the momentum of the electron with the quantum number n. The energy of the state of the electron with the quantum number n is $E(n) \equiv p_n^2/2m_e = \hbar^2/2m_e\lambda^2 = \hbar^2n^2/2m_eL^2$. If the well is three-dimensional, n becomes a vector, with components n_x, n_y, n_z so that $n^2 \rightarrow n_x^2 + n_y^2 + n_z^2$.

The spacing between the different states of the potential scales with $L^{-2} \propto n^{-2/3}$ where n is the cluster size and this works well not only for nanodots of metallic atoms, which is expected, but also for dots of semiconductors (e.g. CdS).

When we treat the electrons as independent, the energy of a many-electron state of a cluster of size n is a sum of the energies of the individual electrons, $E(n) = \Sigma_n E(n)$ where the summation is over all quantum numbers of occupied states. Since the energy of the orbitals depends only on the value of n^2, a "shell" is completed when we have enough electrons to fill all states of given n_x, n_y and n_z where the quantum numbers are such that $n_x^2 + n_y^2 + n_z^2 =$ constant. There is a place for two electrons (of opposite spins) in the ground state shell. Next, six electrons can be placed with a quantum number of unity in the x-, y- or z- direction. For a cluster with one valence electron per monomer (e.g. Na_n), the second shell is completed for $n = 8$.[115,220]

A square confining well is necessarily a simplistic approximation. Other potentials which have been used for metallic clusters include a spherical

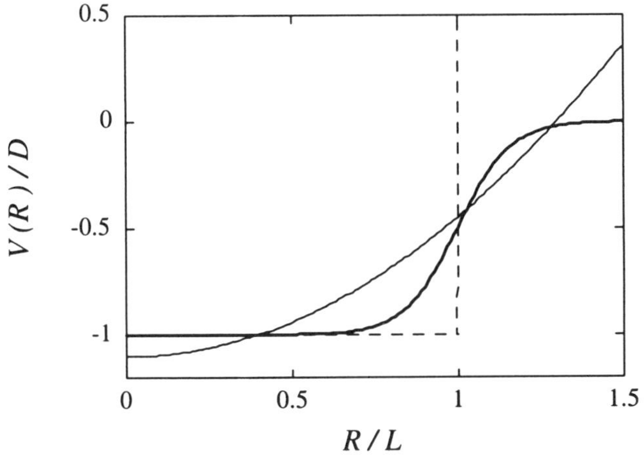

Fig. 9. Three confining potentials for the electrons in a cluster of radius L. One expects that L scales with the size n of the cluster as $L = L_0 n^{1/3}$. D is the well depth. The Woods–Saxon potential $V(r) = -D/(1 + \exp((R - L)d))$, where d is the thickness of the boundary layer, is shown as a heavy solid line and R is the distance from the center of the cluster. The equivalent spherical harmonic potential, $-D' + kR^2/2$ is shown as a light line. Its depth and force constant are given by a fit to the Woods–Saxon potential. The square well potential is shown as a dashed line.

harmonic oscillator potential and a spherical Woods–Saxon potential, both of which come from the nuclear shell model.[221] These three choices are compared in Fig. 9.

When the energy levels of the occupied orbitals in the spherical potentials are added up, one obtains the electronic energy of the cluster of size n, $E(n) = \Sigma_n E(n)$. When this energy is plotted as a function of n, it shows particularly strong binding when a shell is completed. However, on the average it has a monotonic behavior that is often approximated as

$$\overline{E}(n) = -an + bn^{2/3} \tag{5}$$

where a and b are positive energies. This clearly suggests the interpretation that a is the binding energy of the monomer in the bulk (e.g. about 1 eV per Na atom in a sodium cluster) while b is a measure of the reduction in the binding due to monomers which are at the surface and have fewer than typical neighbors. The image is then that of a liquid drop and the dependence expressed in Eq. (5) is therefore referred to as the liquid drop model. Magic numbers are those values of n for which $E(n) - \overline{E}(n)$ is negative, indicating preferential binding.

The computed values of $E(n) - \overline{E}(n)$ are found to be periodic when plotted as a function of $n^{1/3}$. This is understandable from the model because magic numbers of spherical potentials tend to scale as $\nu^3, \nu = 1, 2, \ldots$ The familiar case of a Coulomb potential is perhaps the simplest example. For each principal quantum number n, all the angular momentum states are degenerate. The range of the angular momentum l is from 0 (s states) to $n - 1$. The degeneracy of each level of a given l is $2l + 1$. The number of electrons (of either spin) in a shell of a given n is therefore $\Sigma_{l=0}^{n-1} 2(2l+1) = 2n^2$. The magic numbers n_0 occur when there are exactly enough electrons to complete and fill a shell. so that $n_0 = \Sigma_{n=1}^{\nu} 2n^2 = (2/3)\nu(\nu+1/2)(\nu+1) \approx (2/3)\nu^3$, $\nu = 1, 2, \ldots$ where ν is the principal quantum number of the highest fully occupied shell. For one valence electron per monomer, the number of electrons is the cluster size n, so the magic numbers n_0 scale as ν^3.

The scaling of magic numbers as ν^3 is found for other spherical potentials. There are however grounds for expecting that this cannot be the universal behavior. The reason is the assumption that the cluster is spherical. A familiar example of a broken spherical symmetry is that of a hydrogen atom in an external electrical field. The system still has a cylindrical symmetry but the different angular momentum states of given n are no longer degenerate. (This is the, so called, first order Stark effect). This splitting is found in other potentials even when they are spherical. The reason is the repulsion due to the angular momentum barrier $\hbar^2 l^2 / 2m_e r^2$ (where r is the radial coordinate of the electron in coordinates centered at the proton), which contributes to shifting the energy levels due to its large value at low values of the distance R. The deeply attractive and long-range Coulomb potential is exceptional among all potentials in its relation to the centrifugal term.

The analogy between an electron in an isolated cluster and the Stark splitting due to an external field is less far fetched than it might at first appear. There are no external fields acting on the electron but, on the other hand, the assumption that the core ions provide a uniform spherical background is not quite valid. The cold cluster has definite positions for the atoms and this lattice, even if perfect, has a symmetry that is lower than spherical. The same physical situation obtains when we place an atom with low-lying electronic states in a host lattice or in a coordination compound. The field of the surrounding ligands splits the hydrogenic energy levels. Crystal Field Theory is the highly developed formalism set up to deal with these splittings which give rise, *inter alia*, to the color of gems and of coordination compounds of the transition metals. Our problem is even more complex because the atoms of the cluster need not have a perfectly symmetrical (e.g. octahedral) disposition so that the splittings can be quite extensive.

Non-spherical potentials are often mimicked as three-dimensional harmonic oscillators. In the general case, with no equivalent directions, the expression for the energy is $E(\mathbf{n}) = \hbar\omega_x(n_x+1/2)+\hbar\omega_y(n_y+1/2)+\hbar\omega_z(n_z+1/2)$. For a spherical potential all three frequencies are the same. There are then $(n+1)(n+2)$ degenerate states corresponding to all the combinations of quantum numbers giving rise to the same value of n. The magic numbers in this case are $(1/3)(\nu+1)(\nu+2)(\nu+3)$, namely $2, 8, 20, 40, 70\ldots$ Note that in this case there are fewer distinct levels than for a Coulomb potential. A spherical Woods–Saxon or square well potential has an intermediate number of levels because some of the levels will be split according to the value of the angular momentum. The sequence for either potential begins in the same manner, namely 2, 8 but continues as well 18, 20, 34, 40, 58, 68 at which point these two potentials begin to differ. Beyond the dependence on the exact shape of the potential, the magic numbers can differ due to deviations from spherical symmetry. Quite different sets of magic numbers can be obtained for a non-spherical three-dimensional harmonic oscillator with integer ratios of its frequencies.

5.1.1. *Density Set as Functional Methods*

The density functional method[222–224] is, in principle, able to provide exact results. In practice it is applied in an approximate form which, in more than one way, can be regarded as a refinement of the point of view introduced in the last section. In other words, density functional method centers attention on the density of one electron. It differs from the jellium and other such simple models in that it seeks to follow a more rigorous course to determine the effective potential for the motion of the electron. It does however retain an essential simplicity in that it can handle large systems.[225,226]

The promise of the density functional theory is based on a theorem that, unfortunately, is non-constructive. The theorem states that the exact ground state properties of a many-electron system can be uniquely determined from the knowledge of the electron density alone. The theorem is non-constructive because it provides no instruction on how to go about finding that electron density. It just proves that if it is known, the density will determine the properties. Fortunately, there is a growing number of schemes that provide useful approximations for implementing this program.

What one seeks is that function (or, more correctly stated, functional) of the one electron density which is the ground state energy. Approximations begin with the step of writing the energy as $E[\rho] = T[\rho] + \int dr\rho(r)V(r) + E_{ex}[\rho]$ where $\rho(r)$ is the electron density, T is the kinetic energy term, V is the Coulomb (or other) local potential seen by the electron and the third term is the contribution to the energy due to exchange of electrons. While practical and accurate forms of the kinetic and local potential energy functionals are available, there is a problem with the exchange term. One knows quite well how to write the exchange energy in terms of the distribution of two electrons. The theorem says that it should be possible to write it as a functional of the one electron density and much effort has gone into providing reasonable forms.

The theorem further states that for any density ρ it is the case that

$$E[\rho] \geq E_{\text{ground state}}$$

and equality holds for the exact ground state density. Hence can use a variational principle by varying the density (in the approximate but explicit expression above) $\delta E[\rho]/\delta\rho(r) = \delta T[\rho]/\delta\rho(r) + V(r) + \delta E_{ex}[\rho]/\delta\rho(r)$. The problem, as we said, is to know how to express the exchange energy as a function of the density. Suppose that we have a suitable approximation. If we can define an effective potential such that $U(r) = V(r) + \delta E_{ex}[\rho]/\delta\rho(r)$; then by solving the one electron Schrödinger equation

$$(-(\hbar^2/2m_e)\nabla^2 + U(r))\psi_i(r) = \varepsilon_i\psi_i(r) \tag{6}$$

we can compute the density $\rho(r) = \Sigma_i|\psi_i(\mathbf{r})|^2$. Given the density, one determines the exchange energy and hence its variation with density. One can now recompute the effective potential $U(r)$, solve again the one-electron Schrödinger equation, recompute the density, etc. until the process converges in a self-consistent manner.

As in the simpler jellium model, we retain the simple description of independent electrons, each moving in a confining potential $U(\mathbf{r})$. Here however, that potential is not an arbitrary, made-up model potential chosen to fit data or to make a calculation convenient; this potential includes such effects as the exchange interaction with the other electrons. In this, the present approach is quite reminiscent of the Hartree–Fock self-consistent procedure, which will be described next. There is one essential difference. Unlike the Hartree–Fock procedure, here the exchange term is approximated as a local function, depending only on the one-electron density. This approximation yields fast convergence to a self-consistent density. As in the Hartree–Fock procedure, one may usefully simplify the solution of the one-electron differential Schrödinger equation, which is a non-trivial task by itself, by expanding the one-electron cluster orbital $\psi_i(r)$ as a linear combination of n known site orbitals $\phi(r) : \psi_i(r) = \Sigma_{j=1}^{n} \phi_j(r)c_{ji}$. This converts the one-electron differential Schrödinger equation into an algebraic equation for the unknown coefficients c_{ij}. These coefficients form an $n \times n$ matrix, where n is the number of elements comprising the cluster; the procedure to determine this matrix is one example illustrating why quantum chemical computations scale with a power of n.

Even this procedure still contains an element of arbitrariness and reveals an unsolved problem of quantum chemistry. The representation of the exchange interaction by the local electron probability density (multiplied by a suitable numerical factor, of order 0.5) seems a plausible way to approximate a complex interaction that actually depends on the locations of pairs of electrons. However as yet, there has been no derivation from the many-electron Schrödinger equation of any systematic series of successive approximations to its solution, of which the local density approximation would be the first approximation. There is a belief and a hope that such a derivation will be achieved, but it remains a tantalizing challenge now.

The one-electron matrix Schrödinger equation itself is a quite practical approach[227] and so we outline its derivation. We need two sets of one-electron matrix elements. The first set consists of those of the Hamiltonian: $H_{jk} = \int dr\phi_j^*(r)(-(\hbar^2/2m_e)\nabla^2 + U(r))\phi_k(r)$. The second is the elements of the overlap matrix $S : S_{jk} = \int dr\phi_j^*(r)\phi_k(r)$. Often one assumes, not quite correctly, that the overlap matrix is diagonal and so is the identity matrix. The unknown coefficients in the expression for the wave function $\psi_i(r)$ are the solution of the matrix eigenvalue equation

$$Hc = \varepsilon Sc \,. \tag{7}$$

The eigenvalues are the roots of the $n \times n$ determinental equation det $|H - \varepsilon S| = 0$. There are n roots, not necessarily all different, and n eigenvectors which can be determined, one per eigenvalue, from the matrix equation. It may be obvious, and if so we ask to be excused for pointing it out, but the matrix equation cannot determine a unique set of eigenvectors because the equation is singular $(H - \varepsilon S)c = 0$. One can multiply all the elements of an eigenvector by the same constant, and it remains an eigenvector. To make it unique we must impose an additional condition. To choose that condition, we return to the interpretation of $|\psi_i(r)|^2$ as a probability density, so that if that orbital is occupied, its integral over all space should be unity. This means that we should require the orbital to be normalized: $\int d\mathbf{r}|\psi_i(r)|^2 = 1 = \Sigma_{j=1}^n c_{ij}^* \Sigma_{k=1}^n c_{ki} S_{jk}$. In the conventional notation of quantum mechanics, we write this as $c_i^\dagger S c_i = 1$ and if the overlap matrix is diagonal, normalization of the ith eigenvector has an even a simpler form, $c_i^\dagger c_i = 1$.

Other quantities are also usefully expressed in terms of the eigenvectors. For example, the contribution of the ith orbital to the charge density is just the integrand of the normalization integral $|\psi_i(r)|^2 = \Sigma_{j=1}^n \Sigma_{k=1}^n c_{ij}^* c_{kj} \phi_j^*(r)\phi_k(r)$. The contribution of the jth site to the total charge density is, therefore $\Sigma_{i=1}^n |c_{ij}|^2$ while the charge between two neighboring sites, say j and k, is determined by $\Sigma_{i=1}^n c_{ij}^* c_{ki}$. (This quantity is *not* a probability because it need not be positive; nevertheless it is a measure of the mutual contribution of two sites to the total, which is non-negative, as a probability should be.) In other words, the matrix analog of the one-particle density $\rho(r)$ is the one-particle "density matrix" $\rho_{jk} = \Sigma_{i=1}^n c_{ij}^* c_{ki}$. The analogy, while very useful, is nonetheless imperfect in one sense. The one-electron density is defined in terms of the orbitals that supposedly solve the one-electron differential Schrödinger equation. The density matrix above uses the approximation that the orbital is a linear combination of *given* site orbitals, $\psi_i(r) = \Sigma_{j=1}^n \phi_j(r) c_{ji}$, designated "LCSO". It is not obvious and, in general, it is not quite true, that the exact cluster orbital can be accurately represented as a simple linear combination with one orbital per site. The LCSO representation is a further approximation. It is often very useful and convenient but it is not part and parcel of the density functional method.

Have we finished? No. Everything so far is for a given configuration of the nuclei. We next need to repeat the process for a new configuration, and then another and another. Of course, the electron density that was the solution for each configuration we determine is a good starting point for the next iteration to self consistency, provided we move the nuclei only a bit with each new start. Once this is done for configurations representing all the structures we might expect our system to assume, we have the electronic energy as a function of structure. If we add to it the nuclear repulsions, we can use their sum as a potential energy function governing the motion of the nuclei. One approach, a very expensive one for all but very small systems, is to construct the full potential surface (or at least to find so many points on that surface that we can construct a map to represent it. When we have found such a map of the energy as a function of the positions, we can proceed to the next stage. We let classical mechanics tell how the nuclei must move from any initial geometry. The force that tells us that is precisely the negative gradient (steepest slope) of the potential energy we have just constructed: $F = -\nabla V(r)$. Then we let the nuclei move under these forces to a new configuration, and follow their motion on the complicated potential surface represented by $V(r)$.

A more efficient method is based on the idea of solving the electronic problem step by step as the motion of the nuclei evolves, sometimes known as solving the electronic problem "on the fly". This method has recently become popular and we will return to it when we discuss the dynamics of nuclear motion. In effect, one determines both the potential energy and its gradient, the force, at the initial point, and moves the nuclei a bit in the direction the force pushes them. Then one solves the electronic Schrödinger equation again and finds the potential energy and force at the new configuration, and moves the nuclei a bit more, as the new force pushes them. In this way, one constructs a pathway and a map of the potential energy surface just along that pathway. If one follows the pathway or trajectory very far, and if the system is "quasi-ergodic", i.e. if every trajectory comes arbitrarily close to every accessible point, then the result of the calculation will be a representation of the full surface, and of the system's motion on that surface.

5.1.2. *Hartree–Fock and Beyond*

Now we move to a level a step more advanced than that of the previous discussion. The Hartree–Fock level of description is the highest level approximation at which it is still possible to assign electrons independently to given orbitals. Beyond it one must allow explicitly for correlations between the electrons. Yet one sometimes must allow for such correlations, in order to represent the behavior of a real system. An example is the question of the transition from a localized to a metallic behavior, the so-called Mott insulator to metal transition. This is well known in solid state theory.[228] It arises because an electron that transfers from one site to another must occasionally arrive at a site where an electron (of an opposite spin) is already present. These two electrons have a strong Coulomb repulsion which we can estimate as the energy difference for the change site A + site B $\rightarrow$ site A^++ site B^-. The effective force is a repulsion because ionization potentials are typically much higher than electron affinities, so energy is required to produce the ion-pair from the two neutrals. If the electron is to move relatively freely, the coupling between adjacent sites must be sufficiently strong so as to compensate for this so-called "Coulomb blocking". This effect and possibly also the polarization of an adjacent site by an electron on a given site are important for the understanding of electronic excitations of clusters, yet cannot be described by a one-electron Hamiltonian.

The failure of the orbital picture, in which each electron moves in its own orbital, to properly describe ionic states, is well known. One of the earliest illustration is the electronic structure of H_2. To make the example more like a cluster, let us call it Na_2. We have two valence electrons and in the ground state they occupy the same spatial orbital with opposite spins. Representing this orbital as a linear combination of two atomic orbitals, the space part of the (singlet) wave function looks like $(\phi_a(r_1)+\phi_b(r_1))(\phi_a(r_2)+\phi_b(r_2))$. Opening the brackets, we see that there are four states that contribute with equal weight. Two are covalent, with one electron per atom, e.g. $\phi_a(r_1)\phi_b(r_2))$ and two are ionic, Na^-Na^+, (i.e. $\phi_a(r_1)\phi_a(r_2))$ and Na^+Na^-. We need to go beyond the molecular orbital picture if we want to decrease the importance of the ionic states. The valence bond (VB) approach is one way to do so and generalized valence bond procedures are being increasingly applied to clusters.[229,230] Alternatively, we need to go beyond the Hartree–Fock orbital picture to configuration mixing, i.e. to combining Hartree–Fock bits into more complex wave functions for many electrons.

The Hartree–Fock theory is the starting point for many high level quantum chemical computations of molecules. Hence it is rather well presented in standard sources for quantum chemistry.[202,227] We shall therefore discuss only the bare essence of the theory but we shall do so in such a manner that will allow us to rapidly arrive at a Hamiltonian that contains the essence of the physics we want to describe.

The technical discussion begins with a definition of N (spin)orbitals $\varphi_i(x_i), i = 1, 2, \ldots, N$, where N is the number of electrons and x denotes both the spatial coordinate of the electron and its spin (α or β). For this basis we define one-electron matrix elements

$$h_{ij} = \langle\varphi_i(x_1)|\hat{h}(x_1)|\varphi_j(x_1)\rangle \tag{8}$$

and two-electron matrix elements

$$[ij|kl] = \langle\varphi_i(x_1)\varphi_j(x_2)|\hat{V}(\mathbf{x}_1, x_2)|\varphi_k(x_1)\varphi_l(x_2)\rangle\,. \tag{9}$$

The carat denotes the explicit operators which are the kinetic energy or Coulomb attraction to the nuclei for a one-electron operator or electron–electron repulsion for a two-electron operator. The form of the Hamiltonian operator for the basis is

$$H = \sum_{ij} h_{ij}\hat{a}_i^\dagger \hat{a}_j + \frac{1}{2}\sum_{ij}[ij|kl]\hat{a}_i^\dagger \hat{a}_j^\dagger \hat{a}_k^\dagger \hat{a}_l^\dagger$$

$$= \sum_{ij} h_{ij}\hat{E}_{ij} + \frac{1}{2}\sum_{ij}[ij|kl](\hat{E}_{ik}\hat{E}_{jl} - \delta_{jk}\hat{E}_{il})\,. \tag{10}$$

To write the Hamiltonian in a compact from we have used a, so-called, second quantization notation,[206,231] $\hat{a}_i^\dagger$ and $\hat{a}_i$, respectively, create and annihilate an electron in orbital i and the requirements of the Pauli exclusion principle are satisfied by imposing the (anti)commutation relation

$$\hat{a}_i^\dagger \hat{a}_j + \hat{a}_j \hat{a}_i^\dagger = \delta_{ij}\,. \tag{11}$$

Using this notation, it is very convenient to define a new set of operators, which move an electron from one orbital to another

$$\hat{E}_{ij} \equiv \hat{a}_i^\dagger \hat{a}_i\,. \tag{12}$$

Those familiar with the Dirac bracket notation may want to think of these shift operators (for orbitals which are orthogonal) as $\hat{E}_{ij} = |\varphi_i \times \varphi_i|$. The expectation value of the Hamiltonian in any wave function, made up from the N orbitals, is

$$E = \langle\psi|H|\psi\rangle = \sum_{ij} h_{ij}\langle\psi|\hat{E}_{ij}|\psi\rangle + \frac{1}{2}\sum_{ijkl}[ij|kl]\langle\psi|(\hat{E}_{ik}\hat{E}_{jl} - \delta_{jk}\hat{E}_{il})|\psi\rangle\,. \tag{13}$$

The variational principle of quantum mechanics allows us to chose the best wave function by minimizing this energy subject to such variations in the form of the wave function that leave it normalized. The Hartree–Fock approximation seeks to chose the most flexible form of the wave function that still puts two electrons (of opposite spins) in one space orbital. Such a wave function that is antisymmetrized under the exchange of any two electrons is obtained by writing it as a so-called "Slater determinant". This is an antisymmetrized form of the product (or Hartree) wave function that is, for N electrons,

$$|\psi\rangle = |\varphi_{1\alpha}\varphi_{1\beta}\varphi_{2\alpha}\varphi_{2\beta}\cdots\varphi_{N/2\beta}\rangle \tag{14}$$

The other orbitals, which are empty in the Hartree product, can be used to describe excited states.

The Hartree–Fock single Slater determinant is the best wave function that still allows an orbital description. In this sense it is like the density functional approximation. But now we determine the orbitals from the Schrödinger equation via the variational principle. That is, we chose those orbitals that minimize the energy expression. The restriction of the variation to a single determinant wave function leads to a considerable simplification. Firstly, there are no empty orbitals so an electron can only be moved back to where it came from:

$$\langle\psi|\hat{E}_{ij}|\psi\rangle = \langle\psi|\hat{E}_{ii}|\psi\rangle\delta_{ij} = \begin{cases} 1, i & \text{is a (spin)orbital contained in} \\ & \text{the wave function} \\ & 0 \text{ otherwise}. \end{cases} \tag{15}$$

The other simplification is that the restriction of the variation to a single determinant wave function is equivalent to replacing two-electron operators by one-electron counterparts, e.g. $\hat{E}_{ik}\hat{E}_{jl} \to \langle\hat{E}_{ik}\rangle\hat{E}_{jl} + \hat{E}_{ik}\langle\hat{E}_{jl}\rangle$. Then the energy expression is considerably simplified and becomes a sum over the occupied orbitals only

$$\begin{aligned} E &= \sum_i \langle\psi|\hat{E}_{ii}|\psi\rangle \left(h_{ii} + \frac{1}{2}\sum_j \langle\psi|\hat{E}_{jj}|\psi\rangle([ij|ij] - [ij|ji]) \right) \\ &\equiv \sum_i \left\langle i \middle| \hat{h} + \frac{1}{2}\sum_j (\hat{J}_j - \hat{K}_j) \middle| i \right\rangle . \end{aligned} \tag{16}$$

In the second line we defined the one-particle operators such that $[ij|ij] = \langle i|\hat{J}_j|i\rangle$. To know these operators for orbital i we must know the other orbitals and so we have a self-consistency problem, as in the density functional scheme. Here the problem is a shade more complex. We need to determine all $N/2$ space orbitals in one go, by solving the one-electron Schrödinger equation for electron i,

$$\left\{ \hat{h} + \frac{1}{2}\sum_j (\hat{J}_j - \hat{K}_j) \right\} |\varphi_i\rangle = \varepsilon_i|\varphi_i\rangle \tag{17}$$

consistently with that for the other orbitals. The advantage is that the exchange part of the Hamiltonian, the $\hat{K}$ term, is correctly handled. The kinetic energy part of the one-electron operator $\hat{h}$ means that (17) is a differential equation. As before, it affords a considerable simplification if this equation is converted into a matrix Schrödinger equation by expanding the orbitals φ as a linear combination of some fixed basis functions and solving for the coefficients of the linear combination, $Fc = \varepsilon c$.

The Hartree–Fock approximation allows electron of opposite spins to be put in the same space orbital. As already discussed this is not always adequate. In principle, all that we need to do is to relax the condition that the wave function is given as a single Slater determinant. An often-used generalization is to retain the advantage of the Slater determinant form but to make the wave function a linear combination of determinants, each of which is a "configuration". The variational principle is then used to also determine the coefficients in this linear combination. This method of "configuration interaction" or "configuration mixing" is often used for smaller molecules. However as the number of electrons increases, the number of possible configurations increases so rapidly that, for clusters, one often needs to proceed by a different route. In other words, we want electrons of opposite spins to avoid one another. (The Pauli exclusion principle, which each configuration obeys, serves to do so for electrons of the same spin). So we need to go beyond the Hartree–Fock level, but we cannot do so exactly. So we need first to simplify the Hamiltonian.

We center attention on what is called the Complete Neglect of Differential Overlap or CNDO approximation because it leads to one of the commonly-used procedures, particularly to describe the metal-insulator transition. We shall present both weak and strong versions of this approximation. To motivate the approximation, we begin by taking as given that we need to solve the Schrödinger equation in matrix form so that we have a fixed, finite basis set whose functions are usually centered on the different atoms. Consider then a two-electron integral of the type $[ij|kl]$ as defined by Eq. (9). The indices i and k designate the orbitals for electron 1. If these indices correspond to orbitals centered at different nuclei, then the integrand in (9) contains the "differential" overlap $\varphi_i(\mathbf{x}_1)\varphi_k(\mathbf{x}_1)$. The CNDO approximation is to set such overlaps to zero unless orbitals i and k are on the same atom. We do likewise for electron 2. Then only integrals of the type $[ij|ij]$ remain and the Hamiltonian of Eq. (10) is approximated by

$$H_{\mathrm{CNDO}} = \sum_{ij} h_{ij}\hat{E}_{ij} + \frac{1}{2}\sum_{ij}[ij|ij](\hat{E}_{ii}\hat{E}_{jj} - \delta_{ij}\hat{E}_{ii})\,. \tag{18}$$

For consistency, it makes sense to restrict the one-electron part, the first sum in (18), to orbitals that are on the same atom or that are on near-neighbor atoms. In practice, that consideration is very frequently limited to the valence electrons only and the core electrons are frozen. We shall assume that this is the case.

A further approximation to (18), yielding the "Extended Hückel" Hamiltonian, retains the one-electron part only:

$$H_{\text{Extended Hückel}} = \Sigma_{\text{all valence electrons}} h_{ij}\hat{E}_{ij}\,. \tag{19}$$

In the hands of Hoffmann[204] this Hamiltonian has been very successfully applied to the understanding of bonding in molecules and in extended structures. A prominent version of this form in the literature of organic chemistry dealing with π electrons is the very useful Hückel approximation. In this form, the summation in (19) is even further restricted, to only the $2p_z$ carbon orbitals that participate in the π bonding. The Hückel approximation bears very considerable similarity to what is known as the "tight binding" approximation of solid state physics.[232] In this formulation, each atom in an array of Na atoms contributes one valence electron; this makes the tight binding Hamiltonian formally identical to the Hückel Hamiltonian. However we often want to deal with clusters with more than one valence orbital per atom. Then the appropriate next level for handling this complexity is the extended Hückel approximation. In his book, Hoffmann provides many illustrations of the transition from sharp, separated levels to a band as the size of the system is increased. Our discussion of the metal-insulator transition below gives a simple criterion indicating when this occurs: the coupling h_{ij} between valence orbitals i and j centered on two near neighbors needs to exceed the difference in energy, $h_{ii'}$, between valence orbitals i and i' that are centered on the same atom. When this is the case, orbitals i and i' participate in the same band of electronic states. An example in which such considerations are important are clusters of atoms (e.g. Hg) with two electrons filling an s orbital. When the atoms are far from each other, their interactions are weak and well-described by van der Waals forces. When the atoms are near and therefore coupled, these s orbitals overlap and therefore split in their energy so as to form a band, Fig. 10. The band, like the orbitals that comprise it, must be fully occupied and hence non-conducting. The atoms, however, contribute higher-lying empty p orbitals. (They are higher in energy because in any atom more complex than hydrogen, the p orbitals are not as penetrating as the s orbitals and are better screened from the nucleus). If however the coupling is strong enough, the empty band of p orbitals can overlap the full band of s orbitals, resulting in a partially-filled band.

The Hückel and extended Hückel approximations are one-electron approximations. They are quite useful but they do not take us beyond Hartree–Fock. To do so we need either to use the weak version of the CNDO approximation, as given in Eq. (19) above or to make a further simplification, the strong CNDO approximation, in which we allow only electrons (of opposite spins) which are on the same site to repel. In other words, in the two-center Coulombic repulsion integrals $[ij|ij]$, we allow only $i = j$ and

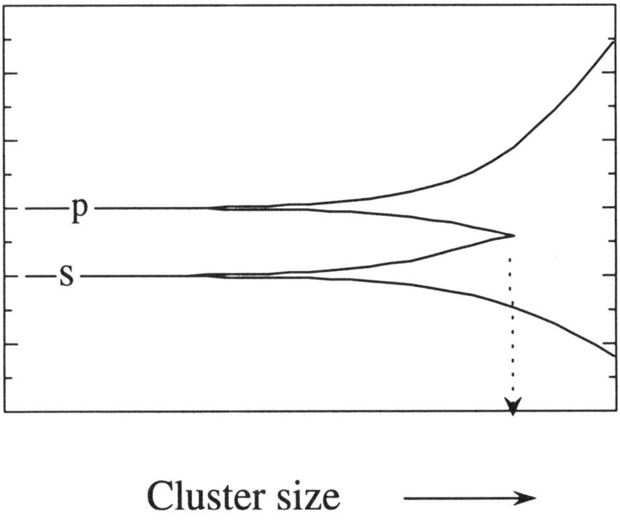

Fig. 10. The evolution of overlapping bands of electronic states as the cluster size increases. The arrow marks the critical size for a non metal–metal transition. For smaller sized clusters of atoms with two valence s electrons the ground state band is full. For larger sizes, mixing with the empty p orbitals provides a half full band. See Refs. 204 and 233 for more technical details. Observing the non-metal to metal transition is still an active research topic.

obtain the "strong CNDO Hamiltonian":

$$H_{\text{Strong CNDO}} = \sum_{ij} h_{ij}\hat{E}_{ij} + \frac{1}{2}\sum_{i}[ii|ii]\hat{E}_{ii}(\hat{E}_{ii} - 1)\,. \tag{20}$$

This is a very popular form of the Hamiltonian that in the solid state physics literature is known as the Hubbard model. It has all the convenience of the Hückel approximation plus the most important part of the electron correlation known as the Coulomb blockade; electrons (of opposite spins) which are on the same site i repel one another with an energy $[ii|ii]$. If all the atoms are the same and each contributes one valence orbital, then (20) has the even simpler form of the "simple Hubbard" Hamiltonian:

$$H_{\text{Simple Hubbard}} = \varepsilon\sum_{i}\hat{E}_{ii} + t\sum_{i \text{ and } j_{\text{neighbors}}}\hat{E}_{ij} + \frac{1}{2}U\sum_{i}\hat{E}_{ii}(\hat{E}_{ii} - 1)\,. \tag{21}$$

The simple Hubbard Hamiltonian is often used. One of the reasons it is so popular is that the Hamiltonian (21) lends itself to an easy physical interpretation: when the atoms are far apart they are uncoupled, $t \to 0$.

In the ground state there is one-electron in the valence orbital, or $\langle \hat{E}_{ii} \rangle = 1$. If the atom is ionized the valence orbital is empty, $\langle \hat{E}_{ii} \rangle = 0$. So ε is the ionization potential of the isolated atom.

Consider next the process, atom A + atom B $\rightarrow$ atom A^++ atom B^-. The change in energy is U, the ionization potential minus the electron affinity (the energy to make the ion pair when the atoms are far apart), minus the Coulomb attraction between the two atoms a distance R_{AB} apart, $U = I.P.(A) - E.A.(B) - e^2/R_{AB}$. Here, t is the strength of coupling of neighboring atoms when they are close. That's it; there are no other parameters.

To determine the energy levels of a cluster one needs to diagonalize the Hamiltonian. Terms like $\hat{E}_{ii}\hat{E}_{ii}$ (or $\hat{E}_{ii}\hat{E}_{jj}$ for the weak CNDO approximation) require knowing where two electrons are with respect to one another. For a single-determinant wave function, these terms contribute only on the average, cf. (13). Therefore one needs to go beyond Hartree–Fock in order to solve the problem. In our experience,[229] it is then not so very advantageous to make the strong CNDO approximation. Making only the weak CNDO approximation, Eq. (18), does not make the computation much more difficult but allows us to include the polarization effects of an electron on site i at the neighboring sites.

To discuss the metal-insulator transition[233] we go back to the beginning to recall that the Hamiltonian does depend on the positions of the atoms. In particular, the coupling, t, between adjacent atoms must depend on the interatomic spacing. Since the properties of a metal require orbitals on adjacent atoms to overlap, and, at large distances, amplitudes of atomic orbitals drop exponentially with distance from their nuclei, we expect t to decrease exponentially as the atoms are pulled apart. So consider a linear array of equally spaced atoms, each contributing one valence electron, and all at large distances from their neighbors. The atoms are noninteracting and the ground state of the array has one electron per atom because in any ionic configuration we are penalized by at least the energy U. The ground state is degenerate because there are many ways of assigning the spins even if we require that the state is a singlet (for an even number N of atoms). Hence the ground "state" consists of a degenerate band of states of energy ε. Above this in energy is a band of excited states corresponding to one site having no electron and another site having two electrons. The next higher are energy bands in which two electrons have moved, and so forth. These bands are all separated approximately by the energy U. Next, we decrease the spacing between the atoms of the array. The parameter t is now

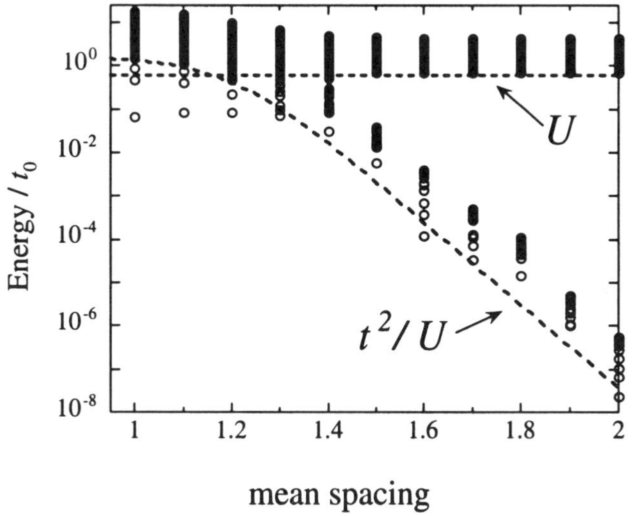

Fig. 11. The Mott non-metal to metal transition as a function of the average separation between the atoms of a cluster of 91 atoms. Shown are the computed energies in units of t (logarithmic scale) of the excited electronic states relative to the ground state. Metallic behavior requires that there is a quasi-continuum of states. In the lowest approximation, shown as dashed lines, states where an electron has moved have excess energies of U above the covalent states. t measures the strength of the exchange coupling between adjacent atoms and hence decreases exponentially with the spacing between atoms. When $t \approx U$, the exchange coupling can overcome the Coulomb repulsion. See Refs. 229 and 334 for details of the computational method.

non-zero, but exponentially small. The operator $\hat{E}_{ij}$ cannot couple different states of the ground (covalent) band because an electron occupies each site. The action of $\hat{E}_{ij}$ on a state of the ground band produces a state of the first excited band. Thus, in the first order of perturbation theory, in which the shift in energy is the expectation value of the perturbation, the site–site coupling has no effect. We need to go to second-order in perturbation theory. Then the shift in energy will be t^2/U because U is the separation between the two bands. This shift is seen in Fig. 11, that shows the energy of the states versus the spacing. When $t^2/U \approx U$ the covalent and ionic bands merge. This is the celebrated Mott transition. The site–site coupling is strong enough to compensate for the energy required for the electron to move to an already occupied site.

There is another type of transition between a localized and a delocalized, band-type wave function. We like to think of it as an Anderson transition.

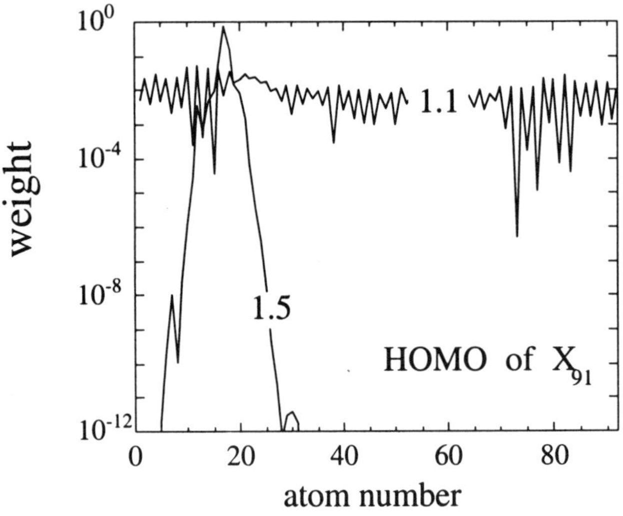

Fig. 12. The transition from a localized state (mean spacing 1.5, as indicated) to a delocalized state (mean spacing 1.1) for a cluster of 91 atoms with one-valence-electron each. Shown is the weight (logarithmic scale) of the highest occupied molecular orbital (HOMO) of the cluster on each one of the 91 atoms. The localized state decreases exponentially as one moves away from the atom on which it is centered. The delocalized state has, on the average, the same weight on all the atoms.

It must however be admitted that the labels "Mott" and "Anderson" refer to the roles played by different parts of the Hamiltonian and reality has everything in the Hamiltonian. The labels refer to limiting behavior in which one or another physical interaction, corresponding to one or another part of the Hamiltonian, dominates the process; in any real case, all those interactions are present to some degree. The Anderson transition describes a limit where U is already smaller than t so we can work with a one-electron Hückel Hamiltonian. Consider a lattice comprised of two types on one-valence-electron atoms. These atoms differ in their ionization potential by $\Delta\varepsilon$. As long as $t < \Delta\varepsilon$, neighboring atoms of different kinds cannot be strongly coupled. The wave functions do not delocalize over the entire system. Rather, electrons remain localized. If the lattice spacing is made so small that a band forms, the wave function becomes very delocalized, as shown in Fig. 12.

One can elaborate on the simple ideas we used to describe the Mott and Anderson transitions. For example, we can extend the models to a lattice of identical atoms, each with two valence electrons, and so forth. The general point at issue is that when the coupling, t, between the atoms becomes strong enough, there can be a transition from a localized to a delocalized state. Different scenarios correspond to different answers to the question "Strong enough compared to what?" and different physical systems fall into different regimes. The possibly counterintuitive result is that if the coupling is weak, the electrons are localized. This is so even if the system consists of identical atoms; if we were to describe the system with a simple molecular orbital (i.e. Hückel type) or tight binding model, we would conclude that the wave function is delocalized. We are taught these approximations early on and so it sometimes comes as a surprise that there can be limiting physical situations where they are qualitatively misleading.

6. Kinetics of Free Clusters

The kinetic behavior of clusters has become known to us largely through simulations, especially through simulations done by the method known as "molecular dynamics".[235] (There is one alternative and widely used way to do simulations, by a random-search procedure known as the Monte Carlo method.[3,236] This is a very efficient way to study many problems associated with clusters, especially thermodynamic properties, but is not particularly suited to the specific topic of kinetics.) Molecular dynamics simulations are nothing more than the development of a history of the system as it evolves according to the laws of mechanics. Those laws may be the classical mechanical laws of Newton, i.e. the law of inertia, the law connecting force to acceleration, and the law of equal and opposite action and reaction, otherwise known now as the law of conservation of momentum. Alternatively, the laws one invokes may be the laws of quantum mechanics, usually taken to mean that the system evolves according to the Schrödinger equation for time evolution.[237] Most simulations of clusters have been done in the classical picture, because many kinds of clusters can be well-described at all temperatures or energies but the very lowest, and even clusters of atoms of the first row of the Periodic Table, Li through Ne, have classical and quantum descriptions that differ quantitatively but not qualitatively.[238–240]

In the classical procedure, one begins by knowing the force law for the interactions among the particles, and by assigning a position and momentum to each particle in the cluster. Then, by solving, over and over, the equations of motion for all the particles simultaneously, as each exerts its force on all the others, we obtain a history of the behavior of all the particles in the cluster.[4–6] We typically carry out the solution process by turning the continuous differential equations for Newton's second law into discrete difference equations, with very small intervals of time and distance, usually only a small fraction of a vibrational period and amplitude. We can study properties of virtually any sort, step by step through this evolution. For example, we may construct pictures of the cluster at each step or, more commonly, at every nth step, to make an animated movie of the cluster's evolution. We can compute positions, velocities, distances traveled, frequencies of vibration and rotation, and correlations of any of these and other properties. The consequences are that we can learn in great detail how clusters would behave if they were composed of classical particles, and, to a fair-to-good approximation, how real quantum particles behave. The limits of such calculations are the limits of computers: at present, it is entirely possible to simulate a cluster of hundreds or even thousands of atoms, for tens or hundreds of millions of time steps, corresponding to tens of nanoseconds of real time. Some calculations have now even approached microsecond time intervals. (However a caution is in order here: although the simulations may go on as long as one can run the computer, in fact the roundoff error of any computer introduces a kind of chaotic noise that prevents any molecular dynamics calculation from being a true solution of the differential equations for intervals of more than about 5000 or 10000 time steps. After about that many steps, enough roundoff error creeps in to make it impossible to retrieve precisely the initial conditions by reversing the direction of the computation. If the procedure were truly producing a solution to the equations of Newton's laws, then the process would be precisely reversible for any time interval, and would always take the system back to the initial state if the direction of time is reversed[241]).

While many of the earlier studies of clusters by molecular dynamics focused on phase transitions,[19,21,22,24,25,158,242–252] more recent investigations have examined other aspects of relaxation and kinetics. For example, the ways water molecules relax and rearrange in clusters has been the object of several molecular dynamics investigations, by both classical and quantum calculations.[253–262] A question becoming more and more important in understanding the kinetics and relaxation of clusters is this: "If a cluster or nanoscale particle forms in some arbitrary structure and then comes to equilibrium, how does it achieve that equilibrium?" This is a question of what pathways does a system take as it relaxes, and has begun to be studied in a variety of cases, especially Lennard–Jones and similar systems, and in alkali halide clusters.[263–265] One generalization that has emerged is that if the multidimensional potential surfrace of the cluster has very few drops in energy, from one local minimum to the next lower, that are large compared with the energy barrier between those minima, then the system is unlikely to find its way to minima that are deep on the surface. Such a potential surface looks much like a sawtooth surface. On the other hand, if the surface has a moderate number of large energy drops, from one minimum to the next, then the system can relax efficiently to its deep minima. This latter kind of surface is something of a rough staircase. Sawtooth surfaces are consequences of short-range interparticle forces, while staircase potentials come from long-range interparticle forces, or from the effective long-range forces imposed by links in polymer chains.

Monte Carlo simulations have been used in some cases to study kinetics of clusters. For example, evaporation kinetics lends itself to this method.[266] The method has been used to study kinetics of passage between pairs of adjacent minima,[267] a subject discussed in a bit more detail later. A Monte Carlo study has described how a cluster of CO_2 molecules containing a single I_2^- molecule relaxes when the extra electron is suddenly removed from the foreign molecule.[268]

Other approaches to kinetic behavior of clusters have been analytic,[141,269] and "reductionist" in the sense of replacing a full description of all the motions of all the particles with a description analogous to that of chemical kinetics. That is, in a kinetic approach, one asks only about the rates of passage of a system among its possible states. The term "states" here means all the locally-stable forms in which the system may exist for periods long enough to be observed with clear signatures of the identity of that form. For example, a cluster with two very stable geometric structures, such as Ar_6 with its low-energy octahedral and higher-energy incomplete pentagonal bipyramidal structures, has what we may well call two states, because if the system has energy enough to be in either of these forms, it may well remain there long enough to establish a clear identity by vibrating for many periods around that structure. More generally, then, a kinetic description is one that yields the rates of passage of a system between each of the regions of local minima of potential energy.

In much of traditional chemical kinetics, we want only the forward and backward rates between a single "reactant" state and a single "product" state. In the case of clusters, as in the situation of many coupled chemical reactions, we typically want to know all the forward and backward rates between every pair of minima linked by a single potential barrier. In the case of rearrangements of clusters among the many minima they have on their potential surfaces, the rates depend only on the concentration or number of clusters in the initial state, and on a rate coefficient for that well-to-well passage. These reactions are, in short, simple first-order reactions. However they are coupled; the rate of clusters passing from the region of one minimum depends on how many clusters are residing in that minimum, which depends in turn on the rate at which clusters have entered that minimum from other minima. It is straightforward in this case to write what is called a "master equation", really the set of all the coupled first-order rate equations, in the form of a matrix equation for the time rate of change of the populations of all the states of the system.

Master equations have been used to describe relaxation and kinetics of clusters. The first approaches were extremely approximate, and served primarily as proof-of-principle.[270,271] Master equations had been used to describe relaxation in models of proteins somewhat earlier[272,273] and continue to be used in that context.[208,274–276] More elaborate master-equation descriptions of cluster behavior have now appeared. These have focused on how accurate the rate coefficients must be in order that the master equation's solutions reproduce the results of molecular dynamics simulations[277,278] and then on what constitutes a robust statistical sample of a large master equation system, again based on both agreement with molecular dynamics simulations and on the results of a full master equation.[279] These are only indications now of how master equations may be used in the future as a way to describe and even control the behavior of clusters and nanoscale systems of great complexity.[280]

7. Chemical Reactions of Clusters

The chemical reactivity of clusters and also the chemistry within and on the surface of clusters provides opportunities for unusual types of changes. Many of these are still not fully explored or even barely imagined. Understanding the reactivity provides challenges for the conceptual and computational tools that were honed for simpler systems. As has been the case throughout this chapter, the prime reasons for this rich chemistry are the variations of cluster properties with size, and the great variety of their bonding types. The chemistry of clusters is as rich as chemistry itself. It was a hope, early in the history of the subject, that clusters would provide us with a bridge between gas phase chemistry and chemistry in solution, at surfaces and in solids. To a large extent, that did happen but what was assumed at the beginning to be a causeway turned out to be a broad landscape with its own distinct ecology. We begin with an example of the dissociation of molecules on clusters where the atoms remain bound on the surface of the cluster.[281] Doing so we hope to learn more about surface chemistry. But of course, atoms can also be incorporated into the cluster and this opens up new directions, etc.

Metals play a key role in catalysis.[282] Diatomic molecules do dissociate on metallic surfaces, e.g. H_2 on Pt or N_2 on Fe. This is taken to be a key step in catalytic processes such as the synthesis of ammonia or the onset of the (potentially explosive) $H_2 + O_2$ reaction. However, directing a beam of diatomic molecules at a clean metal surface often does not lead to reaction, unless the molecules impinge on the surface with quite high energies. On the other hand, exposing a clean surface to the bulk gas does show evidence that dissociation occurs, for example, by taking the gas to be a mixture of H_2 and D_2 and observing that HD molecules are eventually present. Detailed experiments suggest that there are possibly precursor states of the absorbed but undissociated molecule and also that different forms of the energy of the diatomic reactant (e.g. its initial vibrational excitation versus its kinetic energy at impact) can have a marked effect on the yield of dissociation. Do

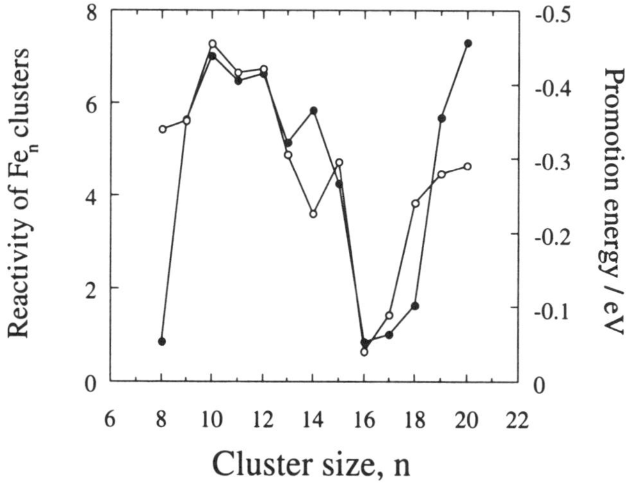

Fig. 13. The reactivity of small clusters of Fe with H_2, as a function of cluster size (solid circle). It is seen that the reactivity is not a simple function of size. Adapted from Ref. 283. See also Refs. 281, 284–286 for further discussion. The right hand ordinate shows that reactivity is well correlated with the energy required to promote the higher most electron of the cluster (open circle). The idea here is that the initial step is electron transfer from the cluster to H_2. Since the bonding σ orbital is full, the electron must go into the anti bonding orbital thereby weakening the bond.

diatomic molecules dissociate on finite-sized metallic clusters? Yes, they do, but, as seen in Fig. 13, the reactivity is not a simple function of cluster size.

Geometric effects are known from earlier work on catalysis. One expects that such surface imperfections as steps or terraces would be sites of preferential reactivity. Even well-ordered planes of a single crystal of a metal differ in their reactivity; for example, different crystal faces of Pt, which have different arrangements of their surface atoms, catalyze the formation of quite different chemical products. It is therefore reasonable that clusters of different sizes, which are packed differently, differ in their reactivity. We can hope that at some point these differences in reactivity of clusters can be used to design better catalysts. However we must recognize that the atoms comprising small clusters usually do not pack in arrangements characteristic of the bulk.

In addition there is the whole topic of "reconstruction": a surface produced to have a specific arrangement of its atoms may begin with the desired structure, but then rearrange to a different structure. The importance of this for surface chemistry is becoming increasingly recognized and there is no reason to think that a similar change does not occur in clusters. Consider first the packing in a bulk metal crystal. Here each atom has optimized its position with respect to its neighbors. When we form a new surface, each atom is no longer at equilibrium with its surroundings. In chemical language one can say that the atom has "dangling bonds" i.e. directions in space where it could be bonded but, being on the surface, it is not. Such a language is, of course, particularly suitable for a covalently bonded solid, e.g. Si, but the essence of the idea is more general. Hence we may expect that the surface atoms of a newly formed surface may rearrange their positions, possibly only slightly because they are constrained by the layers of the bulk under them, so as to change the number of their near neighbors. So far, and qualitatively speaking, this is consistent with clusters, where so many atoms are on the surface, having packings different from that of the bulk. Now let a diatomic molecule dissociate on the new and clean surface. The bond energy of the molecule is high (about 105 kcal mol^{-1} for H_2 and about 225 kcal mol^{-1} for N_2). The energy required to break this bond must come from the chemical interaction of the two new atoms with the atoms of the surface. At the site of the dissociation there must be a significant electronic reorganization. "Free" valences of surface atoms are no longer free and an adsorbate may induce reconstruction of the surface. This is easier to achieve for a cluster, where isomerization can be quite facile and may be energetically easier to achieve than in bulk because the gain in energy due

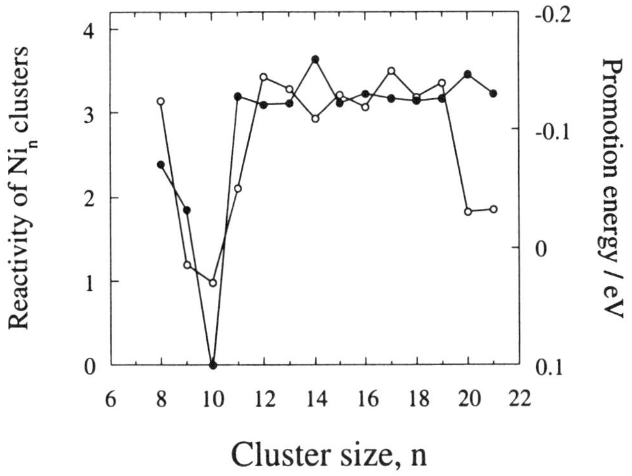

Fig. 14. The reactivity of small clusters of Ni with H_2, as a function of cluster size (solid circle). Compare with Fig. 13 for Fe clusters. Adapted from Ref. 283. See also Refs. 281, 284, 285 for further discussion. The right hand ordinate shows that for this case too, reactivity is well-correlated with the energy required to promote the higher most electron of the cluster (open circle).

to the formation of the two new bonds is shared by fewer atoms. Indeed, one sometimes uses the graphic language that when exoergic absorption occurs on a cluster, enough energy may be released to melt the cluster.

Figure 14 shows that Ni clusters react somehow differently than Fe clusters. It is suggestive to relate this difference to the different catalytic activities of Ni and Fe surfaces in the reduction of CO by H_2, the process known as Fischer–Tropsch synthesis. However, one cannot, on the basis of this evidence alone, rule our another explanation: namely, that the larger sizes of Ni and Fe clusters simply have different structures. Indeed, for Co clusters, one tends to favor this alternative explanation.[281,284,285,287]

Another area that brings forth strong evidence for the role of packing is enzyme catalysis. The selectivity of biochemical recognition has long been described using a "lock and key" metaphor, that is that the substrate must "fit in" the active region of the enzyme. The current evidence is that this too is a dynamic process. The enzyme, rather than being a stiff, static object, is flexible and accommodates its structure towards its substrate. The process is an induced fitting rather than a static one. The fluxional character of clusters makes the protein a better analog than the fairly rigid metal surface where a motion of one surface atom disturbs its entire surroundings.

Electronic effects are equally known from earlier work on catalysis. Even without any detailed theory there must be a reason why transition metals, which readily assume different oxidation states, are effective catalysts. Or why, along a long row of the periodic table, different metals are so different in their catalytic activity. Smaller clusters offer the opportunity to explore these lines further because electronic properties, for a given monomer, can differ considerably with cluster size. As we discussed in Sec. 5, changing the size of the cluster is, in some sense, like selecting "atoms" with different electronic properties: shells may be filled or can be mostly open, etc. Of course the electronic effect is superimposed on (and intimately connected with) the differences in packing for different size clusters. Nonetheless physical properties such as the ionization potential, which are closely tied to chemical reactivity, do vary with the size of smaller clusters in a non-monotonic fashion, as illustrated in Figs. 13 and 14.

It is possible to make simple models that correlate reactivity with electronic properties of the cluster. For example, to induce dissociative adsorption, the cluster must first weaken the H–H bond. This can happen in either of two ways. If charge flows from the cluster into H_2, it must go into a strongly antibonding orbital; if charge flows from H_2 into the cluster, it must come from the bonding orbital. The direction will, of course, depend on the relative heights of the highest occupied (so-called HOMO) and lowest unoccupied (LUMO) orbitals of the two reagents. In either case, the H–H bond must weaken. At the forefront of computational chemistry, *ab initio* theory is being applied to interpret the reactivity of clusters. We discuss the dissociation of H_2 on Si clusters[226,288] as an example.

Two clusters were examined, Si_9 and Si_{10}. These differ in that the former has two exposed Si atoms on its surface (the so-called surface dimer) while the latter has an isolated surface atom. In the simulation, the other surface sites where the dissociation could occur were passivated by H atoms. The electronic structure for each configuration of the atoms was computed at the Hartree–Fock level. Thereby the transition state(s) can be identified; three were found. Of these, two correspond to H–H binding to a single Si atom, in a symmetric and an asymmetric fashion, Fig. 15. As can be seen in Fig. 15, the symmetric transition state, which has a high barrier, occurs early, while the asymmetric structure requires a close approach of one hydrogen to the Si atom through a very constrained geometry.

The idea that the process involves a dihydride intermediate that can be accessed via two different transition states is clearly of interest to the surface passivation of Si by hydrogen. Similar studies have been carried out

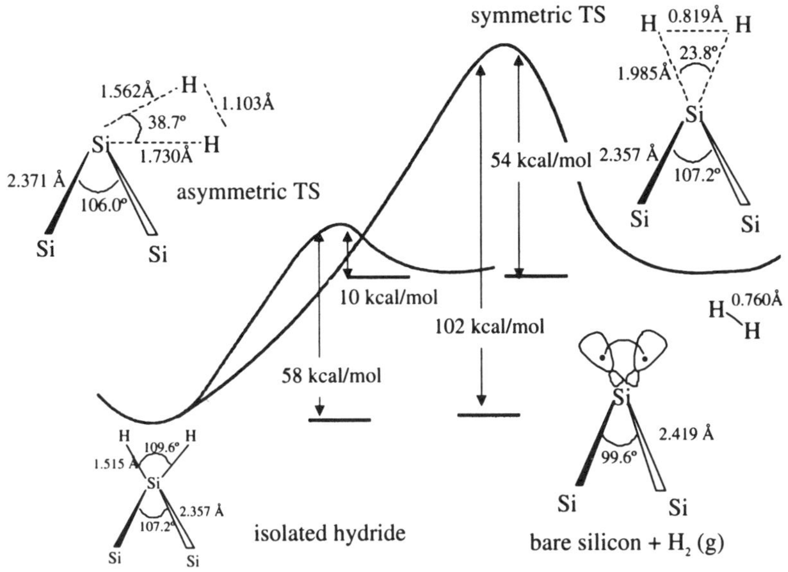

Fig. 15. The potential energy profile along two reaction coordinates for the desorption of H_2 from a Si surface dihydride via a symmetric and an asymmetric transition state. The experimental results are consistent with a lower barrier, suggesting that desorption occurs via the asymmetric transition state. Adapted from Ref. 288. See also Ref. 226.

for the reactions of these clusters with halogens, processes technologically important in the etching of silicon surfaces.

The two examples we just examined used the surface reaction as a reference for the study of clusters. Equally important are the implications of cluster chemistry to reactions in solution. Here, the chemistry of cluster ions has received much attention; this is particularly important because reactions of charged reactants are often quite different in the gas phase and in solution. One canonical case is that of anion-neutral S_N2 reactions such as

$$OH^-(H_2O)_n + CH_3Cl(H_2O)_m \rightarrow Cl^-(H_2O)_{n'}$$
$$+ CH_3OH(H_2O)_{m'} + (H_2O)_{m+n-m'-n'} \cdots \quad (I)$$

Other classes of reactions that have received much attention are proton transfer and electron transfer reactions. These processes can often be initiated by light. This characteristic has made it possible to monitor their time evolution by ultrafast pump-probe techniques. An example is the acid-base reaction of 1-naphthol with ammonia. On the ground electronic surface

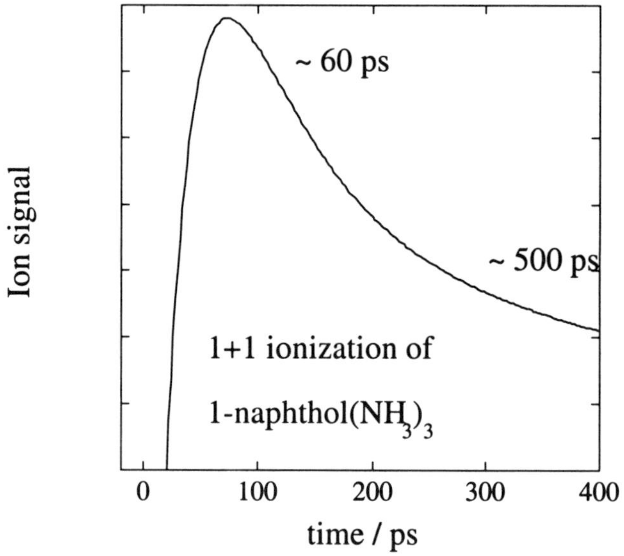

Fig. 16. The biexponential fit to the experimental time resolved 1 + 1 ionization signal of 1 − naphthol(NH3)$_3$ cluster. Adapted from Ref. 289. See text and also Refs. 290–292. The rise time is instrument limited. The short decay of ca. 60ps is the rate of proton transfer. The slow decay, τ ca. 500 ps, is probably due to reorganization of the cluster. The reaction does not occur for fewer ammonia molecules solvating the chromophore.[293,294]

(S_0), the equilibrium constant (K_a) for proton transfer is about 10^{-10}; in the first excited singlet state (S_1) it goes up by ten orders of magnitude. The proton transfer in the clustered molecule can therefore be initiated by photons of energy sufficient to excite low- or higher-lying vibrational states of S_1. Two such photons are sufficient to ionize the cluster. If one assumes that proton translocation changes the cross section for ionization, one can delay the second photon and thereby monitor the progress of the reaction, as shown in Fig. 16.

Solvation plays a key role because the magnitude of the electrostatic ion-polar molecule interaction is large, say 10–20 kcal mol^{-1}, comparable to the barriers for concerted bond-switching reactions. Consider the S_N2 reaction[295–297] $OH^- + CH_3Cl \rightarrow Cl^- + CH_3OH$ in the gas phase. The reaction is concerted so the new (CH_3–OH) bond forms as the old (CH_3–Cl) bond is stretched. This process has an activation barrier but it must be significantly lower than the energy needed to break the old bond. On the other hand, as the reactants approach, the first interaction they experience

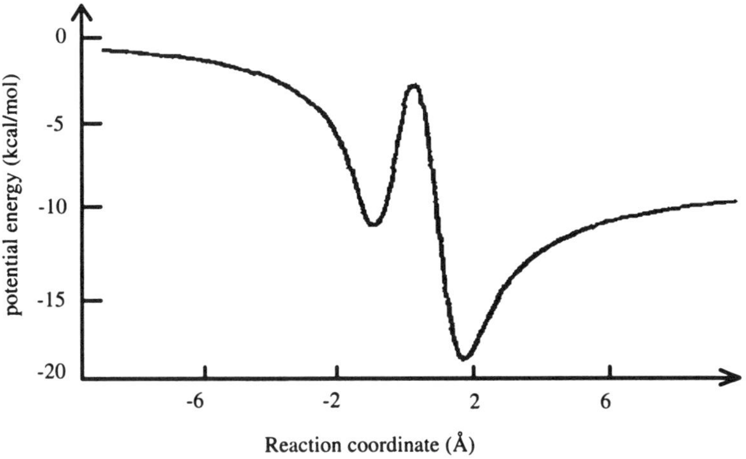

Fig. 17. The potential energy along the reaction coordinate for the exoergic S_N2 ion-molecule reaction $Cl^- + CH_3Br \rightarrow Br^- + CH_3Cl$. Adapted from Refs. 295–297.

is not the chemical one but the physical long-range (R^{-4}) polarization potential of an ion interacting with a polarizable neutral. The reactants are "captured" into the polarization well with a large cross section. (If there were no inner repulsive core, this cross section would be the classical cross section for this process computed long ago by Langevin, and given his name.) But their energy is well above that of the bottom of the well because the colliding partners came from outside and retain their energy. As the reactants move closer in they encounter the chemical barrier, as shown in Fig. 17.

This can be a non-negligible barrier, but it is high relative to the bottom of the potential well corresponding to the chemical interaction, which is nearly zero at the capture distance. In other words, the barrier to bond switching is to be measured from the bottom of the polarization well, not the chemical well, in which the products of the reaction eventually reside. The energy of the colliding gaseous reactants may therefore be sufficient to overcome the barrier to bond switching. The physical attraction in the entrance region thereby provides the energy needed to overcome the barrier for the chemical rearrangement; consequently even reactants with quite low initial translational energy can still proceed to react. With this perspective, it is no wonder that those ion-molecule reactions play a key role in the chemistry of the cold regions of space. If the barrier to reaction is higher than the polarization well, the reaction presents a net barrier to overcome, but it will be much reduced from what it would have been for neutral reactants.

The picture discussed above and depicted in Fig. 17 is quite different when the reaction takes place in a polar environment. The ionic reactant is solvated and stabilized. The ionic product is similarly stabilized. An immediate caveat: because of the transfer of charge, the solvent configuration about the reactants is quite different from the solvent configuration about the products. Solvent motion must take place before the system returns to equilibrium. Clusters, with a variable (and not necessarily large) number of solvent molecules can allow us to probe the dynamics of the solvent motion that must accompany charge translocation. Back to the reaction: in the transition state region the charge is delocalized in a polar solvent. The transition state is not as stabilized by solvation as the reactants or products. If in our mind we move the system slowly along the reaction coordinate, so slowly that at every point the solvent has enough time to equilibrate with the instantaneous charge distribution, we find a barrier to reaction that is significantly higher than in the gas phase. Computations[298–300] for the $Cl^- + CH_3Cl \rightarrow CH_3Cl + Cl^-$reaction show a barrier height, with respect to the reactants at rest, of 3.1 kcal mol^{-1} in the gas phase, 5.4 kcal mol^{-1} when a single water molecule is attached to the anion; this increases to 10.7 kcal mol^{-1} for the dihydrated anion. The activation energy in water solution is about 26 kcal mol^{-1}.

The so-called cage effect is another example of the study of cluster dynamics validating and documenting an important concept often used for portraying reactions in solution. Unlike processes in the gas phase, when a bond is being broken within a cluster or in a solvent, the surrounding molecules can prevent (or delay) the separation of the dissociation products by draining the energy into other modes. Here too, because the process can be initiated by a pulse of light which excites a molecule from its ground to a repulsive electronic state, it is possible to monitor the process in time and also to probe the long-term distribution of products. Figure 18 shows experimental results for yield of photodissociation of $I_2{}^-$ in $I_2^-(CO_2)_n$ clusters.

First of all, there is a direct observation that when the number of "solvent" CO_2 molecules is large enough, dissociation can be essentially fully arrested: the bond that was to be broken heals before the fragments separate. The energy with which the two iodine atoms started to recede from one another dissipates fully into the cluster and, of course, some cluster molecules evaporate as a result. The surviving $I_2^-(CO_2)_n$ clusters have fewer CO_2

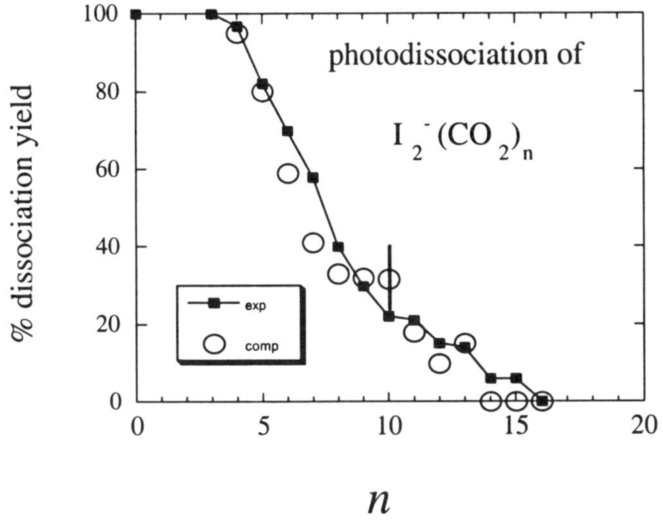

Fig. 18. The Cage effect by a cluster.: The experimental[301,302] and computational[268,303] results for the yield of dissociation versus the cluster size. Once the "solvent shell" is completed, the yield drops to zero. See also Refs. 304 and 305. The error bar in the computation is due to the Monte Carlo sampling.

molecules than the parent cluster. Next, observe that even there are only a few solvent CO_2 molecules, dissociation does not occur with a 100% yield. Some I_2^- remain bound.

The cage effect can be studied theoretically by running classical trajectories on a realistic potential energy surface. One needs to run many trajectories so as to capture the many possible outcomes of such a process. If one does not include the photoexcitation process, then the initial conditions for these trajectories are two iodine atoms, with a narrow distribution of their internuclear distances, at the Franck–Condon region on the repulsive potential curve of I_2^-. The energy is that provided by the photon, minus the binding energy of the ground state. There is however also a distribution in the initial condition of the solvent molecules. Note that because of the Franck–Condon nature of the transition, the cluster molecules have their equilibrium distribution about the ground state of I_2^-. It is this distribution which needs to be sampled. By propagating the different initial configurations in time, one generates an ensemble of final outcomes and can compute such observables as the fraction of initial conditions that end with an undissociated molecule.

The potential energy used in these simulations suggests that the initial CO_2 molecules hug the waist of the I_2^- so as to form a cylinder about it. As the iodine atoms move apart, a CO_2 molecule can move between them and thereby hinder recombination. This view is consistent also with another type of experiment[306,307] further discussed below, in which, in a hot cluster, a CO_2 molecule moves so as to slice the bond of the I_2^- in its electronic ground state.

Cage effects have been seen and/or computed even for a single clustering molecule or atom. Typically, a single atom does not suffice to hinder dissociation but it can take up energy (or angular momentum) and thereby alter the energy disposal in the reaction. This is an important way in which reactions in clusters differ from the corresponding reactions of particles isolated in the gas phase. Typically, in the gas phase, energy disposal can be quite specific. This is less so when the reactants are solvated as the larger number of modes that are coupled tends to make the dynamics more statistical.

The geometrical packing constraints that are expressed in the cage effect can be refined and used to advantage in the study of photochemical reactions within clusters. The essential idea is that the cluster is used to orient (favorably or otherwise) the ground state reactants. The photon then initiates the reaction. The simplest examples are for one clustering molecule and we will discuss both electronic orbital steering and more ordinary steric effects.

In the gas phase, Hg atoms with electrons excited to the $6p$ state (the well-known resonance line) react with H_2 or with hydrocarbons to form HgH. Suppose the initial system is a cold HgH_2 van der Waals cluster. Upon electronic excitation, the otherwise degenerate p orbital of Hg^* split in the presence of the close H_2 molecule. This orbital can have its axis either parallel or perpendicular to the H–H bond; these two states will differ in energy and so can be separately accessed by the photon. Experimentally[308] the two states appear on opposite sides of the Hg resonance line and only the lower state leads to facile HgH formation. The higher-energy state typically dissociates to $Hg^* + H_2$.

As an example of orientational selection, we note photodissociation of the van der Waals cluster $HBrI_2$, in which we take the H atom to point away from the nearly collinear BrII system. Due to the relatively high zero point energy of the bending motion, there is some distribution of the orientation of the H–Br axis relative to I_2. Upon electronic excitation of HBr to a repulsive state, the H atom departs rapidly because it is much lighter than the other atoms and so, to satisfy conservation of momentum, its velocity

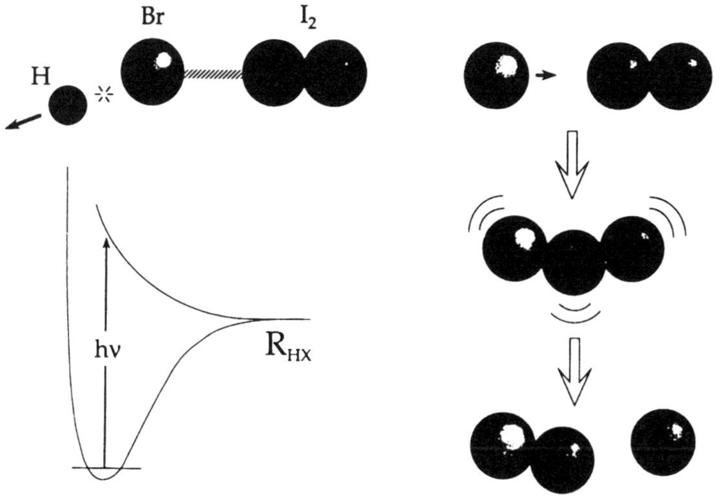

Fig. 19. The ingredients in a time-resolved study of the exchange reaction (II). Ultrafast photolysis of the HBr bond, lower left corner, in the $HBrI_2$ cluster, upper left, defines the zero of time. The Br atom moves slowly towards I_2 and the triatomic system vibrates for a while before exiting as products. Adapted from Ref. 311. See also Refs. 309 and 310.

is high). The Br atom moves slowly towards an I atom and is pulled closer by the van der Waals well. The slightly exoergic reaction

$$Br + I_2 \rightarrow BrI + I \qquad (II)$$

can then take place, presumably via the formation of the BrII intermediate, Fig. 19. This modest well, over which the system is expected to spend some time, is consistent with a rather slow rise in the appearance of the products, measured and computed (using classical trajectories) to be in the range of tens of picoseconds.

Gas phase studies of the reaction (II) have given clear but indirect evidence for the role of the BrII intermediate. By studying the reaction in a cluster, one could assign a "zero" of time to when the fast photolysis laser broke the HBr bond and thereby demonstrate the time delay before the appearance of products.[309,310] The delay is not long and one can wonder if the three-atom system had time to sample all of its available phase space, as is typically assumed in some of the most widely used theories of unimolecular dissociation. (These fall into the category called "transition state theories".)

Cluster experiments have indeed provided early examples of cases of incomplete energy exchange between the different modes of energy storage. The reason is that molecular stretch frequencies are quite different from the much lower frequencies of van der Waals modes. The best-studied example is that of clusters of I_2 with one or more He atoms.[312,313] It does not require a small chromophore. Quite selective chromophore final state distributions have been seen in the dissociation of van der Waals clusters of aromatic molecules with He and other rare gases and with small molecules. Indeed, a not unreasonable estimate of the lifetimes for such dissociations is obtained by restricting the distribution of the available energy to the van der Waals mode alone. Similar non-statistical effects have also been seen in hydrogen-bonded clusters such as dimers and higher monomers of HF, all the way to a distinctly more statistical dynamics whose onset is about at $m > 3$.

We begin the discussion of clustered chromophores with the simple I_2He case. Upon excitation to the lowest electronic excited states, the I_2 molecule can fluoresce with a lifetime of nanoseconds, typical of an allowed transition. The equilibrium distance of the excited state is sufficiently displaced by the excitation process, and the vibrational frequency of I_2 is sufficiently low that the Franck–Condon region corresponds to I_2^* in fairly high vibrational states. (The vibrational quantum numbers are typically of order 20.) Next consider the same experiment but with I_2He. Depending on the excitation wavelength, one observes fluorescence from the cluster or from the bare molecule or both. Here, the experiment provides its built in clock. Without any computations we can conclude that dissociation of the I_2^*He cluster occurs on the same time scale as the fluorescence, a very long time indeed. There is plenty of excess vibrational energy in I_2^*. In fact, one quantum of vibration of I_2 is sufficient to break the van der Waals bond. But the transfer of energy from the chemically bound I_2^* to the very weak (and hence very low frequency) van der Waals bond is very slow. This is only to be expected. For full collisions of atoms with diatomics, at thermal velocities, vibrational deactivation occurs in 1 in $10^4 - 10^6$ collisions; the weaker the bond, the higher the probability of energy transfer. (It was this observation plus the rather efficient energy transfer to rotation that led to the exponential gap rule).[314] The very same is true here. An I atom "collides" with He once per vibrational period, which is about 3×10^{-13} s. If the lifetime of the complex is due to the rate of energy transfer, it should be about 3×10^{-9} s, which is the right magnitude. This runs counter to the central assumption of statistical unimolecular rate theory (RRKM), a theory based on the model that energy equilibrates completely among all the degrees of freedom while the system is in the transition state, and only then does the system cross the barrier to dissociation. Here in I_2He, it is the energy transfer and not the transition state bottleneck to the continuum which is rate determining.

For polyatomic chromophores one can add another diagnostic. By dispersing the fluorescence one can determine the vibrational state of the electronically excited chromophore after the dissociation takes place. Typically, one observes that the vibrational state distribution is quite specific. Vibrational states that are quite allowed energetically are nevertheless not observed.

For larger clusters, selective energy deposition is often, but not always, followed by energy equipartitioning. Evaporation from such a warm cluster is interesting in its own right and also because of its connection with that nucleation which must precede the transition from gas to liquid. Ionization has frequently been used to deposit the initial energy in the cluster.[315] Why and how the excess energy is made available and is then distributed does depend on the bonding type. In the well studied case of hydrogen bonded clusters, the first step after ionization is an exoergic ion molecule reaction, e.g.

$$NH_3^+ + NH_3 \rightarrow NH_4^+ + NH_2\,, \qquad \Delta H^0 = -0.74\ \text{eV}\,. \qquad \text{(III)}$$

Due to this heating and the energy gained from the charge solvation, cluster molecules are lost by "evaporation";

$$(NH_3)_n H^+ \rightarrow (NH_3)_{n-1} H^+ + NH_3\,. \qquad \text{(IV)}$$

In principle, more than one molecule at the time can evaporate, but often this is not the case. As each molecule leaves, the cluster cools down so the next dissociation step is slower.

It is possible to arrange for a fast energy acquisition process in another way, by directing a cold cluster moving at a high speed towards a hard surface. When the front of the cluster reaches the surface, the leading atoms reverse the direction of their motion. But the atoms at the back are still moving in the original directions. This produces a shock wave within the cluster, which forces these opposite camps to collide; *very* shortly, the initially directed velocity of the cluster randomizes. Within the time of a few collisions (tens of fs's) the cluster can be made to reach quite high temperatures. A rough estimate of the temperature rise within the hot and compressed cluster is provided by the equivalence of temperature to the random part of the kinetic energy. If, upon impact, the entire initially directed velocity of the cluster rapidly randomizes, the cluster temperature will reach a value that is v_0^2 times room temperature, where v_0 is the initial velocity in units of the velocity of sound. A cluster striking a surface at 4 km/s, about Mach 10, can therefore be rapidly heated to well above 10^4°K. Molecular dynamics simulations for a cluster of interacting but otherwise structureless particles verify this temperature range is accessible even when energy loss to the surface is allowed. A warm cluster will cool by evaporation but here the energy available is far higher and it is acquired very rapidly. Theory suggested[124] that such a cluster will shatter. Experiments on $(NH_3)_nH^+$ clusters have verified that, unlike evaporation described by reaction (V), the cluster breaks up completely to monomers and some dimers. At higher collision energies only the monomers are seen.

The shattering of hot clusters[316] is interesting in that it shows that the initially directed energy is very rapidly randomized. In other words, that cluster impact produces a very hot cluster. How can the energy be equipartitioned so rapidly? Physical considerations supported by computations[317] suggested that the cluster is able to equilibrate the energy deposited because it is disordered. Additional evidence on this point comes from a simulation of an alkali halide cluster, e.g. $(NaCl)_{32}$, striking the surface of argon.[318] The shock induced by the collision of cluster with surface heats the cluster enough to melt it, but the system cools by extremely rapid evaporation of atoms of the argon substrate. In this case, the melting and subsequent cooling occur in the time of only a few vibrational periods of the cluster, which is fast enough to melt the crystalline cluster and then quench it to an amorphous form. Artificial simulations of rapid annealing-quenching of such clusters had shown that only by cooling at a rate greater than 10^{13} Ks^{-1} would it be possible to quench these systems to amorphous or glassy structures; at cooling rates any slower, they find their way to rocksalt crystalline structures.[263]

Simulations verify that in the absence of strict packing, there is a transient formation of a hot and compressed cluster. This suggests[319] that such a cluster provides an interesting medium within which unusual chemical reactions can be induced. Recent experiments (e.g. Ref. 320) are providing evidence that this is the case, thereby showing that even extreme conditions can be studied using clusters.

One final aspect of chemical reaction kinetics in which clusters have given new insights is the behavior of a reacting system as it passes through a transition state from reactants to products. The rearrangements or isomerizations of the atoms of a cluster of six argon atoms have been used to show that if the "reaction coordinate" or degree of freedom that carries the system across the energy saddle from initial to final state is suitably separated, then the motion in that degree of freedom becomes decoupled from that of all the other degrees of freedom, for transitions even up to rather high energies. That decoupled motion is regular, even if all the other degrees of freedom are coupled together and chaotic. In effect, this result shows that a reacting system approaching its transition state is effectively "committed" by the nature of its trajectory to react, even well before it reaches the mountain pass that separates reactants from products.[319,321,322]

We have shown how clusters provide an interesting and powerful medium for the study of how chemical reactions occur. We no longer view clusters as merely allowing us to link the study of collision dynamics in the gas phase and in solution. The study of reaction dynamics in clusters has opened up new possibilities and uncovered new ways of inducing chemical reactions.

References

1. R. Kubo, *J. Phys. Soc. Japan* **17**, 975 (1962).
2. L. Verlet, *Phys. Rev.* **159**, 98 (1967).
3. N. Metropolis, A. W. Metropolis, M. N. Rosenbluth, A. H. Teller and E. Teller, *J. Chem. Phys.* **21**, 1087 (1953).
4. B. J. Alder and T. E. Wainwright, *J. Chem. Phys.* **27**, 1208 (1957).
5. B. J. Alder and T. E. Wainwright, *J. Chem. Phys.* **31**, 459 (1959).
6. B. J. Alder and T. E. Wainwright, *J. Chem. Phys.* **33**, 1439 (1960).
7. T. L. Hill, *The Thermodynamics of Small Systems*, Part 1 (W. A. Benjamin, New York, 1963).
8. T. L. Hill, *The Thermodynamics of Small Systems*, Part 2 (W. A. Benjamin, New York, 1964).
9. J.-P. Hansen and L. Verlet, *Phys. Rev.* **184**, 151 (1969).
10. J.-P. Hansen, *Phys. Rev.* **A2**, 221 (1970).
11. M. R. Hoare and P. Pal, *Adv. Phys.* **20**, 161 (1971).
12. M. R. Hoare and P. Pal, *J. Cryst. Growth* **17**, 77 (1972).
13. M. R. Hoare and P. Pal, *Nature* **236**, 75 (1972).
14. M. R. Hoare and P. Pal, *Nature* **230**, 5 (1972).
15. M. R. Hoare and P. Pal, *Adv. Phys.* **24**, 645 (1975).
16. M. Hoare, in *The glass transition and the nature of the glassy state*, Vol. 279, edited by M. Goldstein and R. Simha (Annual New York Academic Science, New York, 1976) p. 186.
17. M. R. Hoare, *Adv. Chem. Phys.* **40**, 49 (1979).
18. M. Hoare and J. A. McInnes, *Adv. Phys.* **32**, 791 (1983).
19. D. J. McGinty, *J. Chem. Phys.* **58**, 4733 (1973)
20. J. K. Lee, J. A. Barker and F. F. Abraham, *J. Chem. Phys.* **583**, 166 (1973).
21. R. M. Cotterill, W. Damgaard Kristensen, J. W. Martin, L. B. Peterson and E. J. Jensen, *Comput. Phys. Commun.* **5**, 28 (1973).

22. R. M. J. Cotterill, W. D. Kristensen and E. J. Jensen, *Philos. Mag.* **30**, 245 (1974).
23. R. M. J. Cotterill, *Philos. Mag.* **32**, 1283 (1975).
24. W. Damgaard Kristensen, E. J. Jensen and R. M. J. Cotterill, *J. Chem Phys.* **60**, 4161 (1974).
25. A. Rahman, M. J. Mandell and J. P. McTague, *J. Chem. Phys.* **64**, 1564 (1976).
26. C. L. Briant and J. J. Burton, *J. Chem. Phys.* **63**, 2045 (1975).
27. R. D. Etters and J. B. Kaelberer, *Phys. Rev.* **A11**, 1068 (1975).
28. R. D. Etters and J. B. Kaelberer, *J. Chem. Phys.* **66**, 5112 (1977).
29. R. D. Etters, R. Danilowicz and J. Kaelberer, *J. Chem. Phys.* **67**, 4145 (1977).
30. R. D. Etters, R. Danilowicz and J. Dugan, *J. Chem. Phys.* **67**, 1570 (1977).
31. R. D. Etters and R. Danilowicz, *J. Chem. Phys.* **71**, 4767 (1979).
32. J. B. Kaelberer and R. D. Etters, *J. Chem. Phys.* **66**, 3233 (1977).
33. H. Schröder, *J. Chem. Phys.* **67**, 1953 (1977).
34. K. Hoshino and S. Shimamura, *Philos. Mag.* **A40**, 137 (1979).
35. W. Henkes, *Z. Naturforsch.* **16a**, 842 (1961).
36. W. Henkes, *Z. Naturforsch.* **17a**, 786 (1962).
37. G. C. Lindauer and A. W. Castleman, *Trans. Am. Nucl. Soc.* **12**, 897 (1969).
38. A. W. Castleman, Jr. and I. N. Tang, *Nature* **234**, 129 (1971).
39. A. W. Castleman, Jr., I. N. Tang and H. R. Munkelwitz, *Science* **173**, 1025 (1971).
40. A. W. Castleman, Jr. and I. N. Tang, *J. Chem. Phys.* **57**, 3629 (1972).
41. A. W. Castleman, Jr., M. S. Lian and I. N. Tang, *J. Chem. Phys.* **65**, 4022 (1976).
42. G. Staudenmaier, *Rad. Eff.* **18**, 181 (1973).
43. J. Farges, B. Raoult and G. Torchet, *J. Chem. Phys.* **59**, 3454 (1973).
44. J. Farges, *J. Crystal Growth* **31**, 79 (1975).
45. J. Farges, M. F. DeFeraudy, B. Roualt and G. Torchet, *J. Phys., Paris* **36**, 13 (1975).
46. J. Farges, M. F. DeFeraudy, B. Roualt and G. Torchet, *J. Phys., Paris* **38**, 47 (1977).
47. L. Genzel and T. P. Martin, *Phys. Stat. Sol. (b)* **51**, 101 (1972).
48. J. Gspann and H. Vollmer, in *Rarefied Gas Dynamics*, edited by K. Karamcheti (Academic, New York, 1974) p. 261.
49. O. F. Hagena, in *Molecular Beams and Low Density Gas Dynamics*, edited by P. P. Wegener (Marcel Dekker Inc., New York, 1974) pp. 95–181.
50. G. P. Können, A. Tip and A. E. deVries, *Rad. Eff.* **21**, 269 (1974).
51. A. van Deursen, A. van Leumig and J. Reuss, *Int. J. Mass Spectrom. Ion Phys.* **18**, 129 (1975).
52. A. Yokozeki and G. D. Stein, *J. Appl. Phys.* **49**, 2224 (1978).
53. D. O. Welch, O. W. Lazareth, G. J. Dienes and R. D. Hatcher, *J. Chem. Phys.* **68**, 2159 (1978).
54. C. E. Klots and R. N. Compton, *J. Chem. Phys.* **69**, 1636 (1978).

55. T. P. Martin, *J. Chem. Phys.* **67**, 5207 (1977).
56. T. P. Martin, *J. Chem. Phys.* **69**, 2036 (1978).
57. T. P. Martin, *J. Chem. Phys.* **72**, 3506 (1980).
58. B. G. de Boer and G. D. Stein, *Surf. Sci.* **106**, 84 (1981).
59. J. W. Lee and G. D. Stein, *J. Phys. Chem.* **91**, 2450 (1987).
60. P. Kebarle, *Ann. Rev. Phys. Chem.* **28**, 445 (1977).
61. A. W. Castleman, Jr., in Advances in Colloid and Interface Science, *Nucleation*, edited by A. Zettlemoyer (Elsevier, Oxford, 1979) pp. 73–128.
62. M. Faraday, *Philos. Trans. Roy. Soc.* (London), **147**, 145 (1857).
63. W. Werfelmeier, *Z. Phys.* **107**, 332 (1937).
64. A. Ferdinand and M. E. Fisher, *Phys. Rev.* **185**, 832 (1969).
65. M. E. Fisher, *Rep. Prog. Phys.* **30**, 615 (1967).
66. M. E. Fisher, *Physica* **3**, 255 (1967).
67. M. E. Fisher, in Proceedings of the International School of Physics, Enrico, *Ferm Course LI, Critical Phenomena*, edited by M. S. Green (Academic Press, New York, 1971).
68. S. Kumar, M. Roth, B. Kuchta and R. D. Etters, *J. Chem. Phys.* **97**, 3744 (1992).
69. L. S. Bartell, *Chem. Rev.* **86**, 492 (1986).
70. J. Farges, M. F. DeFeraudy, B. Roualt and G. Torchet, in *Evolution of Size Effects in Chemical Dynamics*, Vol. 2, edited by I. Prigogine and S. A. Rice (Wiley, New York, 1988) p. 45.
71. W. A. de Heer, *Rev. Mod. Phys.* **65**, 611 (1993).
72. O. Chesnovsky, S. H. Yang, C. L. Pettiette, M. J. Craycraft and R. E. Smalley, *Rev. Sci. Instr.* **58**, 2131 (1987).
73. G. Ganteför, M. Gausa, K. H. Meiwes-Broer and H. O. Lutz, *Z. Phys.* **D9**, 253 (1988).
74. G. Ganteför, M. Gausa, K.-H. Meiwes-Broer and H. O. Lutz, *J. Chem. Soc. Faraday Trans.* **86**, 2483 (1990).
75. K.-H. Meiwes-Broer, in *EPS-8 Trends in Physics*, edited by J. Kaczev (Prometheus, Prague, 1991) pp. 523–537.
76. F. Carnovale, J. B. Peel and R. G. Rothwell, *J. Chem. Phys.* **95**, 1473 (1991).
77. C.-Y. Cha, G. Ganteför and W. Eberhardt, *J. Chem. Phys.* **99**, 6308 (1993).
78. X. Zhang, J. M. Smith and J. L. Knee, *J. Chem. Phys.* **97**, 2843 (1992).
79. Y. A. Yang, L. A. Bloomfield, C. Jin, L. S. Wang and R. E. Smalley, *J. Chem. Phys.* **96**, 2453 (1992).
80. D. J. Nesbitt, *Chem. Rev.* **88**, 843 (1988).
81. U. Buck, in *Advances in Atomic, Molecular and Optical Physics*, Vol. 35, edited by B. Bederson and H. Walther (Academic Press, San Diego, 1995) pp. 121–161.
82. M. Knickelbein, *J. Chem. Phys.* **99**, 2377 (1993).
83. P. Ayotte and M. A. Johnson, *J. Chem. Phys.* **106**, 811 (1997).
84. D. L. Freeman and J. D. Doll, *Ann. Rev. Phys. Chem.* **47**, 43 (1996).
85. A. D. Buckingham, P. W. Fowler and J. M. Hutson, *Chem. Rev.* **88**, 963 (1988).

86. S. L. Price and A. J. Stone, *Mol. Phys.* **47**, 1457 (1982).
87. D. J. Wales, P. L. A. Popelier and A. J. Stone, *J. Chem. Phys.* **102** 5551 (1995).
88. T. A. Beu, U. Buck, J. G. Siebers and R. J. Wheatley, *J. Chem. Phys.* **106**, 6795 (1997).
89. E. E. B. Campbell and I. V. Hertel, *Carbon* **30**, 1157 (1992).
90. D. J. Wales, *Chem. Phys. Lett.* **141**, 478 (1987).
91. D. Tománek and M. A. Schluter, *Phys. Rev. Lett.* **67**, 2331 (1991).
92. F. Diederich and R. Whetten, *Angew. Chem. Int.* **30**, 678 (1991).
93. D. E. Manolopoulos and P. W. Fowler, *J. Chem. Phys.* **96**, 7603 (1992).
94. *Clusters and Fullerenes*, edited by V. Kumar, T. P. Martin and E. Tosatti (World Scientific, Singapore, 1993).
95. R. F. Curl and R. E. Smalley, *Sci. Am.* **265**, 54 (1991).
96. W. Krätschmer, L. D. Lamb, K. Fostiropoulos and D. R. Huffman, *Nature* **347**, 354 (1990).
97. H. W. Kroto, A. W. Allaf and S. P. Balm, *Chem. Rev.* **91**, 1213 (1991).
98. B. C. Guo, K. P. Kerns and A. W. Castleman, *Science* **255**, 1411 (1992).
99. C. Bréchignac, M. Broyer, P. Cahuzac, G. Delacretaz, P. Labastie, J. P. Wolf and L. Wöste, *Phys. Rev. Lett.* **60**, 275 (1988).
100. M. E. Garcia, G. M. Pastor and K. H. Bennemann, *Phys. Rev. Lett.* **67**, 1142 (1991).
101. B. Kaiser and K. Rademann, *Phys. Rev. Lett.* **69**, 3204 (1992).
102. R. Kawai and J. H. Weare, *Phys. Rev. Lett.* **65**, 80 (1990).
103. K. J. Taylor, C. L. Pettiette-Hall, O. Chesnovsky and R. E. Smalley, *J. Chem. Phys.* **96**, 3319 (1992).
104. G. M. Pastor and K. H. Bennemanj, in *Clusters of Atoms and Molecules*, Vol. 1, edited by H. Haberland (Springer-Verlag, Berlin, 1994) pp. 86–113.
105. P. Davidovits and D. L. McFadden, *Alkali Halide Vapors* (Academic Press, New York, 1979).
106. T. P. Martin, *Phys. Rep.* **95**, 167 (1983).
107. J. G. Hartley and M. Fink, *J. Chem. Phys.* **87**, 5477 (1987).
108. J. G. Hartley and M. Fink, *J. Chem. Phys.* **89**, 6053 (1988).
109. J. G. Hartley and M. Fink, *J. Chem. Phys.* **89**, 6058 (1988).
110. N. G. Phillips, C. W. S. Conover and L. A. Bloomfield, *J. Chem. Phys.* **94**, 4980 (1991).
111. J. Jortner, D. Scharf, N. Ben-Horin and U. Even, in *The Chemical Physics of Atomic and Molecular Clusters*, Vol. CVII, edited by G. Scoles (North-Holland, Amsterdam, 1990) pp. 43–98.
112. J. Jortner and N. Ben-Horin, *J. Chem. Phys.* **98**, 9346 (1993).
113. H. Müller, H.-G. Fritsche and L. Skala, in *Clusters of Atoms and Molecules*, Vol. 1, edited by H. Haberland (Springer-Verlag, Berlin, 1994) pp. 114–140.
114. W. D. Knight, W. A. de Heer, W. A. Saunders, K. Clemenger and M. L. Cohen, *Chem. Phys. Lett.* **134**, 1 (1987).
115. M. L. Cohen, M. Y. Chou, W. D. Knight and W. A. deHeer, *J. Phys. Chem.* **91**, 3141 (1987).

116. M. M. Kappes, M. Schlär, U. Röthlisberger, C. Yeretsian and E. Schumacher, *Chem. Phys. Lett.* **143**, 251 (1988).
117. D. M. Wood, *Phys. Rev. Lett.* **46**, 749 (1981).
118. C. Bréchignac, P. Cahuzac, F. Carlier and J. Leygnier, *Chem. Phys. Lett.* **164**, 433 (1989).
119. C. Brechignac, in *Clusters of Atoms and Molecules*, edited by H. Haberland (Springer-Verlag, Berlin, 1994) p. 255.
120. J. Kostrowicki, L. Piela, B. J. Cherayil and H. A. Scheraga, *J. Phys. Chem.* **95**, 4113 (1991).
121. C. J. Tsai and K. D. Jordan, *J. Phys. Chem.* **97**, 11227 (1993).
122. S. Schelstraete and H. Verschelde, *J. Chem. Phys.* **108**, 7152 (1998).
123. T. Raz, U. Even and R. D. Levine, *J. Chem. Phys.* **103**, 5394 (1995).
124. T. Raz and R. D. Levine, *J. Chem. Phys.* **105**, 8097 (1996).
125. J. E. Jones, *Proc. Roy. Soc. (London)* **106**, 441 (1924).
126. J. E. Jones, *Proc. Roy. Soc. (London)* **106**, 463 (1924).
127. J. E. Lennard–Jones, *Proc. Roy. Soc.* **A106**, 463 (1924).
128. E. S. Rittner, *J. Chem. Phys.* **19**, 1030 (1951).
129. M. P. Tosi and F. G. Fumi, *J. Phys. Chem. Solids* **25**, 45 (1964).
130. B. M. Axilrod and E. Teller, *J. Chem. Phys.* **11**, 299 (1943).
131. B. M. Axilrod, *J. Chem. Phys.* **17**, 1349 (1949).
132. B. M. Axilrod, *J. Chem. Phys.* **19**, 719 (1951).
133. B. M. Axilrod, *J. Chem. Phys.* **19**, 724 (1951).
134. D. J. Chartrand, R. J. LeRoy, A. Kumar and W. J. Meath, *J. Chem. Phys.* **98**, 5668 (1993).
135. J. P. K. Doye and D. J. Wales, *J. Chem. Soc. Faraday Trans.* **88**, 3295 (1992).
136. T. R. Horn, R. B. Gerber, J. J. Valentini and M. A. Ratner, *J. Chem. Phys.* **94**, 6728 (1991).
137. J. Uppenbrink and D. J. Wales, *J. Chem. Phys.* **96**, 8520 (1992).
138. D. J. Wales, *J. Chem. Soc. Faraday Trans.* **86**, 3505 (1990).
139. A. L. Mackay, *Acta Crystallogr.* **15**, 916 (1962).
140. P. A. Braier, R. S. Berry and D. J. Wales, *J. Chem. Phys.* **93**, 8745 (1990).
141. J. P. K. Doye and D. J. Wales, *J. Chem. Phys.* **105**, 8428 (1996).
142. P. M. Morse, *Phys. Rev.* **34**, 57 (1929).
143. K. P. Huber and G. Herzberg, in *Constants of Diatomic Molecules* (Van Nostrand Reinhold, New York, 1979).
144. D. J. Wales, *J. Chem. Soc. Faraday Trans.* **90**, 1061 (1994).
145. R. S. Berry, J. Jellinek and G. Natanson, *Chem. Phys. Lett.* **107**, 227 (1984).
146. G. Natanson, F. Amar and R. S. Berry, *J. Chem. Phys.* **78**, 399 (1983).
147. N. Quirke and P. Sheng, *Chem. Phys. Lett.* **110**, 63 (1984).
148. M. Bixon and J. Jortner, *J. Chem. Phys.* **91**, 1631 (1989).
149. B. Vekhter and R. S. Berry, *J. Chem. Phys.* **106**, 6456 (1997).
150. F. A. Lindemann, *Phys. Z.* **11**, 609 (1910).
151. V. V. Nauchitel and A. J. Pertsin, *Mol. Phys.* **40**, 1341 (1980).

152. H.-P. Cheng and R. S. Berry, in Symposium on Clusters and Cluster, *Assembled Materials*, Vol. 206, edited by R. Averbach (Materials Research Society, 1991) pp. 241–252.
153. H.-P. Cheng and R. S. Berry, *Phys. Rev.* **A45**, 7969 (1992).
154. S. Sawada and S. Sugano, *Z. Phys.* **D12**, 189 (1989).
155. L. C. Woods, *Thermodynamics of Fluid Systems* (Clarendon, Oxford, 1975).
156. M. E. Kellman and R. S. Berry, *Chem. Phys. Lett.* **42**, 327 (1976).
157. M. E. Kellman, F. Amar and R. S. Berry, *J. Chem. Phys.* **73**, 2387 (1980).
158. F. Amar and R. S. Berry, *J. Chem. Phys.* **85**, 5943 (1986).
159. G. S. Ezra and R. S. Berry, *J. Chem. Phys.* **76**, 3679 (1982).
160. R. S. Berry, T. L. Beck, H. L. Davis and J. Jellinek, in *Evolution of Size Effects in Chemical Dynamics*, Part 2, Vol. 70, edited by I. Prigogine and S. A. Rice (John Wiley and Sons, New York, 1988) pp. 75–138.
161. R. S. Berry, *Sci. Am.* **262**, 68 (1990).
162. R. S. Berry, in *Clusters of Atoms and Molecules*, Vol. 1, edited by H. Haberland (Springer-Verlag, Berlin, 1994) pp. 187–204.
163. R. S. Berry, in *Large Clusters of Atoms and Molecules*, Vol. 313, edited by T. P. Martin (Kluwer, Dordrecht, 1996) pp. 281–297.
164. L. S. Bartell and J. Chen, *J. Phys. Chem.* **96**, 8801 (1992).
165. R. S. Berry and H.-P. Cheng, in Physics and Chemistry of Finite Systems, *From Clusters to Crystals*, Vol. 1, edited by P. Jena, S. N. Khanna and B. K. Rau (Kluwer Academic Publishers, Dordrecht, 1992) pp. 277–286.
166. A. Boutin and A. H. Fuchs, *J. Chem. Phys.* **98**, 3290 (1993).
167. J. Broughton, *Phys. Rev. Lett.* **67**, 2990 (1991).
168. F. Calvo and P. Labastie, *Chem. Phys. Lett.* **258**, 233 (1996).
169. P. R. Couchman and W. A. Jesser, *Philos. Mag.* **35**, 787 (1977).
170. J. P. K. Doye and D. J. Wales, *Z. Phys.* **D40**, 466 (1997).
171. J. C. Earnshaw and C. J. Hughes, *Phys. Rev.* **A46**, R4494 (1992).
172. M. E. Fisher and A. J. Jin, *Phys. Rev. Lett.* **69**, 792 (1992).
173. Z. B. Güvenç and J. Jellinek, *Z. Phys.* **D26**, 304 (1993).
174. J. Jellinek and E. B. Krissinel, in Nanostructural Materials: Clusters, *Composites and Thin Films*, Vol. 679, edited by V. M. Shalaev and M. Moskovits (American Chemical Society, San Francisco, CA, 1997) pp. 239–254.
175. R. E. Kunz and R. S. Berry, *Phys. Rev. Lett.* **71**, 3987 (1993).
176. R. E. Kunz and R. S. Berry, *Phys. Rev.* **E49**, 1895 (1994).
177. T. P. Martin, U. Näher, H. Schaber and U. Zimmerman, *J. Chem. Phys.* **100**, 2322 (1994).
178. C. Rey, J. García-Rodeja, L. J. Gallego and M. J. Grimson, *Phys. Rev.* **E57**, 4420 (1998).
179. S. Saito and I. Ohmine, *Adv. Class. Trajectory Meth.* **4**, 105 (1999).
180. K. Strandburg, *Rev. Mod. Phys.* **60**, 161 (1988).
181. A. TenBosch and F. Celestini, *Z. Phys.* **D28**, 293 (1993).
182. M. Wautelet, *Sol. State Commun.* **74**, 1237 (1990).
183. M. Wautelet, *J. Phys.* **D24**, 343 (1991).
184. H.-P. Cheng, X. Li, R. L. Whetten and R. S. Berry, *Phys. Rev.* **A46**, 791 (1992).

185. R. S. Berry, *Microsc. Thermophys. Eng.* **1**, 1 (1997).
186. R. S. Berry, in *Theory of Atomic and Molecular Clusters*, edited by J. Jellinek (Springer-Verlag, Berlin, 1999) pp. 1–26.
187. A. Proykova and R. S. Berry, *Z. Phys.* **D40**, 215 (1997).
188. A. Proykova, R. Radev, F.-Y. Li and R. S. Berry, *J. Chem. Phys.* **110**, 3887 (1997).
189. R. A. Radev, A. Proykova and R. S. Berry, *Int. J. Chem.* **1**, Article 36, http://ijc.chem.niu.edu/articles/1998v1/36/ (1998).
190. A. Proykova, R. Radev, F.-Y. Li and R. S. Berry, *J. Chem. Phys.* **110**, 3887 (1999).
191. D. J. Wales and R. S. Berry, *Phys. Rev. Lett.* **73**, 2875 (1994).
192. M. A. McMahon, R. N. Barnett and K. B. Whaley, *J. Chem. Phys.* **99**, 8816 (1993).
193. R. N. Barnett and K. B. Whaley, *Phys. Rev.* **A47**, 4082 (1993).
194. R. N. Barnett and K. B. Whaley, *J. Chem. Phys.* **99**, 9730 (1993).
195. E. Cheng, M. A. McMahon and K. B. Whaley, *J. Chem. Phys.* **104**, 2669 (1996).
196. M. Hartmann, R. E. Miller, J. P. Toennies and A. Vilesov, *Phys. Rev. Lett.* **75**, 1566 (1995).
197. M. Hartmann, F. Mielke, J. P. Toennies, A. F. Vilesov and G. Benedek, *Phys. Rev. Lett.* **76**, 4560 (1996).
198. S. Grebenev, J. P. Toennies and A. F. Vilesov, *Science* **279**, 2083 (1998).
199. P. Sindzingre, D. M. Ceperley and M. L. Klein, *Phys. Rev. Lett.* **67**, 1871 (1991).
200. R. L. Whetten, G. S. Ezra and E. R. Grant, *Ann. Rev. Phys. Chem.* **36**, 277 (1985).
201. L. J. Butler, *Ann. Rev. Phys. Chem.* **49**, 125 (1998).
202. R. G. Parr, *Quantum Theory of Molecular Electronic Structure* (Benjamin, New York, 1963).
203. *Free Electron Theory of Conjugated Molecules*, edited by J. R. Platt (Wiley, New York, 1964).
204. R. Hoffmann, in Solids and Surfaces : A Chemist's View of Bonding, *Extended Structures* edited by (VCH, New York, 1988).
205. J. C. Phillips, *Bonds and Bands in Semi Conductors* (Academic Press, New York, 1973).
206. G. C. Schatz and M. A. Ratner, *Quantum Mechanics in Chemistry* (Prentice-Hall, New York, 1993)
207. P. Fulde, *Electron Correlations in Molecules and Solids*, 3rd edn., (Springer, 1999).
208. O. Becker and M. Karplus, *J. Chem. Phys.* **106**, 1495 (1997).
209. J. W. McIver, Jr. and A. Komornicki, *J. Am. Chem. Soc.* **94**, 2625 (1972).
210. R. Shepard, in *Modern Electronic Structure Theory*, edited by D. Yarkony (World Scientific, New York, 1995).
211. B. H. Lengsfield and D. R. Yarkony, *Adv. Chem. Phys.* **82**, 1 (1992)
212. M. C. Payne, M. P. Teter, D. C. Allen, T. A. Arias and J. D. Joannopoulos, *Rev. Mod. Phys.* **64**, 1045 (1992).

213. J.-Q. Sun and K. Ruedenberg, *J. Chem. Phys.* **98**, 9707 (1993).
214. W. Ekardt, *Phys. Rev.* **B31**, 6360 (1985).
215. W. Ekardt, *Phys. Rev.* **B29**, 1558 (1984).
216. M. Brack, *Rev. Mod. Phys.* **65**, 677 (1993).
217. A. P. Alivisatos, *Science* **271**, 933 (1996).
218. M. G. Bawendi, M. L. Steigerwald and L. E. Brus, *Ann. Rev. Phys. Chem.* **41**, 477 (1990).
219. C. P. Collier, T. Vossmeyer and J. R. Heath, *Ann. Rev. Phys. Chem.* **49**, 371 (1998).
220. W. D. Knight, K. Clemenger, W. A. deHeer, W. A. Saunders, M. Y. Chou and M. L. Cohen, *Phys. Rev. Lett.* **52**, 2141 (1984).
221. M. G. Mayer and J. H. D. Hansen, Elementary Theory of Nuclear Shell, *Structure*, (Wiley, New York, 1955).
222. W. Kohn, A. D. Becke and R. G. Parr, *J. Phys. Chem.* **100**, 12974 (1996).
223. R. O. Jones, in *Clusters of Atoms and Molecules*, edited by H. Haberland (Springer-Verlag, Berlin, 1994) p. 67.
224. R. G. Parr and W. Yang, *Density-Functional Theory of Atoms and Molecules* (Oxford University Press, New York, 1989).
225. W. Andreoni, *Ann. Rev. Phys. Chem.* **49**, 405 (1998).
226. M. R. Radeke and E. A. Carter, *Ann. Rev. Phys. Chem.* **48**, 243 (1997).
227. V. Bonacic-Koutecky, P. Fantucci and J. Koutecky, in *Clusters of Atoms and Molecules*, edited by H. Haberland (Springer-Verlag, Berlin, 1994) p. 15.
228. N. F. Mott, *Metal-Insulator Transitions* (Taylor & Francis, London, 1990).
229. F. Remacle and R. D. Levine, *Proc. Natl. Acad. Sci. USA* **97**, 553 (2000).
230. F. Remacle and R. D. Levine, *J. Phys. Chem.* **104**, 2341 (2000).
231. J. Linderberg and Y. Öhrn, *Propagators in Quantum Chemistry* (Academic Press, New York, 1973).
232. D. A. Jelski, T. F. George and J. M. Vienneau, in *Clusters of Atoms and Molecules*, edited by H. Haberland (Springer-Verlag, Berlin, 1994) p. 50.
233. G. M. Pastor and K. H. Bennemann, in *Clusters of Atoms and Molecules*, edited by H. Haberland (Springer-Verlag, Berlin, 1994) p. 86.
234. F. Remacle and R. D. Levine, *J. Am. Chem. Soc.* **122**, 4084–4091 (2000).
235. D. W. Heermann, *Computer Simulation Methods in Theoretical Physics*, Second ed. (Springer-Verlag, Berlin, 1990).
236. *Applications of the Monte Carlo Method in Statistical Physics*, edited by K. Binder (Springer-Verlag, Berlin, 1984).
237. R. Car and M. Parrinello, *Phys. Rev. Lett.* **55**, 2471 (1985).
238. T. L. Beck, J. D. Doll and D. L. Freeman, *J. Chem. Phys.* **90**, 5651 (1989).
239. D. M. Leitner, J. D. Doll and R. M. Whitnell, *J. Chem. Phys.* **94**, 6644 (1991).
240. S. W. Rick, D. M. Leitner, J. D. Doll, D. L. Freeman and D. D. Frantz, *J. Chem. Phys.* **95**, 6658 (1991).
241. R. S. Berry, *J. Phys. Chem.* **98**, 6910 (1994).
242. M. Amini and R. W. Hockney, *J. Non-Cryst. Sol.* **31**, 447 (1979).
243. M. Amini, D. Fincham and R. W. Hockney, *J. Phys.* **C12**, 4707 (1979).

244. F. H. Stillinger and T. A. Weber, *Phys. Rev.* **A28**, 2408 (1983).
245. T. A. Weber and F. H. Stillinger, *J. Chem. Phys.* **80**, 438 (1984).
246. E. Blaisten-Barojas, *Kinam* **6, Series A**, 71 (1984).
247. H. J. C. Berendsen, J. P. M. Postma, W. F. van Gunsteren, A. DiNola and J. R. Haak, *J. Chem. Phys.* **81**, 3684 (1984).
248. J. Jellinek, T. L. Beck and R. S. Berry, *J. Chem. Phys.* **84**, 2783 (1986).
249. H. L. Davis, J. Jellinek and R. S. Berry, *J. Chem. Phys.* **86**, 6456 (1987).
250. T. L. Beck, J. Jellinek and R. S. Berry, *J. Chem. Phys.* **87**, 545 (1987).
251. J. D. Honeycutt and H. C. Andersen, *J. Phys. Chem.* **91**, 4950 (1987).
252. H. L. Davis, T. L. Beck, P. A. Braier and R. S. Berry, in *The Time Domain in Surface and Structural Dynamics*, edited by G. J. Long and F. Grandjean (Kluwer Academic Publishers, 1988) pp. 535–549.
253. D. J. Wales, *J. Am. Chem. Soc.* **115**, 11180 (1993).
254. D. J. Wales and I. Ohmine, *J. Chem. Phys.* **98**, 7257 (1993).
255. M. Schütz, T. Bürgi, S. Leutwyler and H. B. Bürgi, *J. Chem. Phys.* **99**, 5228 (1993).
256. S. Saito and I. Ohmine, *J. Chem. Phys.* **101**, 6063 (1994).
257. M. Schütz, W. Klopper, H.-P. Lüthi and S. Leutwyler, *J. Chem. Phys.* **103**, 6114 (1995).
258. E. Fredj, R. B. Gerber and M. A. Ratner, *J. Chem. Phys.* **105**, 1121 (1996).
259. D. J. Wales and T. R. Walsh, *J. Chem. Phys.* **105**, 6957 (1996).
260. D. J. Wales and T. R. Walsh, *J. Chem. Phys.* **106**, 7193 (1997).
261. A. Baba, Y. Hirata, S. Saito, I. Ohmine and D. J. Wales, *J. Chem. Phys.* **106**, 3329 (1997).
262. L. X. Dang and T.-M. Chang, *J. Chem. Phys.* **106**, 8149 (1997).
263. J. P. Rose and R. S. Berry, *J. Chem. Phys.* **98**, 3262 (1993).
264. K. D. Ball, R. S. Berry, A. Proykova, R. E. Kunz and D. J. Wales, *Science* **271**, 963 (1996).
265. B. Vekhter, K. D. Ball, J. Rose and R. S. Berry, *J. Chem. Phys.* **106**, 4644 (1997).
266. R. S. Dumont and S. Jain, *J. Chem. Phys.* **103**, 6151 (1995).
267. J. E. Hunter III and W. P. Reinhardt, *J. Chem. Phys.* **103**, 8627 (1995).
268. J. M. Papanikolas, P. E. Maslen and R. Parson, *J. Chem. Phys.* **102**, 2452 (1995).
269. C. Seko and K. Takatsuka, *J. Chem. Phys.* **109**, 4768 (1998).
270. R. S. Berry and R. E. Breitengraser-Kunz, *Phys. Rev. Lett.* **74**, 3951 (1995).
271. R. E. Kunz and R. S. Berry, *J. Chem. Phys.* **103**, 1904 (1995).
272. J. Onuchic and P. G. Wolynes, *J. Phys. Chem.* **92**, 6495 (1988).
273. R. Zwanzig, *Proc. Natl. Acad. Sci. USA* **92**, 9801 (1995).
274. A. Fernández and G. Appignanesi, *Phys. Rev. Lett.* **78**, 2668 (1997).
275. K. M. Westerberg and C. A. Floudas, *J. Chem. Phys.* **110**, 9259 (1999).
276. Y.-J. Ye and H. A. Scheraga, in *Slow Dynamics in Complex Systems: Eighth Tohwa Univerity International Symposium*, edited by M. Tokuyama and I. Oppenheim (American Institute of Physics, Tohwa University, Japan, 1999) pp. 4452–475.
277. K. D. Ball and R. S. Berry, *J. Chem. Phys.* **109**, 8541 (1998).

278. K. D. Ball and R. S. Berry, *J. Chem. Phys.* **109** 8557 (1998).
279. K. D. Ball and R. S. Berry, *J. Chem. Phys.* **111**, 2060 (1999).
280. R. E. Kunz, P. Blaudeck, K. H. Hoffmann and R. S. Berry, *J. Chem. Phys.* **108**, 2576 (1998).
281. M. B. Knickelbein, *Ann. Rev. Phys. Chem.* **50**, 79 (1999).
282. G. A. Somorjai, *Introduction to Surface Chemistry and Catalysis* (John Wiley & Sons, New York, 1994).
283. J. Conceição, *et al.*, *Phys. Rev.* **B51**, 4668 (1995).
284. E. K. Parks, L. Zhu, J. Ho and S. J. Riley, *J. Chem. Phys.* **102**, 7377 (1995).
285. E. K. Parks and S. J. Riley, *Z. Phys.* **D33**, 59 (1995).
286. R. L. Whetten, M. R. Zakin, D. M. Cox, D. J. Trevor and A. Kaldor, *J. Chem. Phys.* **85**, 1697 (1986).
287. T. D. Klots, B. J. Winter, E. K. Parks and S. J. Riley, *J. Chem. Phys.* **95**, 8919 (1991).
288. M. R. Radeke and E. A. Carter, *Surf. Sci.* **L289**, 355 (1996).
289. M. F. Hineman, G. A. Brucker, D. F. Kelley and E. R. Bernstein, *J. Chem. Phys.* **97**, 3341 (1992).
290. E. R. Bernstein, *J. Phys. Chem.* **96**, 10105 (1992).
291. E. R. Bernstein, in *Chemical Reactions in Clusters*, edited by E. R. Bernstein (Oxford University Press, 1996) p. 147.
292. J. Castleman, A. W. and J. K. H. Bowen, *J. Phys. Chem.* **100**, 12911 (1996).
293. A. Heikal, S. H. Chong, J. S. Baskin and A. H. Zewail, *Chem. Phys. Lett.* **242**, 380 (1995).
294. J. J. Breen, L. W. Peng, D. M. Willberg, A. Heikal, P. Cong and A. H. Zewail, *J. Chem. Phys.* **92**, 805 (1990).
295. W. L. Hase, *Science* **266**, 998 (1994).
296. A. A. Viggiano, *et al*, *J. Am. Chem. Soc.* **113**, 9404 (1991).
297. A. A. Viggiano, S. T. Arnold and R. A. Morris, *Int. Rev. Phys. Chem.* **17**, 147 (1998).
298. C. S. Pomelli and J. Tomasi, *J. Phys. Chem.* **101**, 3561 (1997).
299. S. C. Tucker and D. G. Truhlar, *J. Am. Chem. Soc.* **112**, 3347 (1990).
300. X. G. Zhao, S. C. Tucker and D. G. Truhlar, *J. Am. Chem. Soc.* **113**, 826 (1991).
301. J. M. Papanikolas, V. Vorsa, M. E. Nadal, P. J. Campagnola, J. R. Gord and W. C. Lineberger, *J. Chem. Phys.* **97**, 7002 (1992).
302. J. M. Papanikolas, J. R. Gord, N. E. Levinger, D. Ray, V. Vorsa and W. C. Lineberger, *J. Phys. Chem.* **95**, 8028 (1991).
303. J. Faeder, N. Delaney, P. E. Maslen and R. Parson, *Chem. Phys.* **239**, 525 (1998).
304. J. K. Wang, Q. Liu and A. H. Zewail, *J. Phys. Chem.* **99**, 11309 (1995).
305. D. M. Willberg, M. Gutmann, J. J. Breen and A. H. Zewail, *J. Chem. Phys.* **96**, 198 (1992).
306. H. Yasumatsu, S. Koizumi, A. Terasaki and T. Kondow, *J. Chem. Phys.* **105**, 9509 (1996).

307. H. Yasumatsu, A. Terasaki and T. Kondow, *J. Chem. Phys.* **106**, 3806 (1997).
308. C. Jouvet, M. Boivineau, M. C. Duval and B. Soep, *J. Phys. Chem.* **91**, 5416 (1987)
309. S. A. Wright, M. F. Tuchler and M. J. D., *Chem. Phys. Lett.* **226**, 570 (1994).
310. I. R. Sims, M. Gruebele, E. D. Potter and A. H. Zewail, *J. Chem. Phys.* **97**, 4127 (1992).
311. C. Wittig and A. H. Zewail, in *Chemical Reactions in Clusters*, edited by E. R. Bernstein (Oxford University Press, 1996).
312. J. Jortner and R. D. Levine, *Adv. Chem. Phys.* **47**, 1 (1981).
313. J. A. Beswick and J. Jortner, *Adv. Chem. Phys.* **47**, 363 (1981).
314. R. D. Levine and R. B. Bernstein, *Molecular Reaction Dynamics and Chemical Reactivity* (Oxford University, New York, 1987).
315. S. Wei and J. A. W. Castleman, in *Chemical Reactions in Clusters*, edited by E. R. Bernstein (Oxford University Press, 1996) p. 197.
316. U. Even, T. Kondow, R. D. Levine and T. Raz, *Commun. Mod. Phys.* **D1**, 14 (1999).
317. T. Raz and R. D. Levine, *Chem. Phys.* **213**, 263 (1996).
318. H.-P. Cheng and U. Landman, *Science* **260**, 1304 (1993).
319. T. Raz and R. D. Levine, *J. Phys. Chem.* **99**, 7495 (1995).
320. W. Christen and U. Even, *J. Phys. Chem.* **102**, 9420 (1998).
321. T. Komatsuzaki and R. S. Berry, *J. Chem. Phys.* **110**, 9160 (1999).
322. T. Komatsuzaki and R. S. Berry, *Phys. Chem. Chem. Phys.* **1**, 1387 (1999).

CHAPTER 2

MOLECULAR AND IONIC CLUSTER SPECTROSCOPY

J. M. Lisy

Department of Chemistry, University of Illinois at Urbana-Champaign, Urbana, IL 61801, USA

E-mail: j-lisy@uiuc.edu

The spectroscopy of neutral and ionic clusters has been a challenging and innovative field for the past three decades. While the first twenty odd years primarily focused on the spectroscopy of binary neutral clusters, the last decade has seen expansion in two areas. Neutral cluster spectroscopy has progressed on to larger systems, both homo- and heterogeneous in composition. Ion cluster spectroscopy has been developed in two areas. High resolution rotationally resolved spectra have been obtained, in general, for binary species. Spectroscopy of larger species has considered basic issues in a variety of condensed phase contexts. The methods, techniques and limitations to these new approaches will be the target of this chapter.

Contents

1. Historical Perspective

While conventional spectroscopic methods had been used to examine weakly bound complexes with Hg_2,[1] Ar–HCl[2] and $(HF)_2$,[3] serving as notable examples, the introduction of molecular beams revolutionized the study of molecular clusters. The adiabatic expansion of a gas or gas mixture through a small orifice or nozzle, generates clusters of atoms and/or molecules at low internal temperatures ($\sim$ 1–20 K). It is also referred as a supersonic expansion, as the local Mach number is greater than 1.[4] Typically a few percent of the species of interest is mixed with a rare-gas carrier, HF in argon as an example, which will in turn generate clusters of HF, argon and mixed clusters of both species. Thus van der Waals and hydrogen-bonded complexes could be synthesized in molecular beams and at internal temperatures, which greatly collapsed the vibrational and rotational partition functions of the clusters. Spectroscopic methods of all types could in principle be applied to these systems heralding a new age in the study of clusters.

The first spectroscopic studies of van der Waals and hydrogen-bonded binary clusters were conducted in the microwave and r.f. regions of the electromagnetic spectrum detailing rotational and hyperfine transitions. At the time, microwave klystrons and r.f. oscillators were the only sources of tunable, coherent radiation. The two principal methods developed were molecular beam electric resonance (MBER) spectroscopy[5] and Fourier transform microwave (FTMW) spectroscopy.[6] MBER utilized the Stark effect to select (by electrostatic focusing) molecules and clusters in specific rotational quantum states in two separate regions of the apparatus. By driving a rotational/hyperfine transition of the cluster in an intervening section of the spectrometer, the quantum state and hence the focusing of the clusters would be altered and the transition was thus detected. By using a mass spectrometer as the detector, MBER spectroscopy was able to initially correlate specific transitions with particular mass peaks. Additionally, the mass spectrometer was quite useful in identifying expansion conditions that led to the optimal formation of clusters, although electron impact ionization of weakly-bound species typically led to extensive fragmentation.[7] As this method was used with continuous nozzle sources, extensive differential pumping of the apparatus was necessary, with large diffusion pumps for the source and skimmer regions, diffusion pump(s) for the resonance region and ion/cryo pumps for the mass spectrometer to eliminate collisions from reducing the cluster beam intensity. There were significant challenges in design and engineering (as well as costs) associated with the assembly of such apparatuses. Among the first structural determinations using this method were two of the most important binary complexes: $(HF)_2$[5] and Ar–HCl.[8]

In contrast to the continuous expansion used in the MBER experiments, FTMW spectroscopy used a pulsed nozzle and a well-timed pulse of microwave radiation to polarize the molecules/clusters within a Fabry–Perot cavity. After the microwave pulse dissipated, coherent emission at the molecular resonance frequency was then detected. The time-domain signal was collected, averaged and then Fourier transformed to give the frequency spectrum.[9,10] Use of a pulsed nozzle *free-jet expansion* eliminated the need for differential pumping and greatly simplified the vacuum design of the FTMW spectrometer to a single (large) chamber. By careful extension of the Fabry–Perot cavity, the spectrometer could be tuned through the microwave region (typically 4–20 GHz). While conditions for producing clusters were generally known, optimal expansion conditions required the detection and assignment of transitions to the cluster of interest. Of the two methods, FTMW spectroscopy is still in use among a number of laboratories around the world, due to its superior sensitivity.

The advent of lasers, in the visible and ultraviolet regions, opened the exciting aspects of vibrational and electronic spectra to the study of weakly bound systems. Laser-induced fluorescence spectroscopy of weakly bound clusters in the visible and UV regions was performed by exciting electronic transitions of binary or larger complexes in a free-jet expansion, using a high resolution tunable laser. The rovibronic fluorescence was collected at right angles to the jet and laser.[11] Structures of aromatic molecules complexed to rare gases or other molecules could be determined from analysis of the rotational structure.

2. Contemporary Methods

In many respects the development of tunable infrared and ultraviolet laser sources when combined with molecular beam expansions, marked the start of the modern or contemporary period of cluster studies. First, it offered the opportunity to selectively excite specific rovibrational or rovibronic levels in a complex. Second, variations in the spectra (linewidth, intensity and frequency) gave insight into dynamical behavior and the presence of nearby perturbing states.[12,13] Finally, the availability of widely tunable sources has enabled the experimentalist to select quantum states that would provide the maximum information content on a cluster system, an impetus that continues to drive the development of new lasers and laser systems. As this is an extremely wide field of research, primary emphasis in this chapter will be placed on vibrational spectroscopic studies of neutral and ionic clusters.

2.1. *Neutral Clusters*

Initially, the spectra of neutral clusters were obtained indirectly. Infrared lasers were used to excite vibrational motion of one molecule within the complex. Following the absorption of a photon, the cluster would acquire sufficient internal energy to become unstable over the lifetime of the experiment and would undergo vibrational predissociation. Two different methods of observing the dissociation process have been developed: mass spectrometric[14,15] and bolometric[16] detection. As was the case for MBER and FTMW methods, the mass spectrometer allowed for detection and optimization of expansion conditions to generate clusters, but the bolometer detector, at 4 K, had greater sensitivity. The vast majority of rotationally-resolved infrared spectra of binary clusters were first obtained using the bolometer beam method in conjunction with a high resolution color center laser.[12] The 2.5 to 4.0 μm region of the infrared was of particular interest since the effects of hydrogen bonding could be readily observed by frequency shifts of the O–H and H–F stretching modes of water and hydrogen fluoride. The magnitude of the shifts and the change in intensity upon hydrogen bond formation became new targets of theoretical calculations,[17] in addition to the determination of the geometrical structure. It should be noted that up to this point, all of the measurements of van der Waals and hydrogen-bonded clusters in this section were performed by taking action spectra, i.e. by observing a change in some detected property of the complex, and not by direct photoabsorption measurements.

The more obvious and direct approach would be to develop methods for direct absorption studies of clusters in molecular beams. This would eliminate the need for an action to take place, for example, should the cluster fail to dissociate on the time scale of the experiment. It would also allow for the measurement of the absorption coefficient, a useful quantity for comparison with theory. The challenge here was to overcome the inherent low density of clusters in order to obtain an absorption with reasonable signal to noise. As in the case of conventional absorption spectroscopy, multipassing the laser through the molecular beam was the obvious first step to take. However, two additional technical innovations were necessary before direction absorption studies could be accomplished: fast differential optical detection and slit nozzles. The combination of these three methods allowed for direct absorption studies in the mid infrared of van der Waals and hydrogen-bonded clusters in pulsed jets.[18] In a short period of time, it became clear that this approach was the most sensitive method yet for measuring weakly bound clusters. The observation of combination bands of intramolecular and van der Waals vibrational modes, and precise measurements of predissociation lifetimes (and their dependence on vibrational-rotational states!) quickly became the newest probes of intermolecular potentials and objectives of high level *ab initio* calculations.[19]

New laser methods allowed extensions into the far infrared where excitation of van der Waals modes in binary complexes could be directly measured

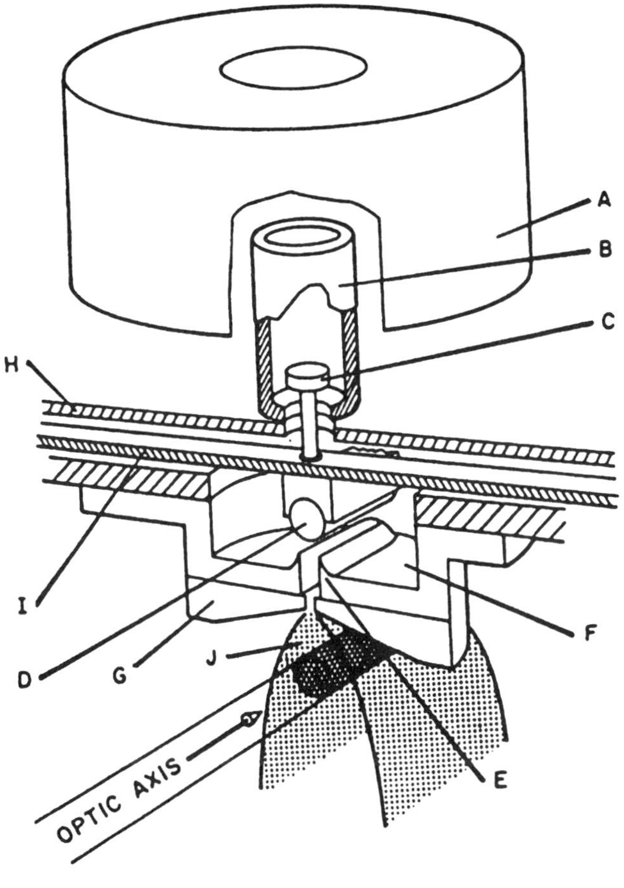

Fig. 1. Pulsed slit nozzle source developed by Nesbitt. A solenoid lifts the gasket seal (D) allowing the gas mixture to expand.[23]

and rotationally resolved.[20] These initial studies were almost immediately superseded by the incorporation of slit nozzles,[21] which dramatically improved the signal to noise. The existence of these modes results directly from the weak intermolecular forces. With resolution of better than 1 MHz, the vibration-rotation-tunneling (VRT) spectroscopy of the van der Waals bond provided an unprecedented probe of intermolecular forces.[22]

The slit nozzle, first introduced by Nesbitt[23] and shown below in Fig. 1, was of particular importance to the development of neutral cluster spectroscopy for three reasons.

First and perhaps foremost, the molecular beam density in a slit source decreases linearly with distance from the slit, as opposed to the quadratic decrease with distance of conventional pinhole/aperture nozzles. The expansion process is confined to only one direction, that perpendicular to the beam axis and the slit direction. Second, the velocity distribution of molecules parallel to the slit direction is narrowed relative to the direction perpendicular to the slit, thus reducing Doppler broadening along the slit axis. Third, the linear drop in density along the direction of expansion makes an ideal setting for multiple pass laser interrogation of the beam. This increase in the effective path length with differential detection methods allowed for direct absorption of the clusters in the beam with unprecedented sensitivity. It is dramatically demonstrated by the detection of extremely weak combination bands between van der Walls or hydrogen bond vibrational modes with intramolecular vibrations of molecules in the complex and in the measurement of trinary and larger clusters. Examples of these spectra and others from the various methods discussed in this section will be presented in the following two sections.

The introduction of the slit jet with direct absorption allowed researchers to go beyond the limitations of action spectra. Laser sources that excited neutral clusters without leading to vibrational predissociation could be utilized. Saykally, who had developed tunable far infrared laser sources, immediately recognized this.[24] These lasers: operated in the region of the infrared, corresponding to the fundamentals and overtones of van der Waals and hydrogen-bonded intermolecular degrees of freedom; were capable of extremely high resolution; and were sensitive to rotational and tunneling motion. Indeed, the Berkeley FIR spectrometer, shown in Fig. 2, was used to measure vibration-rotation-tunneling (VRT) spectra of a wide range of van der Waals and hydrogen-bonded species.[25]

The information content of these spectra has been used to significantly improve and refine intermolecular potentials in important benchmark

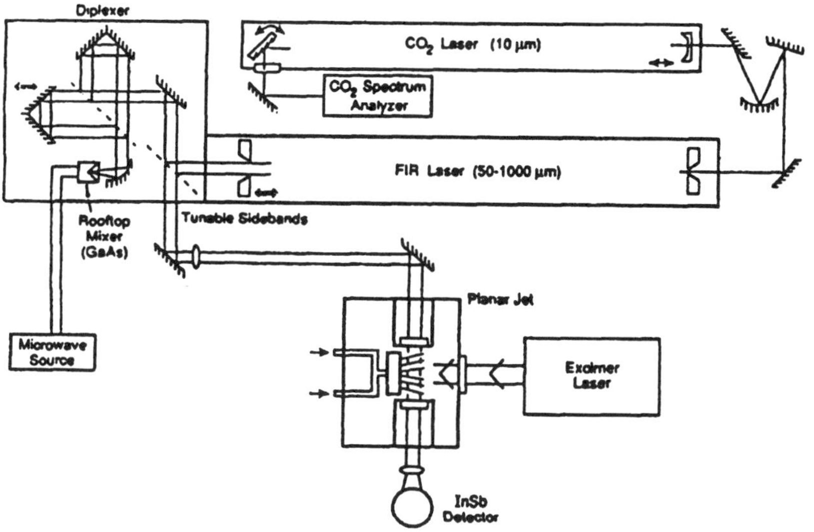

Fig. 2. Berkeley FIR Spectrometer.[25]

systems such as $(H_2O)_3$.[26] The sensitivity has permitted the size-dependent study of larger water clusters, culminating in the determination of a 3-D structure for $(H_2O)_6$.[27,28] These developments have been extremely stimulating for both experimental[29–37] and theoretical[38–45] investigations.

A second independent method of direct absorption spectroscopy has been recently applied to clusters: cavity ring down (CRD) spectroscopy.[46–48] This method, where a sample is introduced into the cavity of a high finesse Fabry–Perot interferometer, and is shown schematically in Fig. 3.

The small amount of radiation that leaks out of the cavity is measured with an appropriate detector. The decay lifetime is affected by the presence of an absorbing species within the cavity. The high finesse, resulting from extremely well fabricated high reflective coatings on the cavity mirrors, translates into an effective path length of tens of meters. If a molecular beam expansion is performed within the cavity, the direct absorption by molecular clusters can be observed. Considerable attention has been paid to the theory associated with the use of both pulsed[49] and cw[50,51] lasers. This method, while still in its infancy, promises to be another useful tool in the characterization of hydrogen-bonded neutral molecular clusters.

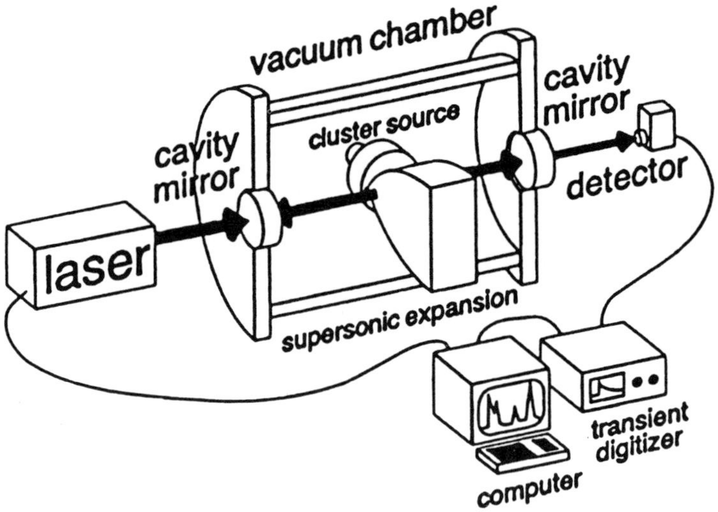

Fig. 3. IR-CRLAS apparatus. Tunable infrared laser radiation is coupled into the ring-down cavity. The light transmitting the output mirror is focused into an InSb detector. The resulting signal is amplified, digitized, and sent to a PC where it is fit to a first-order exponential, which is directly related to the total cavity loss per laser pass. The PC additionally controls the scanning of the dye laser. Base line losses are determined by scanning the laser with the expansion turned off and are then subtracted from the data, yielding the absolute sample absorbance, from which the carrier concentrations are extracted.[47]

While the use of direct absorption methods has grown, indirect action spectroscopic methods continue to be widely and successfully used in the study of neutral molecular clusters. As mentioned earlier, there are two commonly used detection methods, mass spectrometers and bolometers. Because of the variety of mass-spectroscopic methods, there is an equally wide range of techniques used in neutral cluster spectroscopy. One of the oldest among these involves electron-impact mass spectrometry of a cw neutral beam combined with vibrational predissociation spectroscopy using a tunable cw or pulsed laser. The advent of continuously tunable infrared sources (such as color center lasers and $LiNbO_3$ optical parametric oscillators) allowed for detailed studies of size and composition variation in neutral clusters.[52,53] However, fragmentation of the clusters within the ionizer of the mass spectrometer, severely limited the identification of particular clusters with specific masses. Isotopic methods[53,54] were able to mitigate some of the limitations, but only in a few cases.

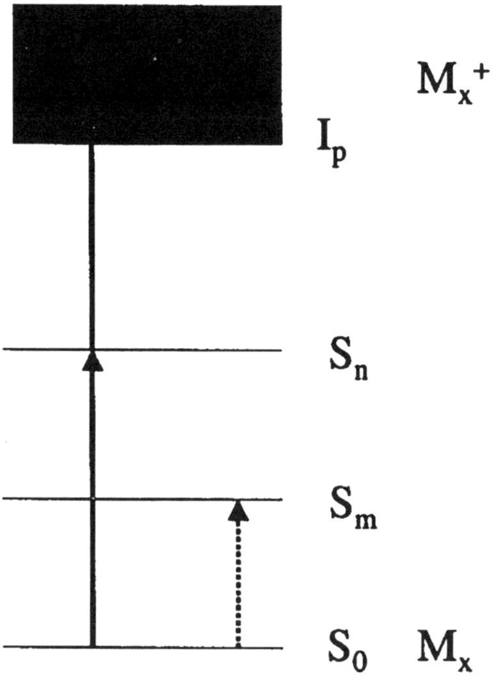

Fig. 4. Ion dip spectroscopy schematic. R2PI is used to ionize and mass-select the cluster, M_x of interest. A second laser pulse, prior or coincident to ionization, excites an intermediate level (S_m), which depletes the ground state and reduces the ion signal of M_x^+.

The use of laser ionization methods offered a more subtle and gentle way to ionize neutral clusters, which could then be mass-analyzed using time-of-flight (TOF) methods. Castleman first introduced these methods to clusters (named ion-dip spectroscopy) and applied to it electronic transitions.[55–57] The method is depicted in Fig. 4.

The mass-selected ion signal is depleted when a second laser pumps a competing transition from the ground state leading to electronic predissociation. As can be easily deduced from this figure, the spectroscopic possibilities were not limited to electronic states alone. As Zwier,[29,58] Mikami[59–61] and others quickly demonstrated, infrared vibrational predissociation processes will work just as well. Although this method does not eliminate all effects of fragmentation, it can be significantly reduced when resonant two photon ionization (R2PI) methods are used to minimize the amount of

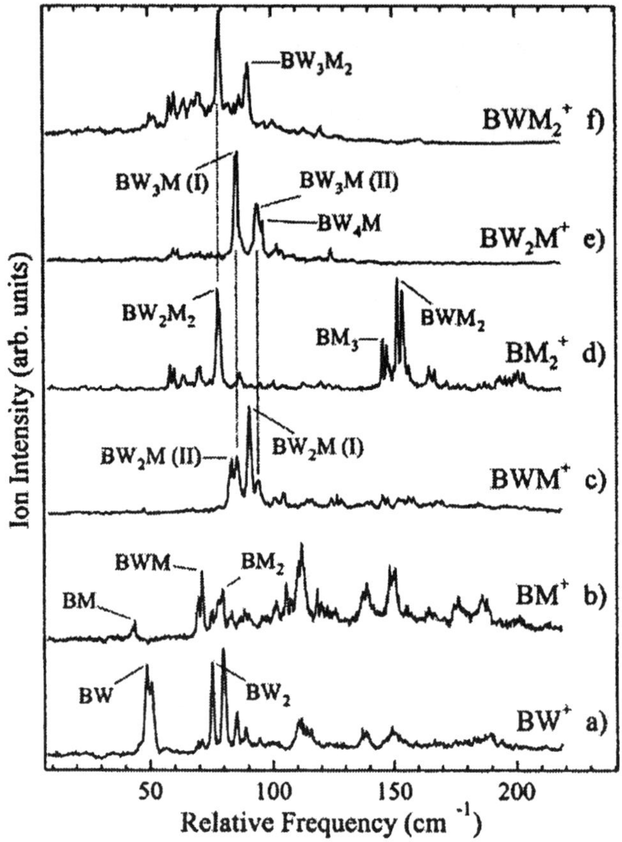

Fig. 5. One-color R2PI spectra recorded in the 610 nm region of the S_0 to S_1 transition of benzene, monitoring the $BW_nM_m{}^+$ mass channels with (a) and (b) $n+m=1$, (c) and (d) $n+m=2$, and (d)–(g) $n+m=3$. The zero of the frequency scale corresponds to the 610 nm transition of the benzene monomer (38609 cm^{-1}). Note that clusters undergo ionization-induced fragmentation with the loss of one water as the primary fragment channel.[216]

internal energy in the ionized species. Clusters containing aromatic species such as benzene or phenol are particularly popular for convenient, accessible and intense electronic transitions. Furthermore, these transitions show a spectral dependence on size and composition that aids in the mass-selection process, as shown in Fig. 5.

The first detection of a vibrational transition in a cluster was performed using a bolometer detector.[62] This device, which is maintained at ~ 4 K, is

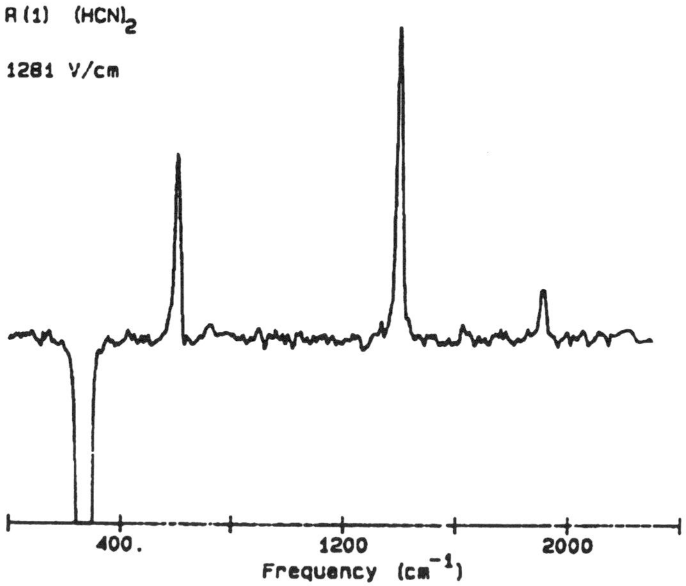

Fig. 6. This spectrum of $(HCN)_2$, was recorded at an electric field strength of 1281 V/cm. The positive (upward) signals are associated with the R(1) transition of the $\nu 1$ band, due to depletion from the molecular beam. The negative (downwards) signal is from the P(1) transition of HCN monomer, due to the increase in energy of the molecular beam striking the bolometer.[215]

sensitive to both translational (imparted when a molecule strikes and sticks to the detector) and internal energy from vibrational excitation. The absorption of an infrared photon by a neutral cluster could then be detected in two possible ways. If the cluster underwent vibrational predissociation, it would scatter out of the path leading to the bolometer resulting in a reduction of the signal recorded by the detector. If the cluster did not undergo predissociation, the internal energy, due to photon absorption, would be retained by the cluster and lead to an increase in signal recorded by the detector. This is depicted in Fig. 6.

Indeed both types of signals have been recorded, with predissociation the most common of the two. The newest innovation to this methodology incorporates the use of superfluid helium clusters (sizes of $\sim$ 1000 – 20000 atoms) as medium in which to grow neutral molecular clusters.[63–66] There is very little interaction between the helium atoms and the cluster of interest resulting in minimal vibrational frequency shifts. The helium clusters maintain an effective temperature of 0.37 K, leading to sharp vibrational spectra of large clusters. The absorption of an infrared photon by the cluster results in a rapid evaporation of helium atoms. This leads to a reduction in signal since the smaller cluster has less mass and thus less translational energy. There is tremendous potential for this approach to van der Waals, hydrogen-bonded and metal clusters.

2.2. *Ion Clusters*

The spectroscopic study of ion clusters has been a more recent development. Mass-spectrometric studies of cluster ions have been extensively conducted and have provided significant information on binding energies and indirect evidence on solvent shell size.[67,68] However, typical ion cluster beam intensities have been too weak for optical investigations using direct absorption methods. The detection of electronic and/or infrared transitions using indirect action methods was also complicated by the strong binding energies of molecular components to the ion core. These strong electrostatic interactions are typically an order of magnitude greater than those present in neutral clusters. At first glance these strong interactions thought to be a curse, but in reality were a blessing for action spectroscopy. Cluster ions could be formed with significant amounts of internal energy without undergoing dissociation due to the strong binding energy. Thus, it was possible to add only an additional increment of energy from the absorption of an infrared photon to induce dissociation on the time scale (typically μs) of the experiment.[69] Since charged species can easily be mass-selected by numerous methods, there has been a wide range of applications developed for cluster ion spectroscopy. The key element in most of these techniques is the use of tandem mass-spectrometric characterization, i.e. a specific cluster ion is chosen for investigation, and the results of the spectroscopic interrogation are evaluated by analyzing the mass fragments.

The use of time-of-flight (TOF) mass-spectrometric methods was ideally suited to the spectroscopic investigation of cluster ions. As demonstrated initially by Lineberger[70–72] and developed by Johnson,[73–75] the use of mass gates and reflection methods allowed for initial selection and analysis by mass of photodissociation products. A typical configuration is shown below in Fig. 7.

Because of the pulsed nature of the mass-selection process, pulsed laser sources are most efficiently used in these applications. Laser dissociation

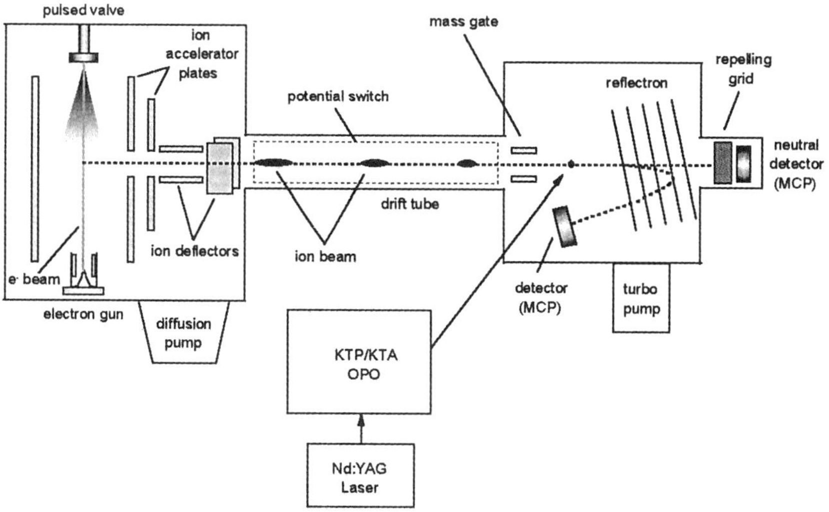

Fig. 7. TOF tandem mass spectrometer. The mass gate selects the ions of interest for photoexcitation by the tunable pulse infrared laser. Photofragment ions are mass-resolved by the reflectron mass spectrometer. (Jude Kelley and Mark A. Johnson, private communication.)

can be induced at two different locations within the apparatus, either after the mass gate or in the reflectron region.

Tandem mass spectrometry has traditionally been accomplished using a combination of magnetic and electric sector devices, but these are not so commonly used for spectroscopic studies. However, both Stace[76–78] and Bowers[79–81] have used these approaches for their electronic and vibrational investigations. The use of tandem quadrupole mass filters with an intermediate quadrupole or octopole ion guide can provide a combination of excellent mass range (up to 3000 amu) with relative compact size. Either in-line configurations[82] or bent (Z- and U-type with quadrupole bending fields) alignments[83–86] have been used as shown in Fig. 8.

Among the newest methods of generating mass-detected cluster ions for spectroscopy involves the use of ion-trap spectrometers such as r.f.[87] and ion cyclotron[88–90] devices. Mass-selected species are confined within a small area for spectral investigation and can be mass-analyzed using a variety of multiple r.f. pulse techniques. In addition, the cluster ions can be kept for long periods of time and thermally equilibrated by collisions with a rare gas such as helium. Alternatively, the cluster ions can be confined in the

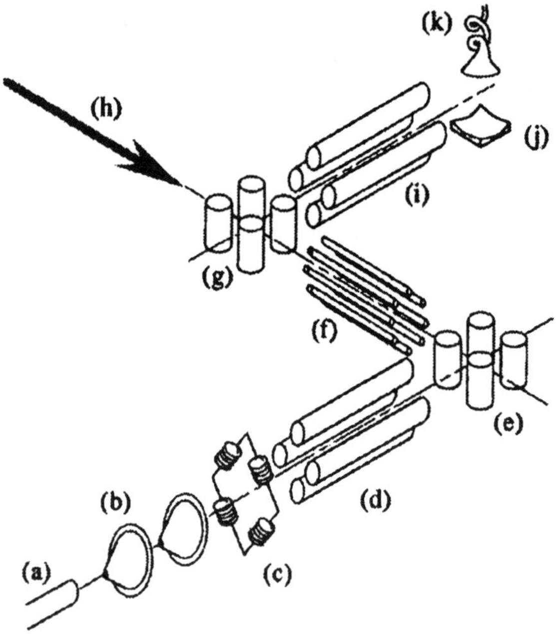

Fig. 8. Schematic drawing tandem mass spectrometer in "Z" configuration for coaxial infrared laser excitation. Aperture lenses have been omitted for clarity. Components: (a) Nozzle, (b) skimmer, (c) electron gun, (d) and (i) quadrupole mass filter, (e) and (g) quadrupole deflector, (f) octopole ion guide, (h) CO_2 laser, (j) ion conversion dynode, and (k) secondary electron multiplier.[86]

absence of collisions for periods of minutes and acquire internal energy via the absorption of blackbody radiation.[91–93]

Clusters ions can be prepared with a wide variety of sources; indeed this is one of the interesting and innovative developments in the expansion of the field. Electron-impact methods have been use to generate both positive and negative ions[94,95] in pulsed expansions. These methods have been among the most successful in generating "cold" cluster ions, marked by containing weakly bound rare gas atoms.[96] Such methods have also been used in continuous (cw) expansions suitable for use with continuous high-resolution tunable lasers.[97] Pulsed discharge sources, where the electric discharge is triggered by the passing pulse of gas, are also capable of producing cluster ions.[98] The continuous version of this source, the corona discharge, has been used for a number of years, not only to generate cluster ions,[99] but originally for a variety of ionic and radical species.[100] A rather different

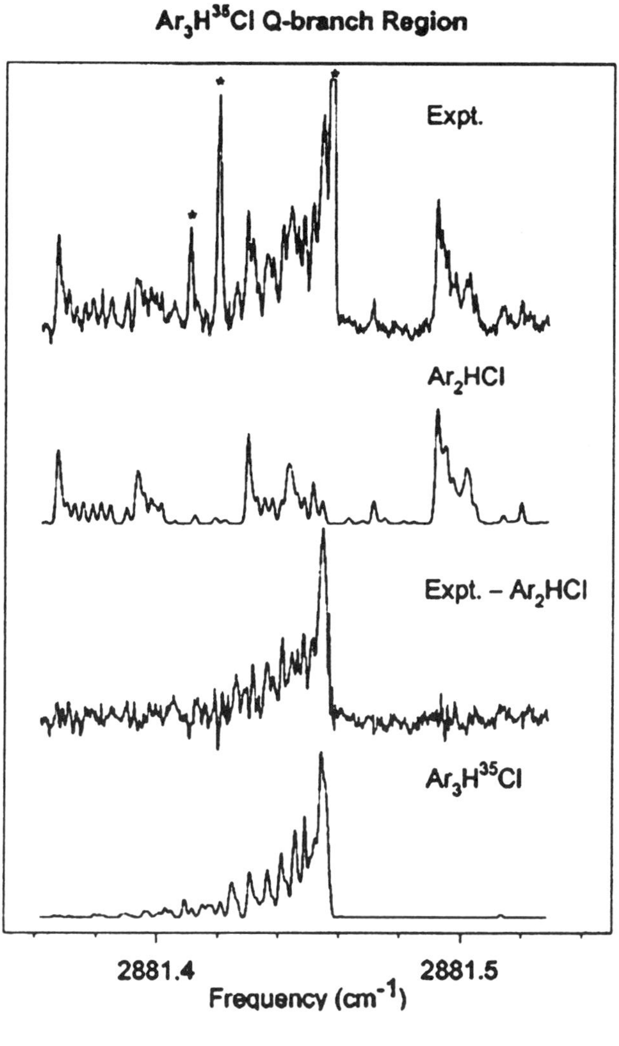

Fig. 9. High-resolution infrared absorption spectra of $Ar_{2,3}HCl$. The overlapping bands can easily be deconvoluted for accurate determination of both the trinary and tertiary complexes.[125]

ion source, which relies directly on ionic precursors as opposed to electrons, employs thermionic cations and anions. These are then directed into an expanding expansion containing neutral clusters.[69,101,102] The nascent cluster ions formed by the collisional process are stabilized by evaporative cooling.[103,104] This continuous source of cluster ions has been used with both cw[105–107] and pulsed lasers.[82,108,109]

3. Neutral Spectra and Characterization

There are numerous examples of spectroscopy of neutral clusters from which to draw examples. This chapter will confine itself to representative spectra related to the methods cited in the preceding section on contemporary techniques. The interested reader can find numerous additional examples in the literature, as well as examples of earlier methods in reviews and monographs from the last decade.[110–118]

While the spectroscopy of neutral complexes is predominantly focussed on binary species[119] the high-resolution spectroscopy of larger complexes has been a more recent development. It was first accomplished in the microwave for Ar_nHF,[120–123] and soon thereafter duplicated in the mid-infrared from rovibrational analysis of the band corresponding to the H–F[124] and H–Cl[125] fundamental stretches. As shown below in Fig. 9, the rotational features of the Ar_3HCl symmetric top are clearly observable, a testament to the high sensitivity and resolution of the beam jet absorption technique.

This experiment provides direct information on the role of many-body interactions in the stabilization of cluster structures, which are displayed in Fig. 10.

These data can be used in conjunction with high-level *ab initio* quantum mechanical calculations to identify the source and magnitude of contributions to two- and higher-body interaction.[126–128] An accurate reflection of these types of interactions in model potentials is essential for computationally efficient methods of determining complex interactions in larger cluster systems.

Further improvements have also enabled the careful measurement of combination bands involving van der Waals or hydrogen-bond bending and stretching modes, depending on the composition of the complex.[52,129–132] These measurements have been particularly useful in characterizing the nature of weak interactions, albeit in combination with intramolecular vibrational degrees of freedom and in excited vibrational states. In a number of cases, it has been possible to measure progressions in the weak modes, in a sense working up the potential well resulting from the intermolecular interaction.[133,134] As in the case of chemically bound species,

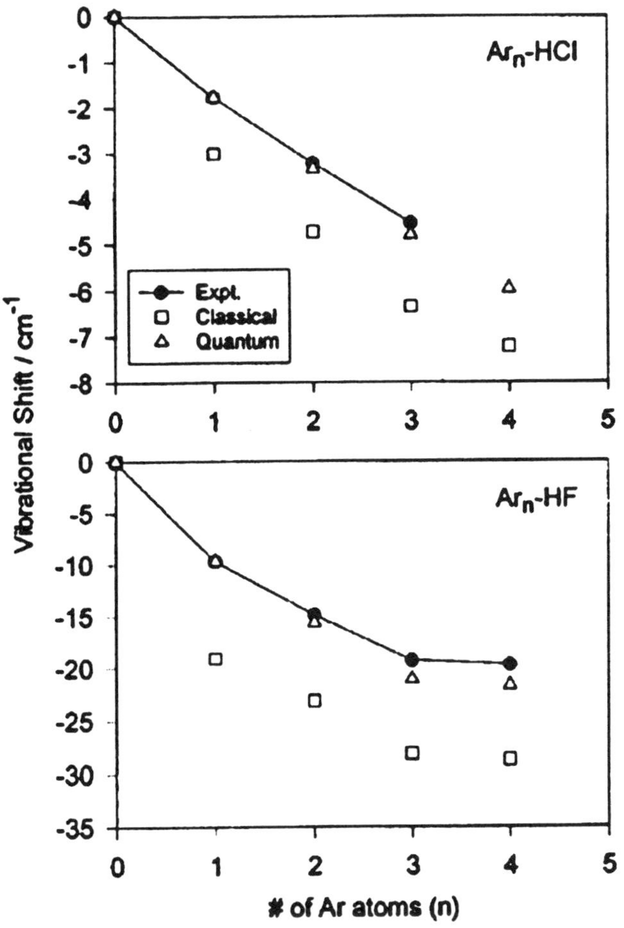

Fig. 10. Frequency shift versus the number of argon atoms in complexes with HCl and HF. The non-linear shifts can be correctly estimated by accurate *ab initio* calculations.[125]

these measurements are quite important and relevant to characterizing the strength of the weak interactions binding neutral clusters together.

In contrast to direct absorption studies of neutral molecular clusters in the mid infrared, the far infrared region offers a number of advantages. At these low energies, clusters do not in general undergo vibrational predissociation, one source of line broadening. The transitions also exhibit less Doppler broadening. The radiation sources, which involve mixing with microwave radiation, also have higher intrinsic resolution (less than 1 MHz) than most mid infrared sources. The additional improvements in linewidth and resolution have allowed for not only measurement of rotational and vibrational features, but also tunneling splittings. It is no surprise that this method has been named vibration-rotation-tunneling (VRT)[21,24] spectroscopy. These advantages have been fully utilized to study the most important of neutral clusters: water. The structures of four, $(H_2O)_{3-6}$, have been determined from VRT spectroscopy in the far IR region from direct absorption measurements by Saykally and coworkers.[27,135–138] A representative spectrum[37] of $(H_2O)_3$ is shown in Fig. 11.

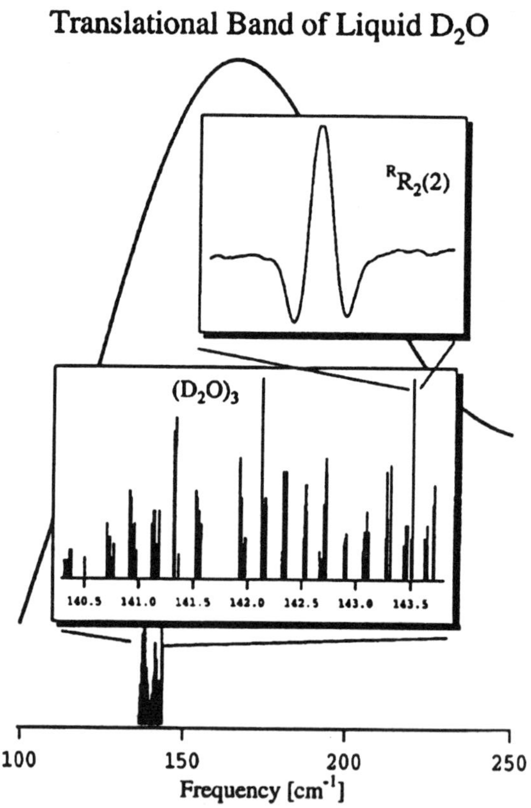

Fig. 11. An overview of the observed D_2O transitions with respect to the translational band of liquid D_2O shows that the 142.8 cm^{-1} band lies well within the translational band of the liquid. The first inset shows a stick spectrum representation of the 142.8 cm^{-1} $(D_2O)_3$ band and the second inset is a scan of the $^RR_2(2)$ transition, representative of the strongest observed rovibrational transitions.[37]

Cyclic ring forms of the trimer, tetramer and pentamer were found to be consistent with the observed VRT transitions. These structures can be viewed as nearly planar oblate rotors with complex tunneling patterns that arise from (a) the flipping of "free", non-hydrogen bonded protons above and below the plane and (b) the tunneling of protons between "free" and hydrogen-bonded positions. The hexamer was of prime interest as calculations predicted three-dimensional structures (prism and cage structures) to have greater stability than the cyclic ring form.[139–141] The cage structure was inferred by comparison of vibrationally-averaged rotational constants from experiment[27,142] and a diffusion quantum Monte Carlo simulation[28] based on a realistic model potential.[143] The three structures are shown in Fig. 12.

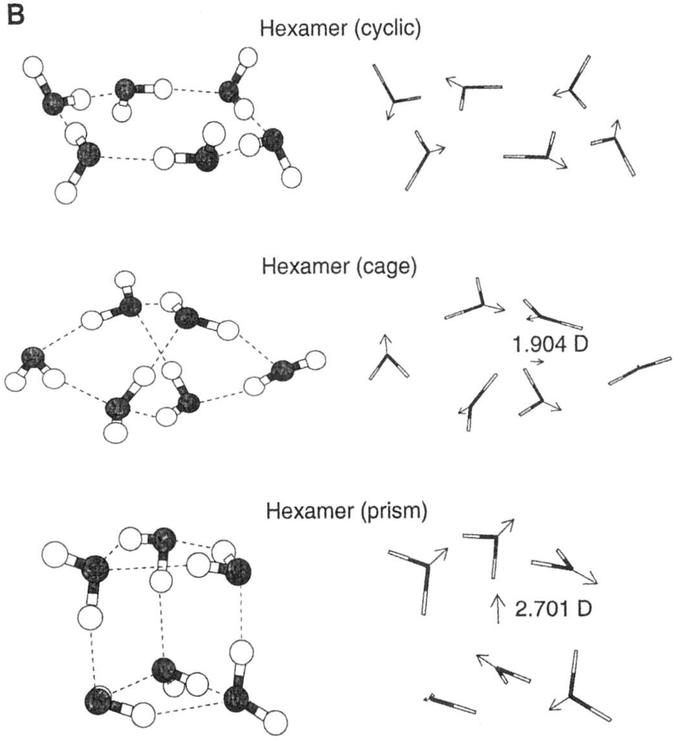

Fig. 12. Three structures of the water hexamer together with their dipole moments. The cyclic structure, like the tetramer, has no net dipole moment because it has S_6 symmetry.[217]

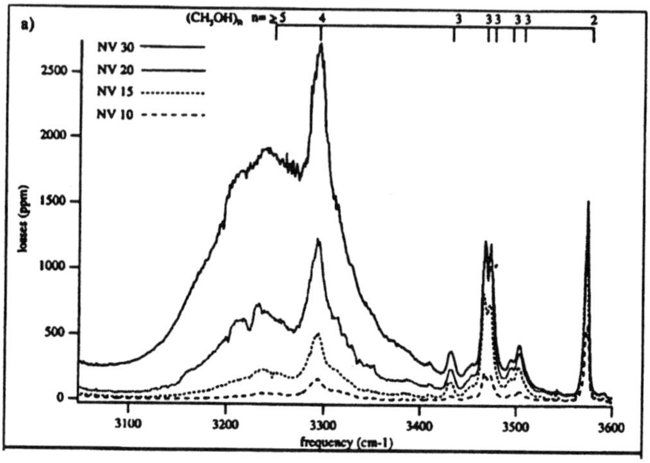

Fig. 13. Typical IR-CRLAS spectra of gas phase methanol clusters. A series of scans taken at various methanol concentrations. The methanol concentration is controlled by use of a needle valve. Larger needle valve (NV) settings correspond to higher methanol concentration.[146]

These experimental observations have been extremely useful in addressing the role of additive and non-additive interactions in the structural contraction of cyclic water clusters as a function of size.[40,44,144,145] There are many other facets of the experimental work that have generated considerable interest including analysis of the dipole moment as a function of size and how tunneling is affected by the form of the potential energy surface. This work has had and will continue to have a dramatic impact on the study of weak interactions and clusters.

Cavity ring down (CRD) spectroscopy, having proven to be a very sensitive method for detecting molecular species in a wide variety of environments, has also been applied to the mid infrared vibrational spectroscopy of hydrogen-bonded clusters of water[47,48] and alcohols.[146,147] As a direct absorption method, it can be used to quantitatively measure important molecular properties, such as absorption cross sections and coefficients. Knowing these properties, as a function of cluster size and structure, is useful in making the connection to the condensed phase. The sensitive detection of methanol clusters, as shown in Fig. 13, is of considerable importance. These particular measurements nicely complement the action spectra of methanol clusters, detected by depletion of mass-detected signal via vibrational predissociation.[148,149]

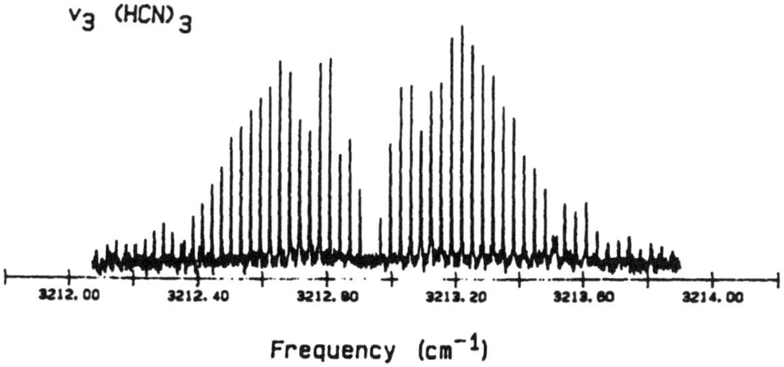

Fig. 14(a). One of the $\nu 3$ band from linear $(HCN)_3$. Predissociation is slow with respect to rotation of the cluster.[151]

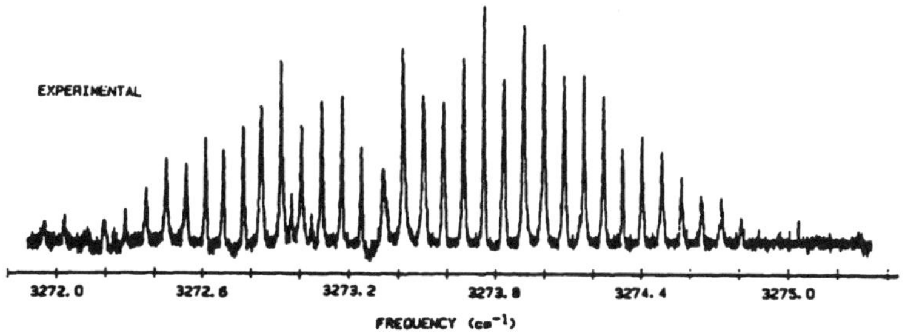

Fig. 14(b). The single infrared active band of cyclic $(HCN)_3$.[151]

It is important to note that while the requirement of predissociation or some other event limits the action spectra, the CRD spectra merely require the absorbance of a photon in order to alter the decay time of the cavity.

While it may appear that direct absorption spectra are superceding the older indirect approaches, such a conclusion would be unwarranted and incorrect. The ingenuity of many workers in the field continues to expand the scope and breadth of these techniques. The first high-resolution study of a molecular trimer was made by the FTMW spectroscopy of $(HCN)_3$.[150] However, only the linear structure was observed, since microwave spectroscopy demands a permanent dipole moment. A subsequent mid infrared study using a high-resolution color-center laser with bolometric detection, observed both linear and cyclic forms,[151] whose spectra are shown in Fig. 14.

The presence of structural isomers in such a small molecular system is of particular interest to both experimentalists and theoreticians, as it allows for testing of *ab initio* molecular structure calculations and model intermolecular potentials.[152–155] Comparisons between such calculations and experiment are shown in Table 1.

Table 1. Values for the frequency shift of the C–H stretch in $(HCN)_n$ clusters.

Cluster	Exp. (cm^{-1})	Computed (cm^{-1})[c]
Dimer	-4^a	-6
Dimer	-70^a	-87
Trimer (linear)	-5^b	-7
Trimer (linear)		-106^d
Trimer (linear)	-99^b	-118
Trimer (cyclic)	-38^b	-36^e

[a] Optothermal measurement.[215]
[b] Optothermal measurement.[151]
[c] SCF-MP2 calculation.[154]
[d] Computed intensity is ~ 0.[154]
[e] Only IR active mode.[154]

Continuing with HCN clusters as a structural motif, it was concluded that the larger clusters beyond the trimer were cyclic and due to rapid predissociation exhibited broad unresolvable spectra.[151] However, using superfluid helium clusters as a growth medium for HCN clusters, linear structures have been observed for $(HCN)_{4-7}$.[65] While lacking in rotational detail, it is possible by merely counting to be able to assign the spectra to specific cluster sizes, as shown below in Fig. 15.

Such particularly simple spectra are a result of the effective 0.37 K temperature maintained by the superfluid droplets. A particularly notable aspect of this work is the ability to synthesize species that are not in their most energetically stable form. Indeed, this approach has proven to be useful in generating the cyclic form of water hexamer as well.[66]

Indirect methods can also use mass spectrometers as detectors. As noted above fragmentation is a potentially severe problem that must be taken into consideration. Both continuous and pulsed beam techniques can be used. A particularly novel approach has been developed using multiphoton ionization via a favorable UV-absorbing chromophore.[156] Neutral clusters are formed containing the readily ionizable molecule, but the weak interactions between the chromophore and the rest of cluster reduces the extent

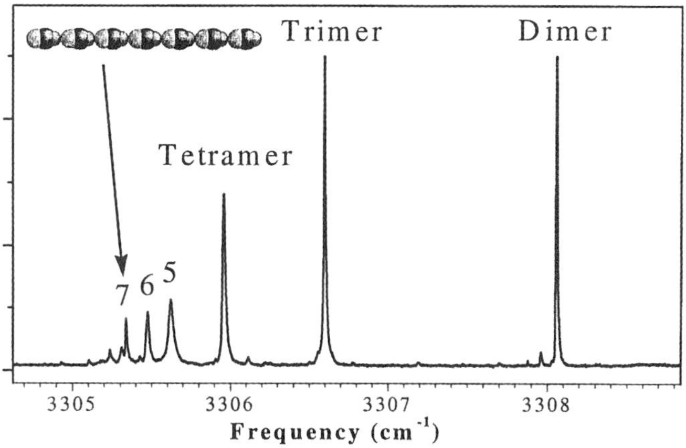

Fig. 15. Spectrum of the free C–H stretching region of the HCN polymer chains, showing clusters up to at least the heptamer. A linear heptamer chain is shown as an inset. This spectrum was obtained in the presence of a large electric field used to orient the polar chains within the helium droplet, thus collapsing the rotational band contours into a single peak for each cluster size.[65]

of fragmentation. In this manner, strongly bound neutral clusters, such as water clusters, can be observed, by ionizing the weakly-attached benzene molecule,[157–159] an approach that can be extended to other aromatic molecules such as phenol[61,160,161] and indole.[162,163] Using ion-dip methods, changes in the mass spectra are observed by exciting vibrational transitions, depleting the ground state of ionizable species. Vibrational spectra and structures of the water hexamer,[29,58] octamer[31,164] and nonomer[165] attached to a benzene molecule have been obtained using this approach. The experimentally observed and computed vibrational spectra for the former two are shown in Fig. 16.

As larger and larger neutral clusters and their associated spectra are pursued, one quickly realizes that even high-resolution spectra are incapable of allowing a completely unambiguous structural determination. Molecular structures are traditionally based on isotopic substitution in order to assign the positions of particular atoms in the molecule. This is necessary since the only experimental data are three rotational constants per isotopomer. However, this becomes an insurmountable task for almost any complex with two or more molecules, even for the smallest cyclic species of $(HF)_3$ and $(H_2O)_3$. As a result, it has been common to make a number of simplifying approximations — the most common — no structural change in the molecular

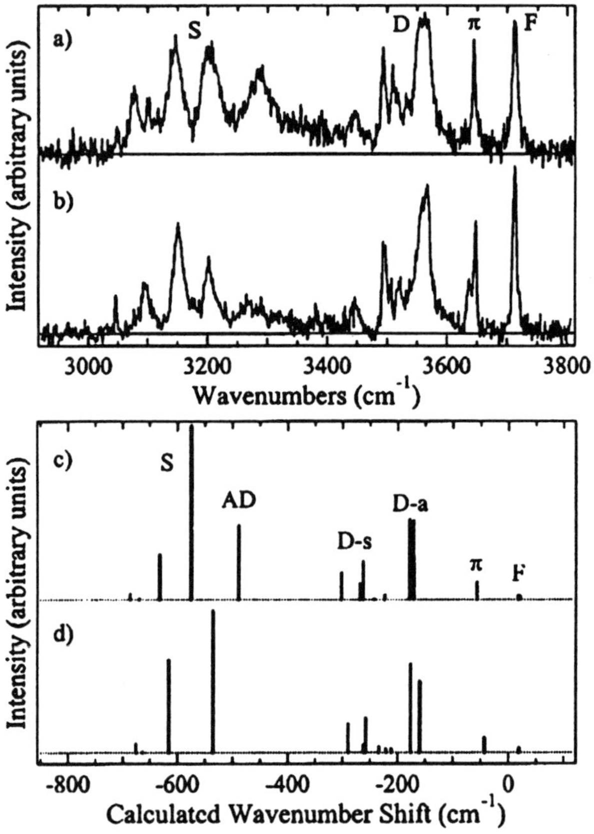

Fig. 16. Comparison of the RIDIR spectra of (a) BW_9 with (b) that for $BW_8(D_{2d})$. (c) and (d) Stick diagram of the calculated OH stretch vibrational frequencies and intensities for (c) one of the isomers of BW_9, and (d) the corresponding $BW_8(D_{2d})$ isomer.[165]

substituents upon complexation. For a binary complex, this is commonly referred to as the pseudo-diatomic approximation. The distance and relative orientation of the two molecules are adjusted to yield rotational constants that are consistent with experimental values and reasonable in regards to chemical intuition.[166,167] As the clusters gain in size, it becomes somewhat problematic whether such structures are accurate representations of reality. Furthermore, such an approach forces one to ignore the role of cooperative intermolecular interactions in modifying the structure of the cluster, modifications that must occur in order to explain sometimes significant changes in vibrational frequencies.[168,169]

An alternative approach is to use high-level *ab initio* calculations to guide the experimental interpretations.[170] This approach allows one to determine not only the relative orientation of the molecular subunits, but changes to the molecules themselves. One must keep in mind that the experiments represent vibrationally averaged structures, while the calculations typically refer to equilibrium structures. For large molecular clusters, such as $C_6H_6(H_2O)_6$, extended basis set + configuration interaction *ab initio* methods are still too computationally intensive. Although reasonable structures have been obtained using DFT methods,[165,171,172] vibrational frequencies must be scaled for agreement with experimental data and questions of vibrational-averaging are unanswered.

An alternative method was developed for the water hexamer, as calculations predicted three-dimensional structures (prism and cage structures) to have greater stability than the cyclic two dimensional forms.[139,140] The cage structure was inferred by comparison of vibrationally averaged rotational constants from experiment and a diffusion quantum Monte Carlo simulation[27] based on a realistic model potential.[143] As can be seen in Table 2, the results of the simulation allow for the selection of one of the many possible structures of the hexamer. At lower resolution and through the use of vibrational frequencies alone, two cage structures for the water octamer subunit of $C_6H_6(H_2O)_8$ have been inferred on the basis of experimental spectra and DFT calculations.[31,164] This is perhaps the largest neutral van der Waals or hydrogen-bonded complex with a reasonably sound structure. The recent analysis of $C_6H_6(H_2O)_9$ has identified three spectroscopically distinct isomers, with supporting DFT calculations aiding in the structural assignment of one isomer.[165]

While one may insist that the use of theory to buttress interpretations of experimental data is not ideal, it is clear that structures of large clusters are not and will not be determined by experiment alone, at least at this point in time. In point of fact, the close cooperation of experimentalists and theoreticians in addressing questions of structure (and dynamics) for clusters is one of the more vibrant synergies that exist in science today. The independence of calculations and experiments, yielding mutually consistent results when performed in an unbiased setting, give additional support to each other. Such close cooperation in this and other areas will spread as the size and complexities of systems grow beyond the ability of experiments and theory to give the highly precise or detailed answers.

Table 2. Rotational Constants (MHz) for Water Hexamer. Theory and Experiment.[27]

Form	A	B	C
Cyclic	1211	1211	598
Book	1798	1078	802
Boat	1733	1225	1054
Prism	1607	1355	1256
Cage	2136	1096	1043
Experiment	2163.61(90)	1131.22(62)	1068.80(62)

4. Ion Spectra and Characterization

High-resolution spectroscopy of cluster ions has largely been limited to binary complexes between a molecular ion and a rare-gas atom or a hydrogen molecule. Much of this work had been done by the group of Maier and Dopfer[173–177] in Switzerland. The spectrum of $Ar–H_2O^+$[178] given below in Fig. 17 is an excellent recent example of the rotationally resolved spectra that are possible using a tandem quadrupole mass spectrometer. Except for, $Rg_2–CH_3^+$ where Rg = Ar and Ne,[174,179] larger clusters have yielded broad vibrational bands without resolvable rotational structure. It is not yet clear why these systems exhibit this particular type of behavior. However, the field of cluster ion spectroscopy is still quite young with respect to its

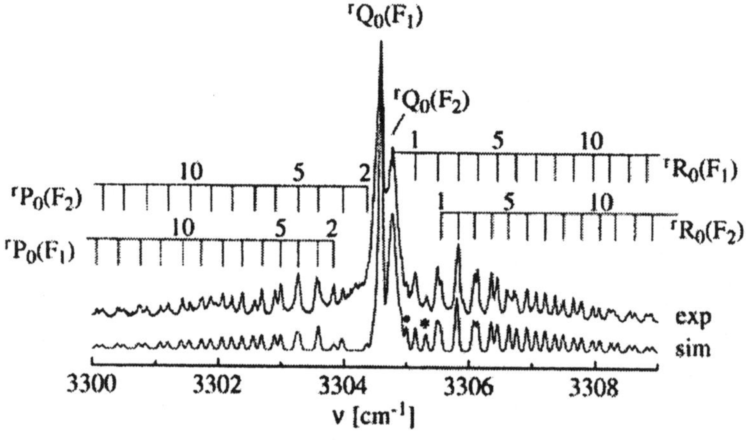

Fig. 17. The K_a (0 to 1) subband of the ν_3 transition of HOH^+–Ar along with assignments and a simulation based on constants from the structural analysis.[178]

neutral counterpart, so future developments are still quite possible. High-resolution spectroscopy of larger molecules bound to atomic ions, such as $Ca^+(C_2H_2)$ have been performed in the near UV,[180,181] and for $Br^-(C_2H_2)$ in the infrared.[182] Indeed, the ability to control and manipulate these systems is still in its infancy.

In contrast to neutral clusters, cluster ions are bound by stronger intermolecular interactions: ion-dipole, ion-quadrupole and ion-induced dipole forces. As a result, considerably more internal energy can be stored in cluster ions without dissociation. When coupled with the charged nature of the species, which permits trapping in ion guides or ICR spectrometers, cluster ions can be confined for periods of even minutes at pressures of 10^{-10} torr. This can cause some interesting and perhaps unexpected "side-effects", such as blackbody radiative heating.[93] As the cluster ion is heated, it undergoes unimolecular dissociation and evaporates an atom or molecule to "cool". The process can be repeated for many cycles as demonstrated

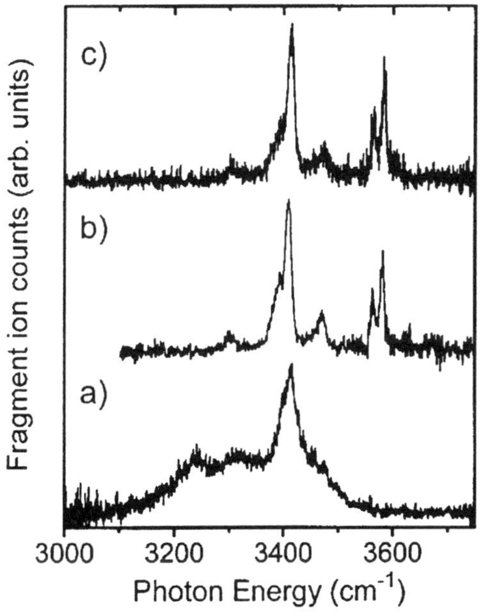

Fig. 18. Vibrational spectra of the $Br^-(H_2O)_3\ Ar_{0,1,3}$ clusters: (a) without Ar, predissociation by loss of one water, (b) with one argon, predissociation by loss of argon, (c) with three argons, predissociation by loss of argon. The cooling of the cluster ions by the presence of argon is quite apparent, but does not depend on the number of argons.[191]

for $I^-(H_2O)_n$ which can lose multiple water molecules as the cluster ion is held in an ICR spectrometer.[183,184] At the other temporal extreme, fsec pump-probe studies of solvated iodide have also been performed.[185] So in principle, one can conduct time-resolved studies of the same cluster ions over a 15 order of magnitude range in time!

The effect of internal energy or "temperature" on cluster ions can be clearly seen in the work on hydrated halide anions. Spectra have been obtained where one or more rare gas atoms have been attached to the cluster ion,[73,186] and can be compared to spectra of the "bare" cluster ion.[187] As can be seen in Fig. 18, the impact of even one argon atom is quite obvious.

The spectrum of the bare cluster ion is much broader and appears to have a substantially different structure than the cluster ions with argon. This has been interpreted as due to additional "cooling" provided by the argon. In the evaporative ensemble treatment,[188–190] the internal energy content of the cluster is determined by the most labile component of the cluster. Since the energy required for evaporating an argon atom is less than that for a water molecule, cluster ions containing argon must have

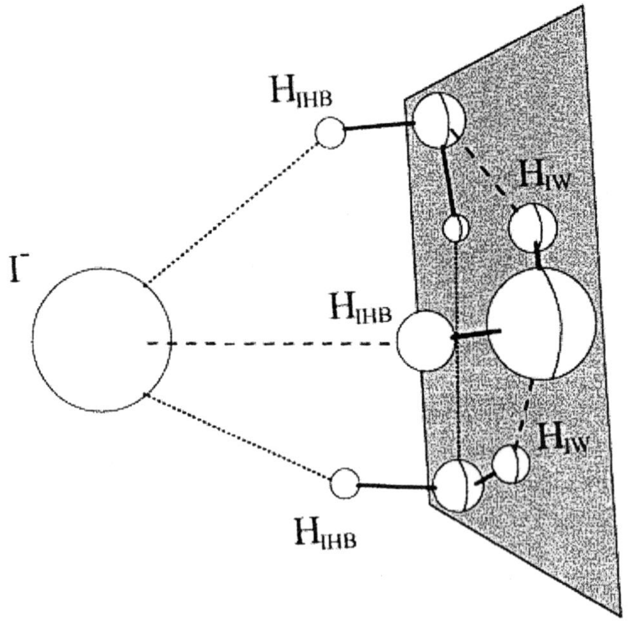

Fig. 19. The *ab initio* structure of $I^-(H_2O)_3$, using 6-31G$^+(2d, p)$ for water and SPV$^+$ for iodide.[218]

lower internal energy or temperature than those without argon. It further appears that in the case of $X^-(H_2O)_3$, X = Cl, Br and I, the presence of argon allows the cluster ion to attain its minimum geometrical configuration where a cyclic water trimer binds on one side of the halide anion.[191] This configuration is shown in Fig. 19.

As the cluster ion gains internal energy, the water–water hydrogen-bonds are broken, while the cluster ion itself remains intact. A detailed study of $Cl^-(H_2O)_2$, using a model potential and molecular dynamics, revealed a considerable change in the vibrational spectrum as a function of temperature.[192] This computational study brought into agreement, two separate experimental spectra that were apparently obtained with different internal energies or temperatures.[98,186,191] It is particularly intriguing that the internal energy or temperature can affect the relative stability or balance of the competing intermolecular forces. Structures favored at temperatures near 0 K may have little or no relation to those near room or biological temperatures. Since cluster ions are apparently intact and stable under such thermal conditions, entropic effects and their role in determining structures may be more important than low temperature energetics.

The hydration of halide anions is of intrinsic interest to the process of solvation. The most important aspect of water, its hydrogen bond network, is directly perturbed by the anion in a simple and direct way. It competes for the protons with its own ionic hydrogen bond. The gas phase studies of the smallest hydrated ions show extremely large shifts in the O–H stretch in the O–H–X^- bond.[98,186,193] This strong interaction must play a role in the bulk solvation process. Interesting implications will be discussed in the final section of this chapter.

Protonated cluster ions are an interesting species to study for a number of reasons. In homogeneous clusters of the form: $H^+(H_2O)_n$, there is the question of the nature of the protonated species. Is it the Zundel ion, $(H_5O_2)^+$, or the Eigen, H_3O^+, form? Infrared experiments with supporting *ab initio* calculations suggest that the larger and more delocalized species is stabilized at clusters sizes of $n > 5$.[194] Spectra supporting this observation are shown below in Fig. 20.

There are also interesting issues associated with heterogeneous protonated cluster ions of the form: $H(CH_3OH)(H_2O)_m{}^+$. Simply put, the question is where is the proton? It can be on either moiety. This system has also been analyzed using vibrational spectroscopy and *ab initio* calculations. Consistent results are obtained that indicate the proton resides on the methanol molecule to form the cation, $CH_3OH_2{}^+$ for clusters with up

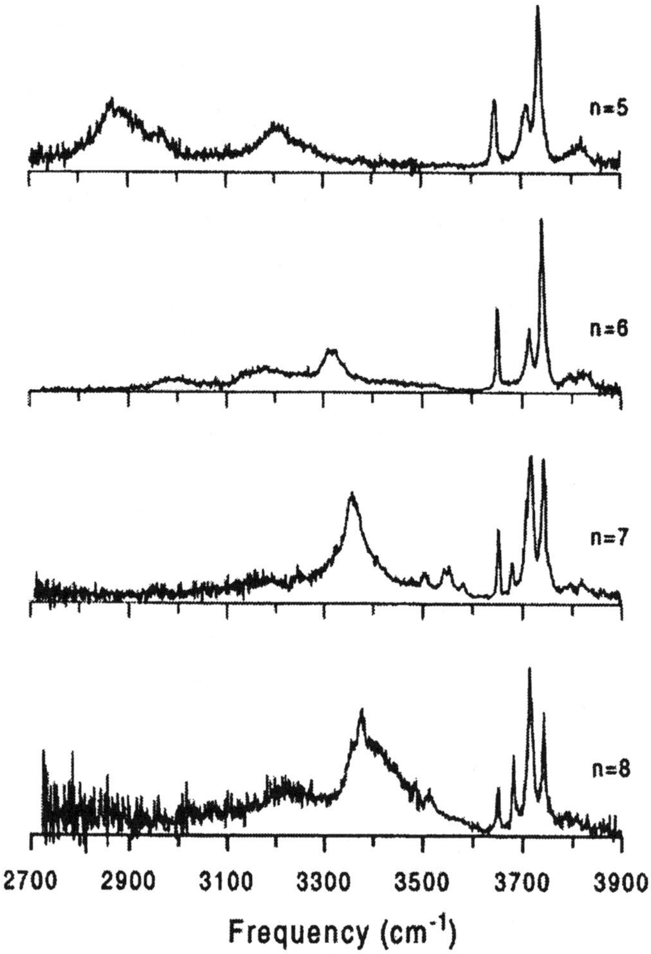

Fig. 20. Vibrational predissociation spectra of $H^+(H_2O)_n$, $n = 5$–8, in the OH stretching region. The band at 3178 cm^{-1} for $n > 5$ was identified with the $H_5O_2^+$ moiety.[194]

to $m = 2$.[195] However, for $m > 2$, the proton appears to be delocalized at the temperatures of the experiment, with little to separate H_3O^+ and $CH_3OH_2^+$ moieties. The competition for the proton in these chemically unique species give an excellent example of how one can utilize ion clusters to model real issues in the solution phase.

While not explicitly stated to this point, it should be clear that issues of competition in heterogeneous ion clusters are of great importance. As noted

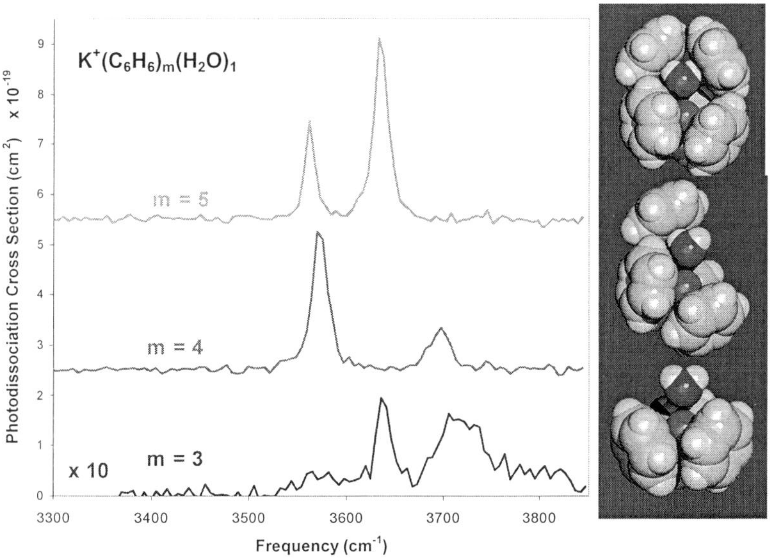

Fig. 21. The vibrational spectrum of $K^+(C_6H_6)_{3-5}(H_2O)_1$. Each additional benzene significantly alters the environment of the water molecule as shown to the right of the spectra.

above, anions in clusters with water or methanol compete with the solvent to form hydrogen bonds. Protons can bind to either water or methanol in clusters of varying size and composition. In even the simple solvation of alkali metal ions, the competition between the electrostatic ion-molecule interaction competes with the solvent–solvent interaction to determine the onset of successive solvent shells.[196] It should come as no surprise that the spectroscopy of cluster ions could also be used to address issues associated with competitive non-covalent interactions that occur in biological systems. The size-selective behavior of certain ion channels shows a strong preference for the passage of potassium ions over that of sodium.[197] It has also been noted that these K^+ channel proteins have a particular Gly-Tyr-Gly peptide sequence in the critical pore region of the channel.[198] Modification of the aromatic Tyr residue (other than to Phe) tends to render the channel useless.[199] It was proposed that the interaction between the aromatic side chain and the potassium ion was unusually favorable enabling the ion to pass through the pore.[200,201] Other ions did not have the proper size to "fit" through, although it was not clear why smaller ions, such as Na^+ could not pass.

This type of competitive non-covalent interaction problem is ideally suited for gas phase cluster ion studies. Benzene could be used to model the aromatic side chain of the protein. The infrared spectra of clusters with the form $M^+(C_6H_6)_n(H_2O)_m$ (M = Na and K) could be used to determine the location of water with respect to the ion. In characterizing the unimolecular dissociation of $K^+(C_6H_6)_{1-5}(H_2O)_{1,2}$ cluster ions,[202] it was observed for cluster ions, with one water and up to three benzene molecules, that water was the most labile species in the cluster. This suggested that benzene was bonded strongly to the potassium ion, an unexpected result. When four or five benzene molecules were present, the water dissociation channel completely closed and only loss of benzene was observed. The vibrational spectra of the cluster ions, $K^+(C_6H_6)_{3-5}(H_2O)_1$ are shown in Fig. 21.

It is clear that there is an abrupt change in the O–H vibrational bands with each additional benzene. With three benzenes, the water O–H bands are very similar to that of uncomplexed water, with the symmetric and asymmetric bands close to their gas phase values of 3650 and 3750 cm^{-1}. With the addition of a fourth benzene, there is a shift to lower frequency

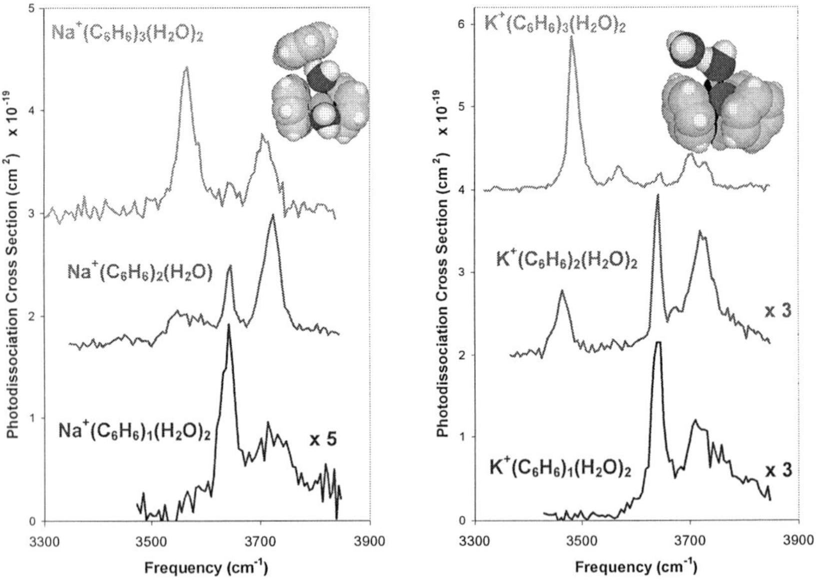

Fig. 22. Vibrational spectra of $Na^+(C_6H_6)_{1-3}$ $(H_2O)_2$ (bottom to top) and $K^+(C_6H_6)_{1-3}(H_2O)_2$. The addition of the third benzene molecule forces a significant difference in the position of the water molecules about the two ions: Na^+ retains both molecules in the first shell, K^+ only one.

and strengthening of the band intensity, both attributes characteristic of hydrogen-bond formation. With only one water molecule present, the band at 3550 cm^{-1} must arise from a hydrogen-bond between the OH group of the water molecule and the π cloud of benzene, a π-hydrogen bond. This was not a surprise as Zwier first identified the π-hydrogen bond in the infrared between benzene and clusters of water.[29,58] It is thus quite easy to characterize the bonding environment of the water molecule in this complex system.

The key question becomes whether the clusters with a potassium ion at the core behave differently, i.e. exhibit different spectroscopic and structural features, than do clusters with a sodium ion. Two sets of spectra directly address this point.[203] The first involve comparisons of $M^+(C_6H_6)_{1-3}(H_2O)_2$, with Na and K as the metal ions. As shown below in Fig. 22, the spectra are different for clusters with three waters. The spectrum with the sodium ion is very similar to one shown in Fig. 21, with π-hydrogen bonding between water in the first shell to benzene in the second shell.

The potassium ion displays a different infrared spectrum. The feature at 3470 cm^{-1} must be due to a stronger hydrogen bond, one between water

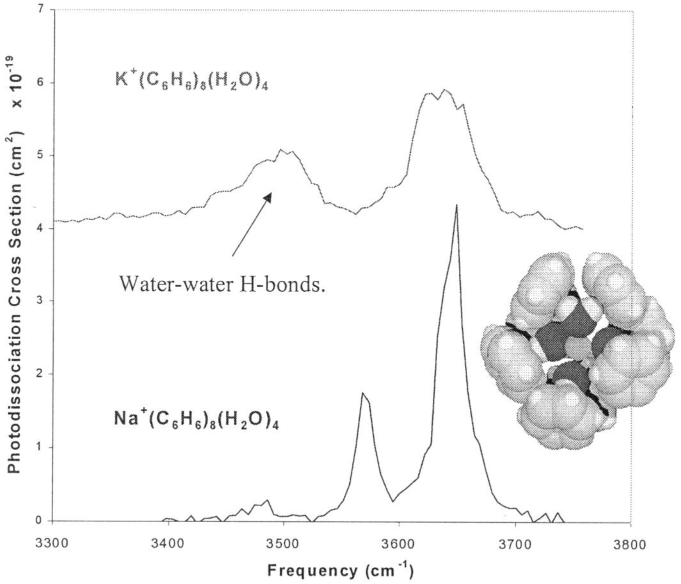

Fig. 23. Vibrational spectra of $K^+(H_2O)_4(C_6H_6)_8$ and $Na^+(H_2O)_4(C_6H_6)_8$, which show the different binding environments for water in the cluster ions. The structure of the solvents about the sodium ion consistent with experiment is shown to the right.

in the first shell and a water in the second shell. This implies that while the potassium ion can be partially dehydrated by benzene, the sodium ion cannot. The cluster ions, containing enough water to fill the first hydration sphere, should provide the crucial test. In Fig. 23, the spectra of $Na^+(H_2O)_4(C_6H_6)_8$ and $K^+(H_2O)_4(C_6H_6)_8$ are compared. As can be clearly seen, the sodium ion retains all of its water in the first solvation shell resulting in symmetric and asymmetric OH stretching motion of the OH groups forming π-hydrogen bonds to benzene molecules in the second solvent shell.

The cluster bearing the potassium ion has spectral features associated with water–water hydrogen bonds. This shows that in this larger system, benzene is still capable of partial dehydration of the ion. Size-selectivity by the aromatic environment can now be explained. The aromatic residues in the channel pore partially dehydrate the potassium ions, allowing them to pass through. The sodium ions retain their first solvent shell and are too large to fit through. A recent X-ray crystal structure of a small inwardly rectifying potassium ion channel suggested that the actual selective mechanism may be due to interactions between the ions and the carbonyls forming the peptide backbone.[204] Regardless of the actual mechanism in the protein, the spectra shown above demonstrate a differentiation between sodium and potassium ions by means of non-covalent interactions. Indeed, solution studies where ions are extracted from an aqueous phase into nitrobenzene support these observations. Determination of the water content of the nitrobenzene indicates that Na^+ carries four waters from the aqueous phase, while K^+ only one.[205] The subtle variations in ion-molecule, conventional and π-hydrogen bonds are responsible in the gas phase and solution studies for the difference between the two ions.

5. Conclusions and Future Directions

When contemplating the progress of cluster research, it is fair to consider its overall goals. There are at least three objectives, on which virtually all researchers would agree, to be identified. The first is to explore new phenomena. There are ample examples contained within this chapter: long time duration cluster ion studies; He microdroplets; and π-hydrogen bonds to name a few. The second is to understand non-covalent interactions at the molecular level. Chemistry occurs between molecules. The orientation and configuration of systems prior to a reaction or other dynamical process (like ion mobility or transfer, are critical to their understanding. The third and perhaps most important is to use the findings in the first two areas to bridge the gap between isolated molecules/clusters and real physical systems.

As can be seen in this and other chapters of the book, the interaction between theory and experiment is stronger than ever. As was stated earlier, the size and complexity of the cluster systems continue to increase. Unambiguous experimental and theoretical results are becoming more and more difficult to achieve. Yet when independent results from both arenas are compared, a strong unified picture can often be obtained. As was discussed earlier in this chapter, such synergy can be found in work on water clusters, benzene bound to water clusters, hydrated ions, and aromatic molecules attached to methanol, as notable examples.

Perhaps for the first time, since cluster studies were proposed, it seems that the bridge to real physical systems may actually exist. In the area of atmospheric chemistry, conventional models of salt-bearing aerosols could not explain the rate of conversion from chloride to chlorine in marine environments. Asymmetric hydration of chloride, resulting in an increase in the surface concentration of chloride was able to bring the observed and modeled rates into agreement.[206] The asymmetric solvation of chloride and the larger halide anions was observed in cluster studies,[109,191,207] supporting the atmospheric work. Solution studies will also benefit from the current areas of research. Between water and fluoride, the interaction is so strong, that the proton may actually transfer to the fluoride upon vibrational excitation, generating ($FH–OH^-$).[208] The hydration of more complex ions such as O_2^- is beginning to reveal how water adapts to solvate non-spherical ions,[209] with potential applications to a molecular understanding of solvent environment about charged groups in molecules and proteins. The understanding of structural and dynamical forces in biological systems is not beyond the reach of cluster studies. As discussed earlier, size-selectivity can be examined at the molecular level for systems such as ion channels and ionophores.[203,210–213] There are many examples of non-covalent interactions in living systems that shape fundamental processes necessary for life. These are all potential targets for molecular level investigation.

While neutral cluster studies have a more extensive history and accumulated information than ion clusters, it has been difficult to devise a general methodology for performing size-selective studies when the clusters have three or more molecular units. It appears that the use of He microdroplets, which cool the clusters to 0.37 K, can simplify very complex spectral patterns allowing identification by size and composition. Miller has shown very exciting results not only for water[66] and HCN[65] clusters, but new studies on metal-bearing species may open the door to the atom by atom understanding of metalization.[214] These are indeed exciting times for cluster science and for spectroscopic studies.

Acknowledgements

The author would like to acknowledge the support of the National Science Foundation (grants CHE-0072178 and INT-9726311) in the preparation of this work. He would also like to thank Prof. Tamotsu Kondow and Dr. Fumitaka Mafune and the members of the Cluster Research Laboratory of the Genesis Research Institute, Inc. for their gracious hospitality, which enabled this work to be completed.

References

1. H. Kuhn, *Proc. Roy. Soc. (London)* **158**, 212 (1937). (see also p. 230).
2. A. W. Miziolek and G. C. Pimentel, *J. Chem. Phys.* **65**, 4462 (1976).
3. A. S. Pine and W. J. Lafferty, *J. Chem. Phys.* **78**, 2154 (1983).
4. D. R. Miller, in *Atomic and Molecular Beam Methods*, Vol. 1 edited by G. Scoles (Oxford, New York, 1988).
5. T. R. Dyke *et al.*, *J. Chem. Phys.* **56**, 2442 (1972).
6. T. J. Balle and W. H. Flygare, *Rev. Sci. Instr.* **52**, 33 (1981).
7. K. C. Janda *et al.*, *J. Am. Chem. Soc.* **100**, 8074 (1978).
8. S. J. Harris *et al.*, *J. Chem. Phys.* **60**, 3208 (1974).
9. T. J. Balle *et al.*, *J. Chem. Phys.* **71**, 2723 (1979).
10. T. J. Balle *et al.*, *J. Chem. Phys.* **72**, 922 (1980).
11. D. H. Levy, *Ann. Rev. Phys. Chem.* **31**, 197 (1980).
12. R. E. Miller, *J. Phys. Chem.* **90**, 3301 (1986).
13. M. D. Schuder *et al.*, *J. Chem. Phys.* **99**, 4346 (1993).
14. J. M. Lisy *et al.*, *Springer Ser. Opt. Sci.* **30**, 324 (1981).
15. J. M. Lisy *et al.*, *J. Chem. Phys.* **75**, 4733 (1981).
16. R. E. Miller, *NATO ASI Ser. Ser.* **C212**, 131 (1987).
17. D. W. Michael *et al.*, *J. Chem. Phys.* **81**, 5998 (1984).
18. C. M. Lovejoy *et al.*, *J. Chem. Phys.* **85**, 4890 (1986).
19. D. J. Nesbitt, *Chem. Rev.* **88**, 843 (1988).
20. R. L. Robinson *et al.*, *J. Chem. Phys.* **87**, 5149 (1987).
21. R. C. Cohen *et al.*, *J. Chem. Phys.* **89**, 4494 (1988).
22. R. C. Cohen and R. J. Saykally, *J. Chem. Phys.* **98**, 6007 (1993).
23. C. M. Lovejoy and D. J. Nesbitt, *Rev. Sci. Instr.* **58**, 807 (1987).
24. G. A. Blake *et al.*, *Rev. Sci. Instr.* **62**, 1693 (1991).
25. G. A. Blake *et al.*, *Rev. Sci. Instr.* **62**, 1701 (1991).
26. K. Liu *et al.*, *Faraday Disc.* **97**, 35 (1994).
27. K. Liu *et al.*, *Nature (London)* **381**, 501 (1996).
28. K. Liu *et al.*, *Science* **271**, 929 (1996).

29. R. N. Pribble and T. S. Zwier, *Science* **265**, 75 (1994).
30. F. Huisken *et al.*, *J. Chem. Phys.* **104**, 17 (1996).
31. C. J. Gruenloh *et al.*, *Science* **276**, 1678 (1997).
32. T. Watanabe *et al.*, *J. Chem. Phys.* **105**, 408 (1996).
33. U. Buck *et al.*, *Phys. Rev. Lett.* **80**, 2578 (1998).
34. L. M. Goss *et al.*, *J. Phys. Chem.* **103**, 8620 (1999).
35. A. Bach *et al.*, *J. Chem. Phys.* **113**, 9032 (2000).
36. F. N. Keutsch *et al.*, *J. Chem. Phys.* **114**, 4005 (2001).
37. F. N. Keutsch *et al.*, *J. Chem. Phys.* **114**, 3994 (2001).
38. W. Klopper *et al.*, *J. Chem. Phys.* **103**, 1085 (1995).
39. D. J. Wales, *J. Chem. Phys.* **111**, 8429 (1999).
40. A. Milet *et al.*, *J. Phys. Chem.* **103**, 6811 (1999).
41. R. S. Fellers *et al.*, *Science* **284**, 945 (1999).
42. H. M. Lee *et al.*, *J. Chem. Phys.* **112**, 9759 (2000).
43. G. C. Groenenboom *et al.*, *Phys. Rev. Lett.* **84**, 4072 (2000).
44. S. S. Xantheas, *Chem. Phys.* **258**, 225 (2000).
45. H. M. Lee *et al.*, *J. Chem. Phys.* **114**, 10749 (2001).
46. J. J. Scherer *et al.*, *Chem. Rev.* **97**, 25 (1997).
47. J. B. Paul *et al.*, *J. Phys. Chem.* **A101**, 5211 (1997).
48. J. B. Paul *et al.*, *J. Chem. Phys.* **109**, 10201 (1998).
49. A. O'Keefe and D. A. G. Deacon, *Rev. Sci. Instr.* **59**, 2544 (1988).
50. C. R. Bucher *et al.*, *Appl. Opt.* **39**, 3154 (2000).
51. B. A. Paldus *et al.*, *Opt. Lett.* **25**, 666 (2000).
52. G. T. Fraser *et al.*, *J. Chem. Phys.* **87**, 1502 (1987).
53. D. W. Michael and J. M. Lisy, *J. Chem. Phys.* **85**, 2528 (1986).
54. K. D. Kolenbrander and J. M. Lisy, *J. Chem. Phys.* **85**, 6227 (1986).
55. R. J. Stanley and A. W. Castleman Jr., *J. Chem. Phys.* **92**, 5770 (1990).
56. R. J. Stanley and A. W. Castleman Jr., *J. Chem. Phys.* **94**, 7744 (1991).
57. R. J. Stanley and A. W. Castleman Jr., *J. Chem. Phys.* **98**, 796 (1993).
58. R. N. Pribble and T. S. Zwier, *Faraday Disc.* **97**, 229 (1994).
59. A. Fujii *et al.*, *Chem. Phys. Lett.* **225**, 104 (1994).
60. T. Ebata *et al.*, *J. Phys. Chem.* **99**, 5761 (1995).
61. T. Ebata *et al.*, *Int. J. Mass Spectrom. Ion Process.* **159**, 111 (1996).
62. T. E. Gough *et al.*, *J. Chem. Phys.* **69**, 1588 (1978).
63. M. Hartmann *et al.*, *Science* **272**, 1631 (1996).
64. M. Hartmann *et al.*, *Phys. Rev. Lett.* **75**, 1566 (1995).
65. K. Nauta and R. E. Miller, *Science* **283**, 1895 (1999).
66. K. Nauta and R. E. Miller, *Science* **287**, 293 (2000).
67. A. W. Castleman Jr. and R. G. Keesee, *Chem. Rev.* **86**, 589 (1986).
68. R. G. Keesee and A. W. Castleman Jr., *J. Phys. Chem. Ref. Data* **15**, 1011 (1986).
69. J. A. Draves *et al.*, *J. Chem. Phys.* **93**, 4589 (1990).
70. M. L. Alexander *et al.*, *Phys. Rev. Lett.* **57**, 976 (1986).
71. M. A. Johnson *et al.*, *Chem. Phys. Lett.* **112**, 285 (1984).
72. M. L. Alexander *et al.*, *J. Chem. Phys.* **82**, 5288 (1985).
73. P. Ayotte *et al.*, *Chem. Phys.* **239**, 485 (1998).

74. P. Ayotte *et al.*, *J. Chem. Phys.* **108**, 444 (1998).
75. P. Ayotte *et al.*, *J. Chem. Phys.* **110**, 6268 (1999).
76. S. Atrill and A. J. Stace, *Int. J. Mass Spectrom* **180**, 253 (1998).
77. J. F. Winkel *et al.*, *J. Chem. Phys.* **103**, 5177 (1995).
78. A. B. Jones *et al.*, *Int. J. Mass Spectrom Ion Process.* **133**, 83 (1994).
79. M. F. Jarrold *et al.*, *J. Chem. Phys.* **79**, 6086 (1983).
80. M. F. Jarrold *et al.*, *J. Chem. Phys.* **81**, 222 (1984).
81. J. T. Snodgrass *et al.*, *J. Phys. Chem.* **94**, 648 (1990).
82. C. J. Weinheimer and J. M. Lisy, *Chem. Phys.* **239**, 357 (1998).
83. K. Ohashi *et al.*, *Chem. Phys. Lett.* **323**, 43 (2000).
84. Y. Inokuchi *et al.*, *Chem. Phys. Lett.* **279**, 73 (1997).
85. M. Ichihashi *et al.*, *J. Phys. Chem.* **100**, 10050 (1996).
86. M. Ichihashi *et al.*, *J. Phys. Chem.* **102**, 8287 (1998).
87. T. Sawamura *et al.*, *J. Phys. Chem.* **100**, 8131 (1996).
88. G. Niednerschatteburg *et al.*, *J. Chem. Phys.* **99**, 9664 (1993).
89. C. Berg *et al.*, *Int. J. Mass Spectrom Ion Process.* **167**, 723 (1997).
90. Y. J. Yang *et al.*, *J. Phys. Chem.* **A101**, 2371 (1997).
91. T. Schindler *et al.*, *Chem. Phys. Lett.* **250**, 301 (1996).
92. S. E. Rodriguez-Cruz *et al.*, *J. Am. Chem. Soc.* **120**, 5842 (1998).
93. R. C. Dunbar and T. B. McMahon, *Science* **279**, 194 (1998).
94. E. J. Bieske *et al.*, *J. Chem. Phys.* **102**, 5152 (1995).
95. M. A. Johnson and W. C. Lineberger, *Tech. Chem. (New York)* **20**, 591 (1988).
96. W. H. Robertson *et al.*, *Rev. Sci. Instr.* **71**, 4431 (2000).
97. J. F. Winkel *et al.*, *J. Chem. Phys.* **101**, 9436 (1994).
98. J.-H. Choi *et al.*, *J. Phys. Chem.* **102**, 503 (1998).
99. Y.-S. Wang *et al.*, *J. Am. Chem. Soc.* **120**, 8777 (1998).
100. P. C. Engelking, *Rev. Sci. Instr.* **57**, 2274 (1986).
101. J. M. Lisy, *Adv. Mol. Vib. Collision Dynam.* **B1**, 281 (1991).
102. W. L. Liu and J. M. Lisy, *J. Chem. Phys.* **89**, 605 (1988).
103. O. M. Cabarcos and J. M. Lisy, *Chem. Phys. Lett.* **257**, 265 (1996).
104. O. M. Cabarcos *et al.*, *J. Phys. Chem.* **A103**, 8777 (1999).
105. T. J. Selegue and J. M. Lisy, *J. Phys. Chem.* **96**, 4143 (1992).
106. T. J. Selegue *et al.*, *J. Chem. Phys.* **96**, 7268 (1992).
107. T. J. Selegue *et al.*, *J. Chem. Phys.* **100**, 4790 (1994).
108. C. J. Weinheimer and J. M. Lisy, *J. Phys. Chem.* **100**, 15305 (1996).
109. O. M. Cabarcos *et al.*, *J. Chem. Phys.* **110**, 9516 (1999).
110. S. E. Novick *et al.*, *Stud. Phys. Theor. Chem.* **68**, 359 (1990).
111. M. R. Topp, *Int. Rev. Phys. Chem.* **12**, 149 (1993).
112. P. M. Felker *et al.*, *Chem. Rev.* **94**, 1787 (1994).
113. U. Buck, *Springer Ser. Chem. Phys.* **52**, 396 (1994).
114. Z. Bacic and R. E. Miller, *J. Phys. Chem.* **100**, 12945 (1996).
115. D. W. Pratt, *Ann. Rev. Phys. Chem.* **49**, 481 (1998).
116. K. Muller-Dethlefs and P. Hobza, *Chem. Rev.* **100**, 143 (2000).
117. H. J. Neusser and K. Siglow, *Chem. Rev.* **100**, 3921 (2000).
118. D. M. Neumark, *Ann. Rev. Phys. Chem.* **52**, 255 (2001).

119. S. E. Novick, *Bibliography of Rotational Spectra of Weakly Bound Complexes.* Electronic updates are available on the web at http://www.wesleyan.edu/chem/faculty/novick/vdw.html (2000).
120. H. S. Gutowsky *et al.*, *J. Am. Chem. Soc.* **107**, 7174 (1985).
121. H. S. Gutowsky *et al.*, *J. Chem. Phys.* **83**, 4817 (1985).
122. H. S. Gutowsky *et al.*, *J. Am. Chem. Soc.* **109**, 5633 (1987).
123. H. S. Gutowsky *et al.*, *J. Chem. Phys.* **88**, 2919 (1988).
124. A. McIlroy *et al.*, *J. Phys. Chem.* **95**, 2636 (1991).
125. D. T. Anderson *et al.*, *J. Chem. Phys.* **107**, 1115 (1997).
126. C. M. Lovejoy *et al.*, *J. Chem. Phys.* **97**, 8009 (1992).
127. A. McLlroy and D. J. Nesbitt, *J. Chem. Phys.* **97**, 6044 (1992).
128. D. J. Nesbitt, *Faraday Disc.* **97**, 1 (1994).
129. C. M. Lovejoy and D. J. Nesbitt, *Chem. Phys. Lett.* **146**, 582 (1988).
130. E. J. Bohac and R. E. Miller, *J. Chem. Phys.* **99**, 1537 (1993).
131. S. Davis *et al.*, *J. Chem. Phys.* **105**, 6645 (1996).
132. R. J. Bemish and R. E. Miller, *Chem. Phys. Lett.* **281**, 272 (1997).
133. D. J. Nesbitt and C. M. Lovejoy, *J. Chem. Phys.* **93**, 7716 (1990).
134. D. J. Nesbitt and C. M. Lovejoy, *J. Chem. Phys.* **96**, 5712 (1992).
135. K. L. Busarow *et al.*, *J. Chem. Phys.* **90**, 3937 (1989).
136. N. Pugliano and R. J. Saykally, *J. Chem. Phys.* **96**, 1832 (1992).
137. J. D. Cruzan *et al.*, *J. Chem. Phys.* **105**, 6634 (1996).
138. K. Liu *et al.*, *Science* **271**, 62 (1996).
139. K. Kim *et al.*, *J. Am. Chem. Soc.* **116**, 11568 (1994).
140. J. K. Gregory and D. C. Clary, *J. Phys. Chem.* **100**, 18014 (1996).
141. J. K. Gregory and D. C. Clary, *J. Phys. Chem.* **A101**, 6813 (1997).
142. K. Liu *et al.*, *J. Phys. Chem.* **101**, 8995 (1997).
143. C. Millot and A. J. Stone, *Mol. Phys.* **77**, 439 (1992).
144. J. W. Caldwell and P. A. Kollman, *J. Phys. Chem.* **99**, 6208 (1995).
145. W. L. Jorgensen *et al.*, *J. Chem. Phys.* **79**, 926 (1983).
146. R. A. Provencal *et al.*, *J. Chem. Phys.* **110**, 4258 (1999).
147. R. A. Provencal *et al.*, *J. Phys. Chem.* **104**, 1423 (2000).
148. F. Huisken *et al.*, *J. Chem. Phys.* **95**, 3924 (1991).
149. U. Buck and F. Huisken, *Chem. Rev.* **100**, 3863 (2000).
150. R. S. Ruoff *et al.*, *J. Chem. Phys.* **89**, 138 (1988).
151. K. W. Jucks and R. E. Miller, *J. Chem. Phys.* **88**, 2196 (1988).
152. I. J. Kurnig *et al.*, *J. Chem. Phys.* **92**, 2469 (1990).
153. B. F. King and F. Weinhold, *J. Chem. Phys.* **103**, 333 (1995).
154. A. Karpfen, *J. Phys. Chem.* **100**, 13474 (1996).
155. A. J. Stone *et al.*, *J. Chem. Phys.* **107**, 1030 (1997).
156. T. S. Zwier, *Ann. Rev. Phys. Chem.* **47**, 205 (1996).
157. A. J. Gotch and T. S. Zwier, *J. Chem. Phys.* **96**, 3388 (1992).
158. A. W. Garrett and T. S. Zwier, *J. Chem. Phys.* **96**, 3402 (1992).
159. J. D. Augspurger *et al.*, *J. Phys. Chem.* **97**, 980 (1993).
160. K. Kleinermanns *et al.*, *J. Phys. Chem.* **A103**, 232 (1999).
161. T. Ebata *et al.*, *J. Phys. Chem.* **104**, 7974 (2000).
162. J. R. Carney *et al.*, *J. Chem. Phys.* **108**, 3379 (1998).

163. J. R. Carney and T. S. Zwier, *J. Phys. Chem.* **103**, 9943 (1999).
164. C. J. Gruenloh *et al.*, *J. Chem. Phys.* **109**, 6601 (1998).
165. C. J. Gruenloh *et al.*, *J. Chem. Phys.* **113**, 2290 (2000).
166. L. W. Buxton *et al.*, *J. Chem. Phys.* **75**, 2681 (1981).
167. P. D. Aldrich *et al.*, *J. Chem. Phys.* **75**, 2126 (1981).
168. J. P. Devlin *et al.*, *J. Phys. Chem.* **104**, 1974 (2000).
169. J. P. Devlin *et al.*, *J. Phys. Chem.* **105**, 974 (2001).
170. C. E. Dykstra and J. M. Lisy, *J. Mol. Struct.* **500**, 375 (2000).
171. S. S. Xantheas, *J. Chem. Phys.* **102**, 4505 (1995).
172. E. R. Batista *et al.*, *J. Chem. Phys.* **112**, 3285 (2000).
173. O. Dopfer *et al.*, *J. Chem. Phys.* **114**, 7081 (2001).
174. O. Dopfer *et al.*, *J. Chem. Phys.* **112**, 2176 (2000).
175. O. Dopfer *et al.*, *J. Chem. Phys.* **113**, 120 (2000).
176. O. Dopfer *et al.*, *J. Phys. Chem.* **A103**, 2982 (1999).
177. O. Dopfer *et al.*, *J. Chem. Phys.* **111**, 10754 (1999).
178. O. Dopfer, *J. Phys. Chem.* **104**, 11702 (2000).
179. R. V. Olkhov *et al.*, *J. Chem. Phys.* **108**, 10046 (1998).
180. M. R. France *et al.*, *J. Chem. Phys.* **108**, 7049 (1998).
181. M. R. France *et al.*, *J. Chem. Phys.* **109**, 8842 (1998).
182. D. A. Wild *et al.*, *J. Chem. Phys.* **113**, 1075 (2000).
183. U. Achatz *et al.*, *Chem. Phys. Lett.* **291**, 459 (1998).
184. U. Achatz *et al.*, *J. Am. Chem. Soc.* **123**, 6151 (2001).
185. A. V. Davis *et al.*, *J. Electron Spectrosc. Relat. Phenom.* **108**, 203 (2000).
186. P. Ayotte *et al.*, *J. Phys. Chem.* **103**, 10665 (1999).
187. P. Ayotte *et al.*, *J. Phys. Chem.* **A102**, 3067 (1998).
188. C. E. Z. Klots, *Phys. D - At. Mol. Clusters* **5**, 83 (1987).
189. C. E. Klots, *J. Phys. Chem.* **92**, 5864 (1988).
190. C. E. Klots, *Z. Phys. D - At. Mol. Clusters* **20**, 105 (1991).
191. P. Ayotte *et al.*, *J. Chem. Phys.* **110**, 7129 (1999).
192. H. E. Dorsett *et al.*, *J. Phys. Chem.* **A103**, 3351 (1999).
193. O. M. Cabarcos *et al.*, *J. Chem. Phys.* **110**, 5 (1999).
194. J. C. Jiang *et al.*, *J. Am. Chem. Soc.* **122**, 1398 (2000).
195. C. C. Wu *et al.*, *J. Chem. Phys.* **112**, 176 (2000).
196. J. M. Lisy, *Proc. SPIE-Int. Soc. Opt. Eng.* **1638**, 92 (1992).
197. C. Miller, *Science* **252**, 1092 (1991).
198. L. Heginbotham *et al.*, *Science* **258**, 1152 (1992).
199. L. Heginbotham *et al.*, *Biophys. J.* **66**, 1061 (1994).
200. R. A. Kumpf and D. A. Dougherty, *Science* **261**, 1708 (1993).
201. D. A. Dougherty, *Science* **271**, 163 (1996).
202. O. M. Cabarcos *et al.*, *J. Chem. Phys.* **108**, 5151 (1998).
203. O. M. Cabarcos *et al.*, *J. Chem. Phys.* **110**, 8429 (1999).
204. D. A. Doyle *et al.*, *Science* **280**, 69 (1998).
205. T. Osakai *et al.*, *J. Phys. Chem.* **B101**, 8341 (1997).
206. E. M. Knipping *et al.*, *Science* **288**, 301 (2000).
207. D. J. Tobias *et al.*, *J. Chem. Phys.* **114**, 7036 (2001).
208. P. Ayotte *et al.*, *Chem. Phys. Lett.* **316**, 455 (2000).

209. J. M. Weber *et al.*, *Science* **287**, 2461 (2000).
210. T. J. Marronne and K. M. Merz, *J. Am. Chem. Soc.* **114**, 7542 (1992).
211. P. S. H. Wong *et al.*, *J. Am. Soc. Mass Spectrom.* **5**, 632 (1994).
212. K. S. Kim *et al.*, *J. Phys. Chem.* **B102**, 461 (1998).
213. H. S. Choi *et al.*, *Proc. Natl. Acad. Sci. USA* **95**, 12094 (1998).
214. K. Nauta K. *et al.*, *Science* **292**, 481 (2001).
215. K. W. Jucks and R. E. Miller, *J. Chem. Phys.* **88**, 6059 (1988).
216. C. J. Gruenloh *et al.*, *J. Phys. Chem.* **103**, 503 (1999).
217. J. K. Gregory *et al.*, *Science* **275**, 814 (1997).
218. P. Ayotte *et al.*, *J. Phys. Chem.* **103**, 443 (1999).

CHAPTER 3

PHYSICAL AND CHEMICAL PROPERTIES OF METAL CLUSTERS IN THE GAS PHASE AND ON SOLID SURFACES

A. Terasaki

Cluster Research Laboratory, Toyota Technological Institute in East Tokyo Laboratory, Genesis Research Institute, Inc., 717-86 Futamata, Ichikawa, Chiba 272-0001, Japan

E-mail: terasaki@toyota-ti.ac.jp

In this chapter, we describe the size-dependent physical and chemical properties of metal clusters and ions isolated in the gas phase and supported on solid surfaces from experimental points of view. At first, the methodologies and procedures employed in the experimental studies are surveyed for the readers who are not familiar with this field. Then we refer to size-specific features of the electronic and geometric structures of isolated metal-cluster ions together with their magnetic properties. The specific heat measurements of isolated metal-cluster ions are mentioned, and the phase transitions manifested in the specific heats of the cluster ions are argued. As a representative of the chemical properties, we pick up the size-dependent reactivity of cluster ions in the gas phase and clusters on solid surfaces. The reaction processes change with the cluster size dramatically.

Contents

1. Introduction

The physical properties and the chemical reactivity of a cluster change dramatically with its size (the number of constituent atoms). A new class of material science attracts much attention by taking advantage of the size-dependent characteristics. Among others, a transition metal-cluster is particularly important because it is expected to show specific magnetic properties and catalytic activity for instance, which are scarcely encountered in conventional solid materials. Manganese is antiferromagnetic as a sold specimen, while shows a ferromagnetic character as a small cluster. Nickel catalyzes various chemical reactions. In the reaction of a nickel cluster ion with a methanol molecule, demethanation (methane is released from methanol) occurs at the size of 4, chemisorption of methanol at the size of 6, and carbide formation at the sizes of 7 and 8. There are many other reports on the size-dependent geometric and electronic structures, specific heats, phase transition, electronic conductivity, superfluidity, etc. These reports demonstrate size-specific behaviors of metal clusters in various physical and chemical phenomena.[1]

Because of a low number density of cluster ions isolated in the gas phase, conventional methodologies available for condensed matters are not applicable to the measurements of the cluster properties. The first step of the measurements is to obtain a cluster ion with a desired size by mass-spectroscopic separation of cluster ions in a cluster source. After the size-selection, the number density of the size-selected cluster ion is typically 10^6 cm^{-3} or lower, which is too low for the conventional optical absorption spectroscopy, for instance. In the measurement of an electronic conductivity, one should attach electrodes to the specimen that you intend to measure.

How we could attach electrodes to a cluster for the conductivity measurement? Similarly, there is no thermometer, which is small enough to be applied to any cluster. Many endeavors have been made to develop new methodologies for the measurements of the physical and chemical properties of clusters in the following sections.

2. Method of Cluster-Size Selection

Typical measurement procedures are given as follows:

(1) Production of clusters by laser ablation, ion sputtering, plasma discharge, etc.
(2) Ionization of the clusters by electron-impact, metastable atom impact, photoionization, etc.
(3) Selection of monosized cluster ions by mass spectrometry.
(4) Interaction of the size-selected cluster ions with photons, atoms and molecules, solid surfaces, etc.
(5) Mass-spectroscopic identification of product ions after the interaction and subsequent theoretical analysis.

As described above, cluster "ions" are employed to elucidate the size-dependent characteristics because of easiness of the size-selection, but neutral clusters are not. Even though, several studies have been reported on size-selective attempts on neutral clusters. For instance, neutral clusters with various sizes are ionized by photons having an energy close to the threshold energy. Expectedly, the size-distribution of the product ions agrees with that of the original neutral clusters, because the nearly threshold energy ionization minimizes fragmentation in the ionization of the neutral clusters. Size-selection is performed by collisional momentum transfer from helium atoms to neutral clusters with various sizes. A beam of the neutral clusters is allowed to collide with that of helium atoms. As the kinetic energies of the neutral clusters and the helium atoms are given, a neutral cluster with a given size is scattered to a predetermined direction with respect to its original beam axis. Besides these examples, various unconventional experimental techniques have been developed.[2]

3. Electronic and Geometric Structures

The physical and chemical properties of a cluster or a cluster ion are closely related to its geometric and electronic structures, which are measured by

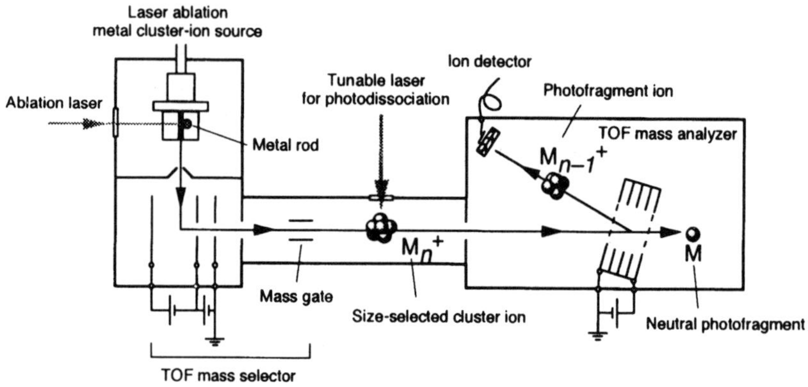

Fig. 1. Experimental setup for the optical absorption measurement of size-selected cluster ions by the photodissociation laser spectrometry.

optical absorption and photoelectron spectroscopies with the aid of theoretical calculations. Several works described below exemplify the relations.

3.1. *Optical Absorption Spectroscopy*

Usually optical absorbance is obtained from the reduction of incoming light intensity due to optical absorption by a sample placed in an absorption cell. However, the reduction by the cluster or its ion is too small to overcome background noise of the measurement system. Therefore, a special methodology is required to obtain the absorbance in the presence of such a low-density sample. In order to overcome this difficulty, fragmented ions accompanied by the optical absorption are detected as a function of the wavelength of the incoming light (photodissociation action spectrum). Figure 1 shows a schematic diagram of the apparatus together with a principle of the measurement. Cluster ions are produced by laser ablation of a metal plate/rod and size-selected by time-of-flight (TOF) mass spectrometry. A cluster ion, M_n^+, thus size-selected is irradiated with a laser. Subsequently, photofragment ions are identified and detected by a reflectron TOF mass spectrometer. If the primary cluster ion, M_n^+, is readily fragmented upon photoabsorption, the action spectrum is equivalent to the optical absorption spectrum. In principle, the optical absorption spectrum can be measured until the wavelength of the incoming light is short enough to fragment the primary cluster ion. In the wavelengths longer than this critical wavelength, one needs a special care. A complex ion, M_n^+Rg, in which a rare-gas atom,

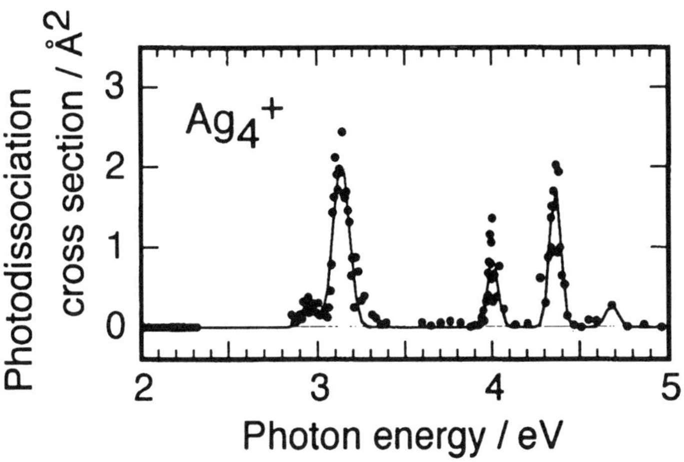

Fig. 2. The optical absorption spectrum of Ag_4^+ obtained by the measurement of photodissociation action spectrum. The dots show experimental data; the solid line is for eye-guide. The temperature of the cluster ion was estimated to be below 150 K. The transitions are attributed to one-electron transition between the ground and excited electronic states.[6]

Rg, is weakly attached to M_n^+, is used instead of M_n^+. As Rg is bound by an energy of $\sim$ 0.1 eV, one can extend the spectrum measurement to the infrared region by using this ion.[3–5]

3.2. *Optical Absorption Spectra of Silver Cluster Ions*

Figure 2 shows the optical absorption spectrum of silver tetramer cation, Ag_4^+, which is produced by the laser ablation method.[6] As shown in the figure, three intense peaks at 3.1, 4.0 and 4.4 eV are discernible. It looks that this ion has a low internal temperature because of their narrow peak widths of $\sim$ 0.1 eV or less, and hence possesses a rigid geometric structure (molecular-like). Spectra of Ag_4^+ calculated by a recent molecular orbital (MO) theory are shown in Fig. 3(a) and (b) for Ag_4^+ having a diamond-shaped structure with the lowest-energy and a T-shaped structure with the second lowest-energy, respectively.[7] The experimental spectrum agrees well with the theoretical spectrum for the most stable Ag_4^+; it is concluded that the silver tetramer cation has a diamond-shaped structure in its ground state. The electronic ground state has $^2B_{1u}$ and the excited states observed experimentally are assigned to 2^2A_g (3.1 eV), 3^2B_{2g} (4.0 eV), and 4^2B_{2g} (4.4 eV), which correspond to one-electron transitions involving the

5*s*-dominant initial state. The geometric and electronic structures of transition metal cluster ions are determined as such.

On the other hand, the spectrum changes drastically if the internal temperature of a cluster ion is raised. Figure 4 shows an optical absorption spectrum of Ag_9^+ having a high internal temperature produced by ion sputtering.[8] The spectrum has a broad and intense peak at ~ 4 eV, which can no longer be interpreted in terms of single-electron excitation, but excitation of the collective motion of electrons. The silver cluster cation is in a liquid-like state, so that the constituent silver atoms move freely in the cluster ion. It turns out that the valence electrons donated from the constituent silver atoms over a uniform background potential built by the atomic ion cores of silver. This picture is called by the "jellium model". The intense peak is assigned to a surface plasma oscillation. In general, the electronic structure of a metal-cluster or its ion depends critically on the cluster temperature (internal energy) whether the cluster is in a solid-like state or in a liquid-like state. Then it is an important question what electronic structure one may imagine in a transition region where the solid-like and liquid-like states appear alternately. The temperature dependence of the electronic structure of alkali-metal cluster ions has been investigated systematically.[9,10]

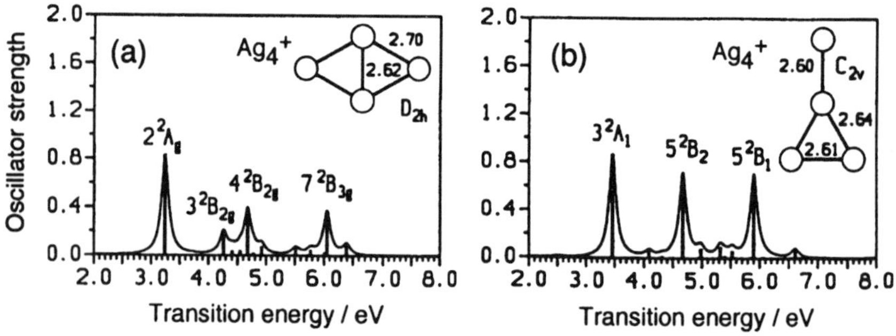

Fig. 3. Quantum chemical calculations of optical absorption spectra of Ag_4^+. (a) The most stable rhombic structure. (b) The second lowest-energy isomer of the T-shape, of which total energy is higher than that of (a) by 0.02 eV. The bars indicate oscillator strengths; solid lines show spectral profiles. Bond lengths in the Å unit are indicated in the insets. The spectrum obtained for the most stable structure (a) best reproduces the experimental spectrum in Fig. 2. (Adapted from Ref. 7.)

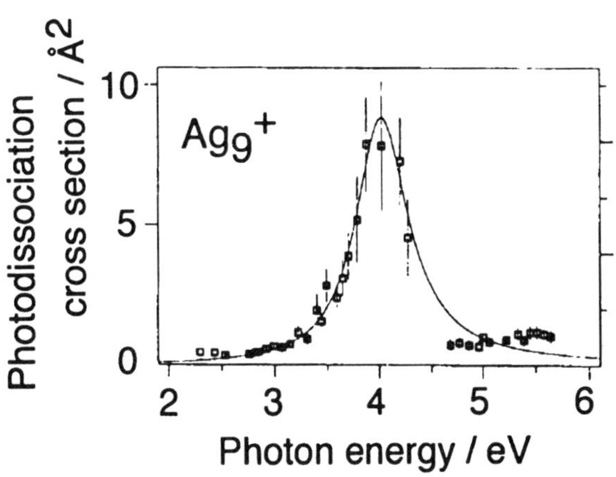

Fig. 4. The optical absorption spectrum of Ag_9^+ produced by ion sputtering. The temperature of the cluster ion is estimated to be above 400 K. Squares show experimental data; the solid line is for eye-guide. The broad spectral feature is attributed to a surface plasmon resonance, where electrons are excited collectively. (Adapted from Ref. 8.)

3.3. *Optical Absorption Spectra of Transition Metal Cluster Ions*

Let us explain optical absorption spectra of cluster ions of transition metals containing 3*d* valence electrons. It is well-known that 3*d* electrons have large electron correlation and exhibit specific magnetic properties. In this section, we describe the magnetic properties of several transition metal-cluster ions derived from their optical absorption spectra and photoelectron spectra in combination with theoretical analyses.[5,11–16]

Figure 5(a) shows a typical optical absorption spectrum of vanadium pentamer cation, V_5^+. The intensity increases from about 1 eV with the photon energy; there appear a peak at 1.2 eV and a broad feature above 1.6 eV. Electrons related to this energy range are three 3*d* electrons and two 4*s* electrons of each vanadium atom; 24 electrons altogether are involved. Because of these many electrons involved in the transitions, the resulting absorption spectrum is congested with many absorption transitions.

The transition energies and the oscillator strengths are calculated by methodology based on a density functional method, which deals well with a many electron system. The density-functional calculation shows that the most stable structure of V_5^+ is a bipyramid with C_{2v} symmetry with the average interatomic distance of 2.4 Å and the triplet spin manifold.

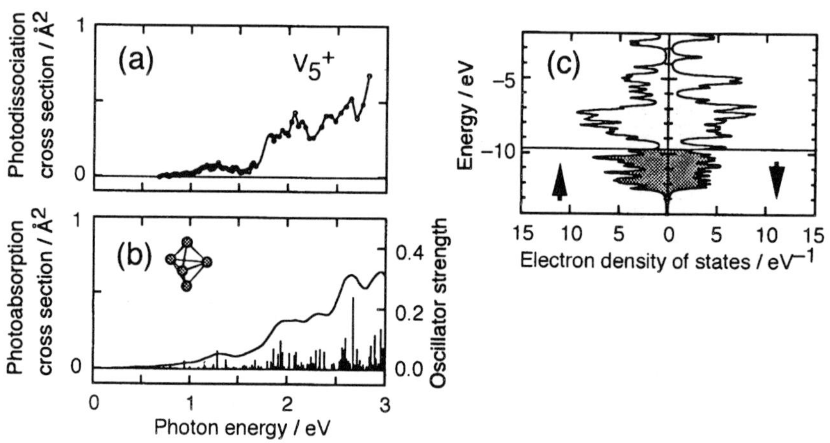

Fig. 5. The optical absorption spectrum and the electronic structure of $V_5{}^+$. (a) Experimental data, where a photodissociation action spectrum of a rare-gas complex, $V_5{}^+Ar$, was measured by observing a photofragment, $V_5{}^+$. (b) Density-functional calculation of the spectrum for the most stable isomer illustrated in the inset. The bars show oscillator strengths; the solid line a spectral profile. (c) Density-of-states profiles of the majority or and the minority-spin electrons obtained by the density-functional calculation. The shadows indicate occupied electronic levels. The vanadium pentamer ion, $V_5{}^+$, was shown to be in the spin triplet state with a trigonal bipyramid structure, where the average bond length was 2.4 Å.[13]

Optical transition energies and their oscillator strengths are calculated on the basis of the geometric structure thus obtained. The optical absorption spectrum is obtained as histograms whose energy positions and heights correspond to the transition energies and the oscillator strengths, respectively. In order to take the effect of vibrational motions of the constituent atoms in a finite temperature, each histogram is widened by a Lorentzian function with the width of 0.2 eV. The spectrum is simulated to sum all the Lorentzian functions. Figure 5(b) shows the histograms and the simulated spectrum. In comparison with the observed and the calculated spectra [see Figs. 5(a) and (b)], one concludes that the experimental spectrum agrees well with the calculated. Figure 5(c) shows the density-of-states profile of the spin-polarized electronic structure of $V_5{}^+$. The difference in the electron occupations in the spin-up and spin-down states is two, that is, $V_5{}^+$ has the spin magnetic moment of 2 μ_B (triplet state). The vanadium cluster ions of other sizes are also treated by a similar procedure. As the cluster size increases, the spin magnetic moment, which is 4 μ_B at the monomer, decreases rapidly and gradually approaches the paramagnetic bulk value.

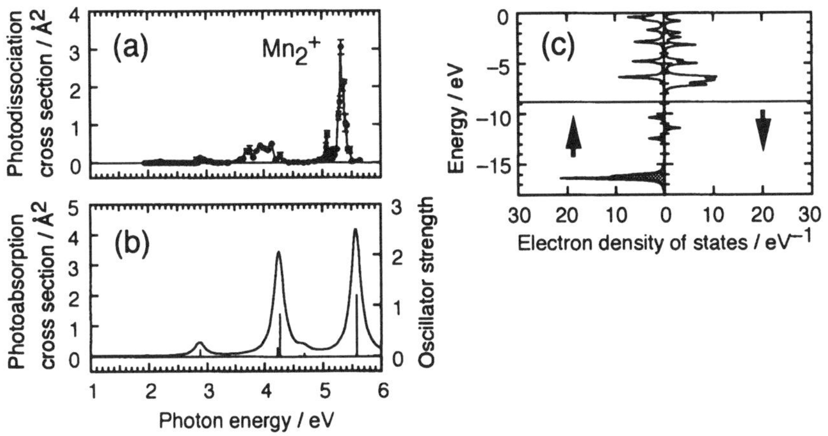

Fig. 6. The optical absorption spectrum and the electronic structure of Mn_2^+. (a) Experimental data, where a photodissociation action spectrum of Mn_2^+ was measured by observing Mn^+ photofragment. (b) The spectrum calculated by a hybrid-type density-functional method. The bars show oscillator strengths; the solid line a spectral profile. (c) Density-of-states profiles of the majority and the minority spin electrons obtained by the same theoretical calculation. The shadows indicate occupied electronic levels. The manganese dimer ion, Mn_2^+, was shown to have a spin multiplicity of twelve with a bond length of 3.01 Å.[14]

However, the spin magnetic moment enhances greatly when the interatomic distance in the cluster ion is elongated by changing its size or its charge state.[12,13,15,17]

Let us show another example, manganese cluster ions. Figure 6(a) represents the optical absorption spectrum of Mn_2^+ experimentally determined and Fig. 6(b) the theoretical spectrum; the two spectra agree well.[14] The dimer ion, Mn_2^+, in the ground state has 12-fold degenerate spin manifolds and the interatomic distance of 3.01 Å. The electron spin resonance of Mn_2^+ in an argon matrix at a low-temperature also demonstrates the spin moment as high as that observed in the gas phase.[18] As shown in Fig. 6(c), Mn_2^+, has all the 3*d* electrons and two of the 4*s* electrons polarized upwardly and one of the 4*s* electrons downwardly. The population difference amounts to the spin magnetic moment of 11 μ_B. This giant magnetic moment of Mn_2^+ originates from its longer interatomic distance, which results from the weak bonding energy of 1.4 eV between the two manganese atoms.[19–21] The bonding energy is much smaller than any typical transition metal dimers and their ions. As the cluster size increases, the manganese cluster or its ion is predicted to change from a ferromagnetic spin-coupling scheme to an antiferromagnetic one at the sizes below 8 in accordance with the increase of the bonding energy.

A solid specimen of cobalt exhibits ferromagnetism with the magnetic moment of 1.72 μ_B per atom. The spin magnetic moment of a cobalt cluster ion is 2.0 and 1.7 μ_B per atom at the sizes of 3 and 4, respectively, which are close to the bulk value. It looks that the magnetic moment per atom does not change appreciably with the size.[5,11,16]

It is worth to note that even vanadium and manganese, whose solid specimens are none ferromagnetic, exhibit ferromagnetic coupling between adjacent spins of the small cluster ions in particular. There is a general trend that two adjacent spins of a cluster ion couple ferromagnetically when the interatomic interaction is weak or the interatomic distance is long and as a result the overlapping of the relevant electron orbitals is not significant.

We focused on the size-selected cluster ions with small sizes in the above. On the other hand, the magnetic moments of neutral transition metal clusters in the size range between several and several hundreds are successfully measured by deflection in an inhomogeneous magnetic field.[22] This method is basically the same as the Stern–Gerlach experiment known by the discovery of the electron spin by using the deflection of an atomic beam of silver.

3.4. *Photoelectron Spectra of Transition Metal Cluster Anions*

The other powerful method for determining the electronic structure of a transition metal-cluster is photoelectron spectroscopy. The cluster anion is irradiated with a laser, which has a sufficiently high energy so that the resulting electrons gain a large kinetic energy containing information on the electronic structure of the cluster anion. Figure 7(a) shows a schematic diagram of the apparatus used. A cluster anion of a given size is selected out of those produced by laser ablation by using a TOF mass spectrometer, and is decelerated down to several tens of electron volts at the laser interaction region of a magnetic bottle-type photoelectron spectrometer. An electron emitted to any direction by laser irradiation on the cluster anion is collected in a magnetic field of ~ 1000 G, and detected after travelling by 2 m in a weak magnetic field of 10 G. Because of the large collection efficiency of the photoelectron, the sensitivity of the spectrometer is raised by 2–3 orders of magnitude in comparison with an ordinary spectrometer. The transit

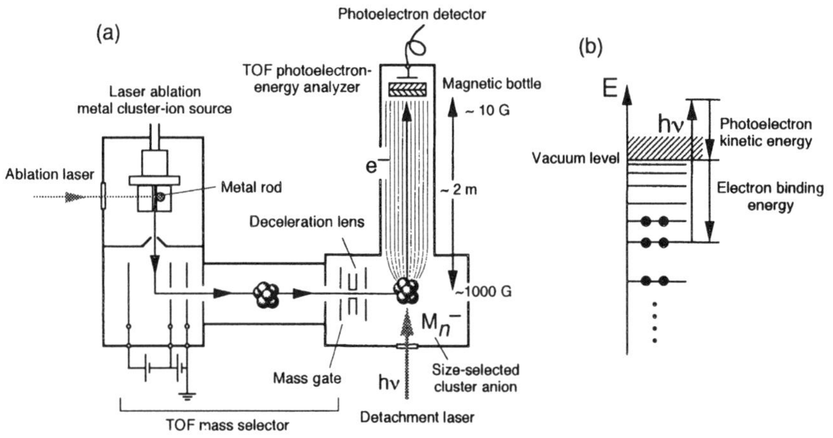

Fig. 7. Photoelectron spectroscopy of size-selected cluster anions. Schemes of (a) the experimental setup and (b) the principle.

time of the photoelectron measured is converted to its kinetic energy. The energy resolution reaches as high as 10 meV by reducing the transnational energy of the incoming cluster anion. As depicted in Fig. 7(b), the electron binding energy is given by the difference between the photon energy and the electron kinetic energy.

In the first place, the photoelectron spectrum of vanadium cluster anions (V_n^-; $3 < n < 100$) is shown in Fig. 8.[15] As the cluster-size increases, the spectral feature changes significantly until the size of ~ 10 and, above it, remains unchanged, but its onset energy gradually increases. The vertical detachment energy of the vanadium cluster anion derived from the onset energy of the photoelectron is plotted as a function of $n^{-1/3}$ (see Fig. 9); the value, $n^{-1/3}$, is roughly proportional to the reciprocal of the diameter of the cluster anion. All the data points above the size of ~ 10 are on the solid line, which is calculated on the assumption that the cluster anion consists of a conductive sphere and one-electron attached by a charge-induced dipole interaction (spherical conducting droplet model). In other words, the vanadium cluster anions with the sizes larger than ~ 10 are regarded to be "metallic" and their valence electrons are delocalized. Below this critical size, vanadium cluster anion is regarded as "molecular" and the valence electrons are localized in the molecular orbitals constituting the covalent bonds. Let us remark that almost the same picture is true to other cluster anions of $3d$-transition metals (except for manganese, which

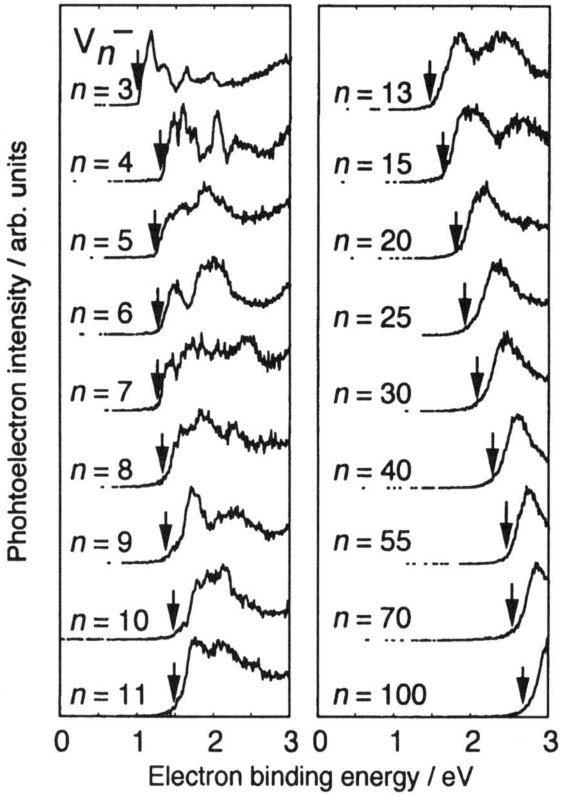

Fig. 8. Photoelectron spectra of vanadium cluster anions, V_n^- (n = 3–100). The arrows show the onsets of the spectra corresponding to the electron affinities.[15]

has not been reported). The deviation of the data points from the linear relationship at the size of $\sim$ 10 is contrasted to the size-dependence of copper cluster anions, where the spherical conducting droplet model is valid even at smaller sizes.[23]

3.5. *Metal-Insulator Transition of Mercury Cluster*

Then, a fundamental question arises; how many metal atoms should be in a cluster, which exhibits a metallic nature? Mercury is the most suitable element to examine such a question. Liquid mercury is a typical metal while two mercury atoms is weakly bonded by a dispersion force, because a mercury atom has a closed electronic structure, where all the orbitals

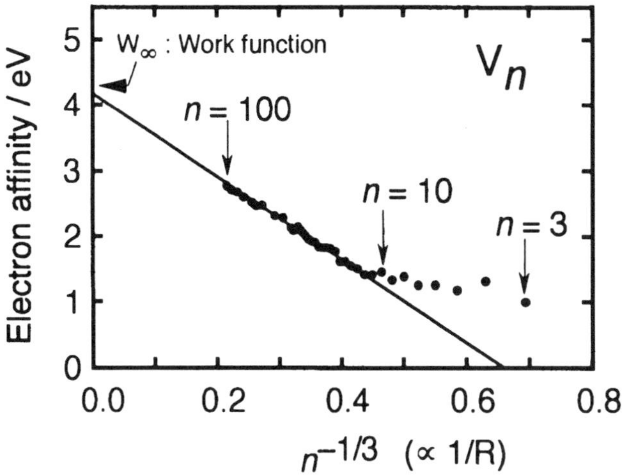

Fig. 9. Size dependence of electron affinities (EA) of vanadium clusters. The abscissa is proportional to the reciprocal of the cluster radius. The solid line shows the predicted values by the spherical conducting drop model; $EA = W_\infty - e^2/(2R)$, where W_∞ is the work function of a bulk metal. The model well explains the experimental data above $n \approx 10$.[15]

are occupied until 6s levels. It follows that a mercury cluster, which is an insulator at small sizes, should change to be metallic above a critical size ("metal-insulator transition"). In a mercury atomic anion, the electrons fully occupy the atomic levels up to 6s-level, while the outermost electron singly occupies 6p-level. As the size increases, the energy bands related to 6s and 6p levels are broadened and finally overlap at the critical size. Accordingly, the metal-insulator transition can be attested by observing the shape of the energy band in the vicinity of the "Fermi energy" of the cluster anion. In fact, inner shell excitation spectroscopy, optical spectroscopy (plasmon excitation), measurements of the cohesive energy and the ionization energy have examined the disappearance of the energy gap. These examinations have shown that the electronic structure begins to change gradually and 6s and 6p bands are mixed at the sizes of 80–100.

Recently a more detailed feature of the energy gap has been observed in the energy region in the vicinity of the Fermi energy by using photoelectron spectroscopy.[24] Figure 10 shows typical photoelectron spectra of mercury cluster anions with different sizes. The spectrum of a mercury cluster anion, Hg_n^-, consists of an isolated peak assigned to the transition of a 6p electron to the continuum and a broad band assigned to the transition of

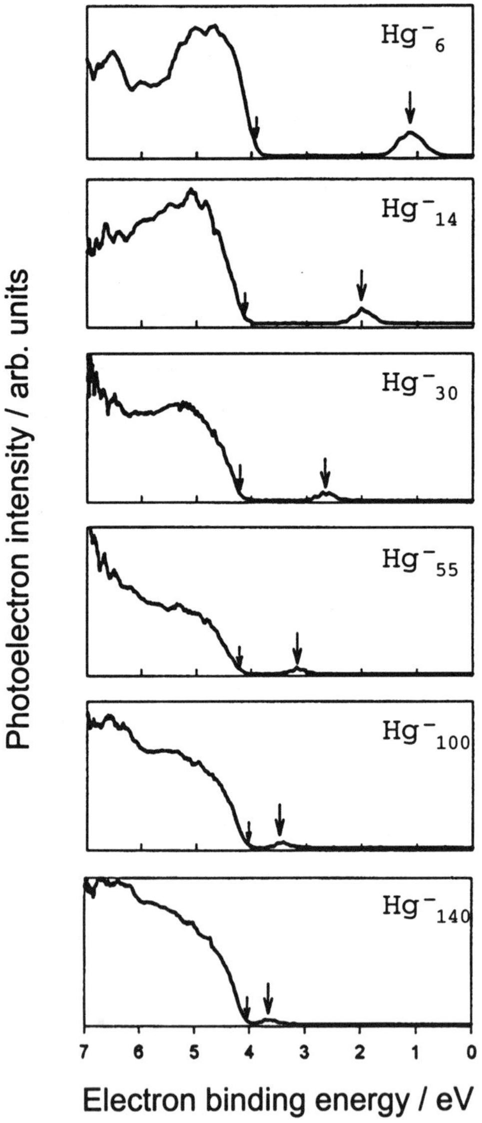

Fig. 10. Photoelectron spectra of mercury cluster anions. The right arrows indicate the bottom of the 6*p* band and the left ones the top of the 6*s* band. The difference between the two values corresponds to the *s*-*p* band gap. (Adapted from Ref. 24.)

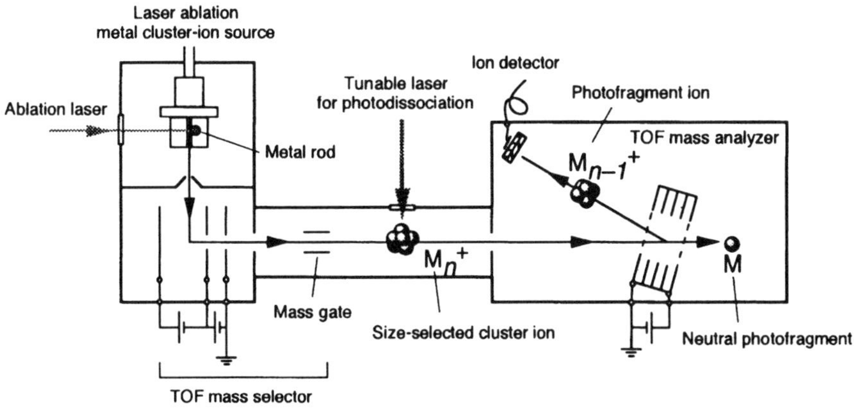

Fig. 11. Size dependence of electron-binding energies of the s and p bands along with the s-p gap energies. The abscissa, $n^{-1/3}$, is proportional to the reciprocal of the cluster radius. The solid line was obtained by fitting the s-p gap data between $n = 50$ and 250. Extrapolation of the fitting predicts that the s-p gap vanishes at $n = 400 \pm 30$. (Adapted from Ref. 24.)

$6s$ electrons to the continuum. The arrows on the $6p$-peak and in the onset region of the $6s$-band represent the energies of the lowest-unoccupied and the highest-occupied orbitals of Hg_n, respectively. Therefore, the difference of the energies indicated by the two arrows corresponds to the energy gap of Hg_n. Figure 11 shows the energies of s and p bands corresponding to the two arrows and their difference (energy gap or s-p gap) as a function of $n^{-1/3}$ proportional to the reciprocal of the cluster diameter. Fitting the data points for the sizes from 55 to 250 makes the solid line. The extrapolation of the solid line to a larger size leads us to conclude that the band gap becomes zero at the size of 400 (± 30). On the other hand, the data points tend to deviate from this solid line in the vicinity of a size of several tens or smaller. The deviation is not related to the metal-insulator transition but a different one. It is likely that former studies misinterpret the latter one to be the metal-insulator transition.

4. Specific Heat and Melting Point

A specific heat is one of the most important observables for understanding the thermodynamical feature of a cluster and its ion. In spite of the importance of the specific heat measurement, no study has been reported until recently because no suitable probe is available for the temperature measurement of a cluster. Recently two methods for the specific heat measurements have been developed, as described in the next subsections.

4.1. *Transition of Magnetic Phases in Ferromagnetic Metal Clusters*

The clusters of ferromagnetic metals exhibit superparamagnetism above 100 K, when the magnetic moments are measured by the Stern–Gerlach method, although such a cluster itself is a single-domain ferromagnet. As the single cluster thermally fluctuates around the magnetic easy axis, it turns out to be superparamagnetic as a result of time average during the interaction of the cluster with an inhomogeneous magnetic field whose interaction time is much longer than the relaxation time for the fluctuation. Therefore, the magnetic moment measured by the Stern–Gerlach method contains the information of the cluster temperature, and can be used as a "thermometer". On the basis of this principle, the specific heats of the clusters of iron, cobalt and nickel have been measured.[25,26] Figure 12 shows a schematic diagram of the apparatus. Clusters of interest are produced in a small cell containing a metal plate by laser ablation. The clusters in the cell are in equilibrium with the temperature of the cell, which is varied from 78–1000 K. A pulsed beam of the clusters having the same temperature as

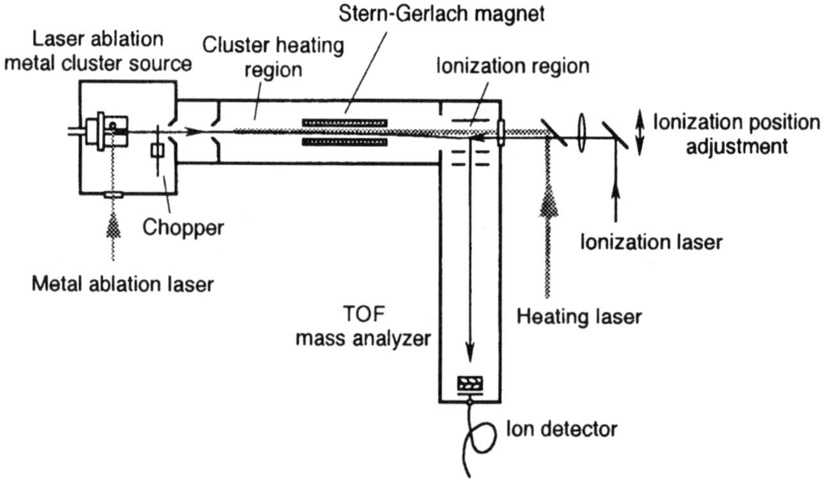

Fig. 12. The Stern–Gerlach type experimental setup for the measurements of magnetic moments and specific heats of metal clusters.

that of the cell are passed through an inhomogeneous magnetic field, and is ionized by an excimer laser (photon energy of $\sim$ 6.4 eV) on the assumption of no fragmentation by the ionization. The cluster ions thus photoionized are detected by a TOF mass spectrometer. Passing through the inhomogeneous magnetic field, each cluster is deflected by an extent proportional to its magnetic moment. The extent of the displacement by the inhomogeneous field is measured by changing the spatial position of the ionization excimer laser. The magnetic moment of the cluster is evaluated from the displacement at which the intensity of the cluster ion is maximized. Figure 13(a) shows the magnetic moments of nickel clusters thus obtained with several different sizes as a function of the temperature (of the cell). As the temperature and the size increase, the magnetic moment of the cluster decreases. In comparison with the temperature dependence of a nickel metal (dotted curve with Curie temperature of 625 K), one concludes that a cluster with a size of 550–600 or larger behaves similarly to the nickel metal.

In the specific heat measurement, a cluster of interest is irradiated with a 532-nm laser (2.33 eV) in order to increase its internal energy, and a temperature rise due to the internal-energy increase is determined from a simultaneous change of its magnetic moment. The specific heat of the cluster is evaluated from the internal-energy increase divided by the temperature rise. The temperature dependence of the specific heat of nickel clusters with the sizes of 200–240 is shown in Fig. 13(b). The specific heat

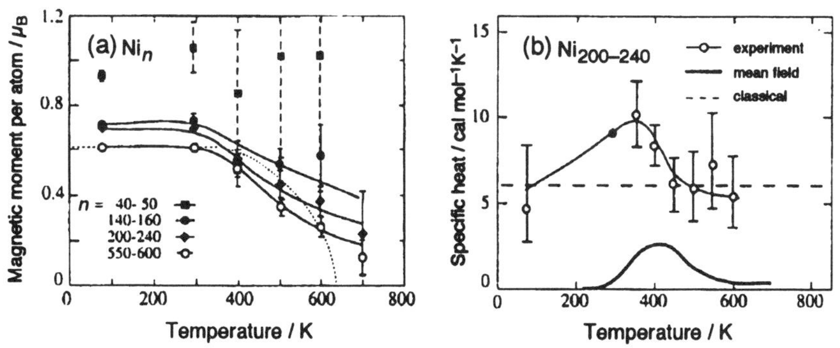

Fig. 13. Magnetic moments and specific heats of nickel clusters as a function of the temperature. (a) Magnetic moment per atom. Solid squares, n = 40–50; solid circles, n = 140–160; solid diamonds, n = 200–240; open circles, n = 550–600. (b) Specific heat obtained by averaging over n = 200–240. Open circles, experimental data; thick line, the mean field theory; dashed line, the classical Dulong–Petit prediction. (Adapted from Ref. 25.)

at 80 K agrees with the Dulong–Petit specific heat based on the classical thermodynamics (dashed line). As the temperature increases, the specific heat increases, reaches the maximum at about 350 K, and then decreases. The temperature dependence above the Dulong–Petit value originates from the magnetic specific heat, which is well reproduced by the mean field theory (thick solid curve). The magnetic phase transition (second-order phase transition) of the nickel clusters is manifested on the temperature dependence of the specific heat. Cobalt clusters exhibit more or less the same behavior, but iron clusters show more complex temperature dependence, which could be related to the fact that a solid iron has bcc and fcc structures in addition to high and low spin states of the fcc structure.

4.2. *Solid-Liquid Phase Transition in Alkali-Metal Clusters*

In the previous subsection, we described the magnetic phase transition as a typical example of the second-order phase transition. In this subsection, the transition from a solid-like to liquid-like states is explained as an example of the first-order phase transition.[27–29] In general, the melting point and the latent heat of a particle decrease with the decrease of its diameter, mainly because the surface energy of the particle is reduced and hence it becomes floppy. A solid material is regarded to be "melted" intuitively when the thermal displacement of the constituent atoms exceeds 10–15% of the lattice constant (Lindemann criterion). The specificity of the clusters is that the transition occurs in a finite temperature range; the melting point does not coincide with the freezing point. Actually this phenomenon is predicted theoretically.

Let us show an example of an alkali-metal cluster ion. In the specific heat measurement, the temperature is probed by detecting a number of alkali atoms evaporating from the alkali-metal cluster ion concerned when it is irradiated with a laser. Figure 14 shows the principle of the measurement. At first, a beam of alkali-metal cluster ions with a given size is equilibrated with a helium gas at a given temperature, T_1, by passing the beam through it. Then, the beam is irradiated with a laser. The cluster ions are fragmented after the absorbed energy is converted to the internal-energy [see Fig. 14(b)]. Note that no dissociation is observed without the laser irradiation [see Fig. 14(a)]. The two profiles shown in Fig. 14(b) originate from photodissociation by absorption of n and $(n+1)$ photons. When the temperature is increased to T_2, the profiles shift toward smaller sizes [see Fig. 14(c)]. A further increment of the temperature from T_2 to T_3 results

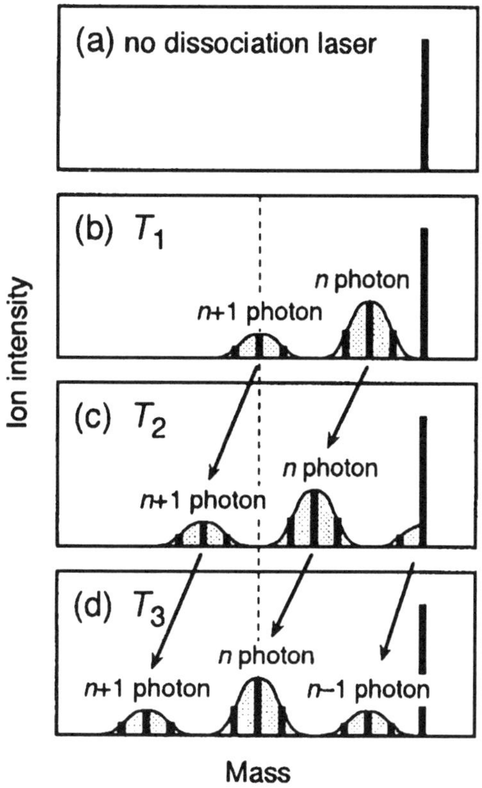

Fig. 14. Principle of the specific heat measurement. Size-distribution of photo fragments is observed as the temperature of the primary cluster ion is varied. As the temperature increases, the sizes of the photofragments generated by the n-photon process shift toward lower sizes. Panels (b) and (d) illustrate that the n-photon process at the temperature, T_3, generates photofragments of the same sizes that produced by the $(n+1)$-photon process at the temperature, $T_1 (< T_3)$. The specific heat, C, is evaluated by $C = \Delta Q/\Delta T$, where the increase in the internal energy of the cluster, ΔQ, is the photon energy, $h\nu$, and the temperature rise, ΔT, is equal to T_3–T_1.

in a further shift toward much smaller sizes [see Fig. 14(d)], where the profile due to $(n + 1)$-photon absorption observed at the temperature of T_1 coincides with that due to n-photon absorption (at T_3). Namely, the one-photon energy introduced into the internal-energy of the cluster ion causes the cluster temperature to increase by T_3–T_1. The specific heat at the mid-point, $T = (T_3 + T_1)/2$, is given by the internal-energy increase (photon energy) divided by T_3–T_1. Figure 15 shows the temperature dependence

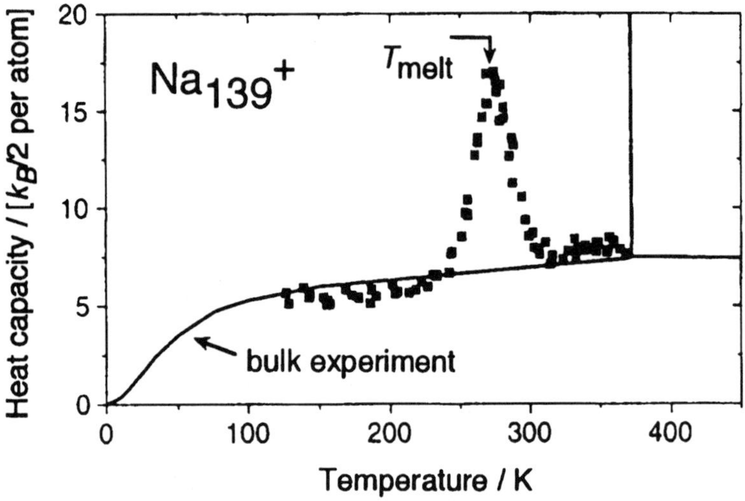

Fig. 15. Temperature dependence of the heat capacity per atom of $Na_{139}{}^{+}$ (solid squares). The peak at $T_{\text{melt}} = 267$ K indicates the melting temperature. The solid line shows the experimental data for the bulk sodium, whose melting point appears at 371 K. (Adapted from Ref. 29.)

of the specific heat (or heat capacity per atom) of $Na_{139}{}^{+}$. The peak of the specific heat curve at 267 K corresponds to the melting temperature of $Na_{139}{}^{+}$. It is lower by 104 K (or 28%) than that of a solid sample of sodium; the specific heat curve of the bulk sodium is given as a solid curve in Fig. 15. On the other hand, the latent heat evaluated to be 1.98 eV by integration of the specific heat curve; the latent heat of $Na_{139}{}^{+}$ is smaller by 48% than that of the bulk sodium (3.69 eV). Figure 16 shows the melting temperature of sodium cluster ions as a function of the size. The two peaks at the sizes of 57 and 142 are discernible, but are not explained by any icosahedral structure and the jellium model. The melting temperature as well as the latent heat does not depend smoothly on the size, and is hardly explained by a simple theory.

5. Reactivity of Isolated Clusters

Size-dependent characteristics are also manifested in the reactivity of clusters, especially in the clusters of transition-metal elements used very often as catalysts. Several examples are shown in the following subsections.

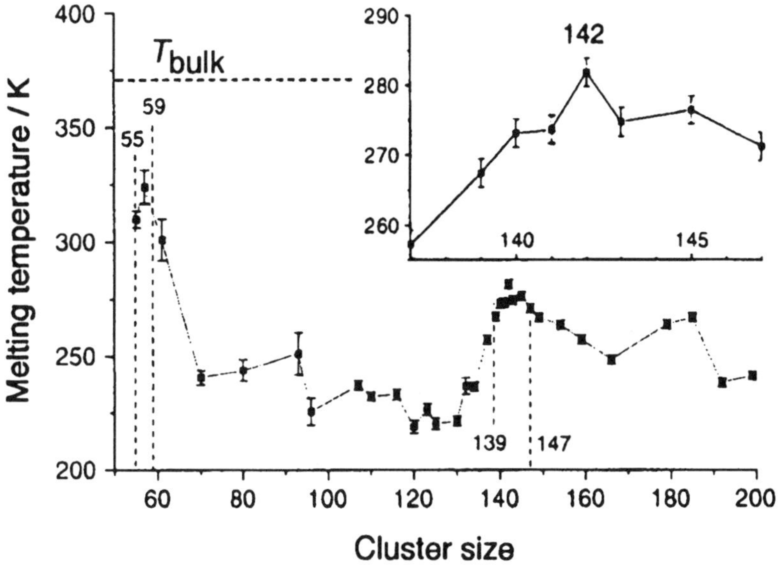

Fig. 16. Size dependence of the melting temperature of sodium cluster ions. (Adapted from Ref. 29.)

5.1. *Adsorption of Simple Molecules on Iron, Cobalt and Nickel Clusters*

Hydrogen adsorption has been examined on transition metal clusters, since this is a bench mark process in the catalysis studies. Clusters produced by the laser ablation method are allowed to collide with a hydrogen gas in a narrow reaction tube and are photoionized for mass-spectroscopic identification. The relative reactivity is estimated from the ratio of the intensities of a given cluster ion with and without a hydrogen gas in the tube.

Figure 17 shows the relative reactivity of iron (top panel), cobalt (middle panel) and nickel (bottom panel) clusters as a function of the cluster size. The relative reactivity changes by 2–3 orders of magnitude in the size range below 20. The size dependence of the relative reactivity has a good correlation with the promotion energy, the energy required for an electron in the highest-occupied molecular orbital to be promoted to the lowest-unoccupied molecular orbital.[30] The promotion energy, E_p, is given in terms of the ionization energy, IP, electron affinity, EA, and the Coulomb energy of an electron to a positively charged sphere with the diameter, R; $E_p = IP - EA - e^2/R$. The good correlation implies that both the electron

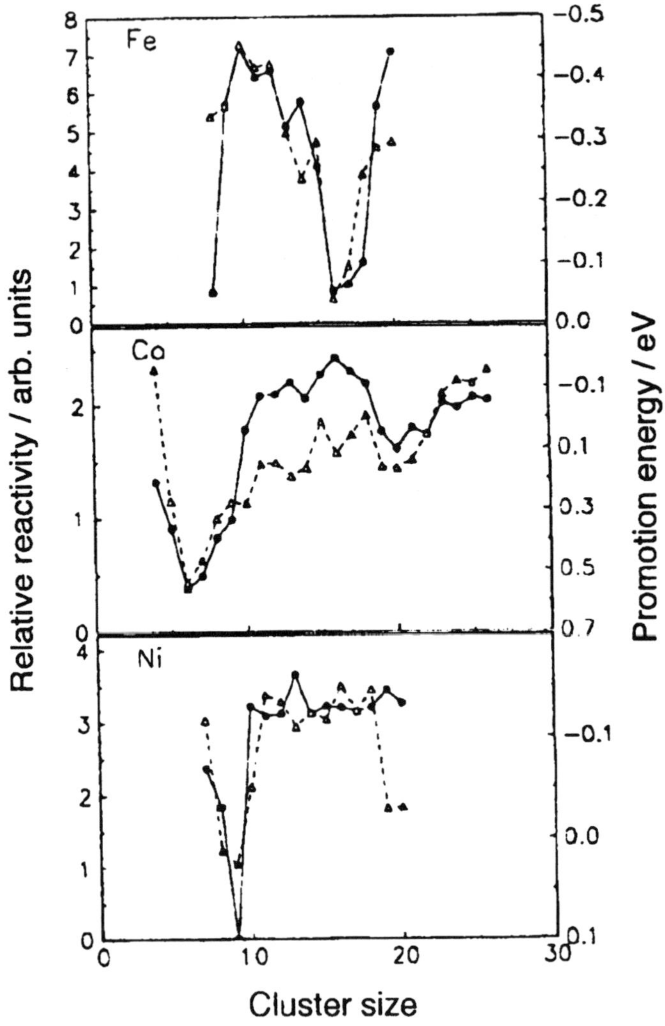

Fig. 17. Size-dependent reactivity of iron, cobalt and nickel clusters. Solid circles connected by the solid lines show reactivity in hydrogen-molecule adsorption. Open triangles connected by the dashed lines show promotion energies of an electron from the highest-occupied molecular orbital to the lowest-unoccupied molecular orbital. The correlation between the promotion energy and the reactivity was found, where the higher the promotion energy , the lower the reactivity. (Adapted from Ref. 30.)

donation and the back donation between the adsorbed molecule and the surface involved contribute to the energy barrier for the adsorption process.

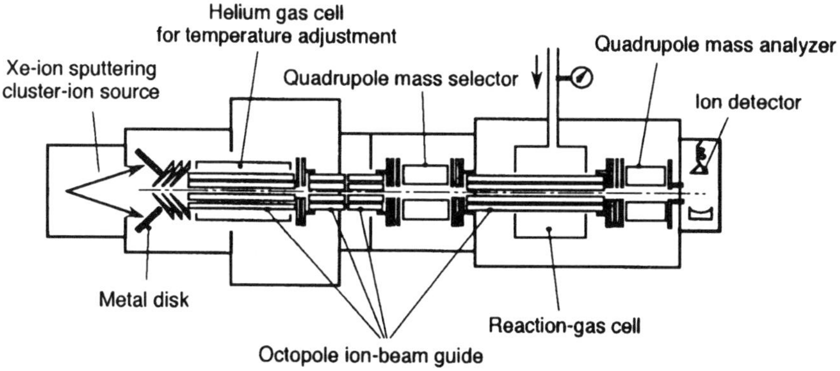

Fig. 18. Experimental setup for the collision-induced reaction of size-selected cluster ions with gas phase molecules.

5.2. *Size-Dependent Reaction of Nickel Cluster Ions with Methanol*

Size-dependent chemical reactions proceed on transition metal clusters and their ions. In this subsection, we describe the reaction of a methanol molecule on a size-selected nickel cluster ion.[31] Figure 18 depicts a schematic drawing of the apparatus employed. Nickel cluster ions are produced by sputtering four targets of nickel metal by xenon ions, are cooled in an octopole ion-beam guide in collision with a helium gas having a controlled temperature, and then are size-selected in a quadrupole mass spectrometer. A size-selected nickel cluster ion is allowed to react with a methanol molecule in a single-collision condition. The product ions are identified and detected by the second quadrupole mass spectrometer. The absolute reaction cross section is determined from the methanol pressure dependence of ions produced by the reaction. The reaction cross section is measured as a function of the collision energy and the size of the primary nickel cluster ion with a given temperature, which is actually the temperature of the helium gas. Figure 19 shows mass spectra of the product ions from the primary nickel cluster ions with the sizes of 4, 6 and 8. The reaction changes drastically with the size of the primary cluster ion; demethanation (formation of oxide by dissociation of methane from methanol) at the size of 4, methanol chemisorption at the size of 6 and carbide formation at the size of 8. Figure 20 demonstrates the size-dependent reaction cross sections for these three processes obtained by the analysis of the mass spectra.

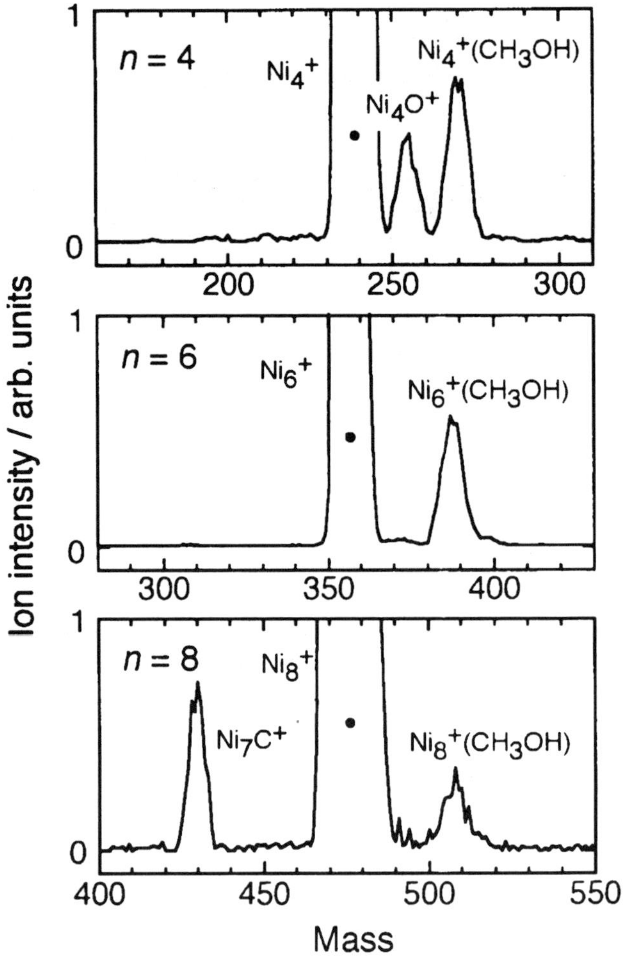

Fig. 19. Mass spectra of product ions generated by the reaction of nickel cluster ions, Ni_n^+, with a methanol molecule. (Adapted from Ref. 31.)

The results are explained in terms of the scheme described as follows: an incoming methanol molecule is bound weakly to a given nickel cluster ion by charge-induced dipole interaction, and is chemisorbed after surpassing the energy barrier between the physisorbed and the chemisorbed states. The demethanation and the carbide formation proceed through the methanol chemisorption. The reaction cross section for the physisorption is regarded to be the Langevin cross section. The reaction cross sections for

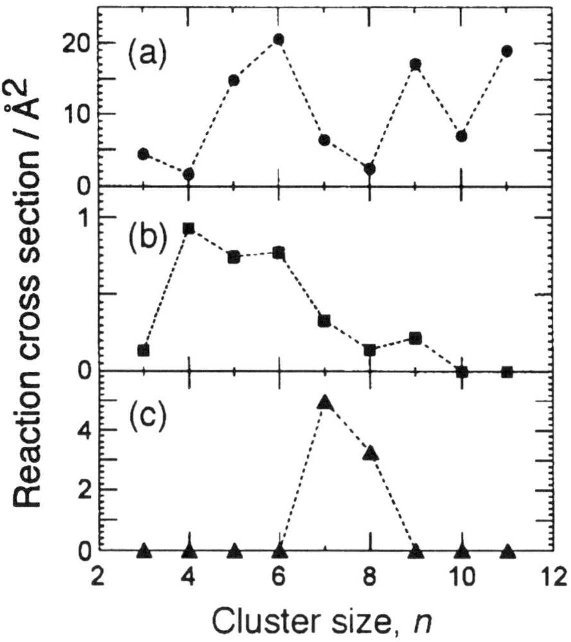

Fig. 20. Partial cross sections of the reaction between Ni_n^+ and a methanol molecule. (a) Adsorption; (b) demethanation; (c) carbide formation. (Adapted from Ref. 31.)

the methanol chemisorption, the demethanation, and the carbide formation are calculated with the aid of the statistical Rice–Ramsperger–Kassel (RRK) theory. The energy barrier is the lowest at the size of 6 and high at the sizes of 4, 7 and 8, in accordance with the largest cross section for the chemisorption at the size of 6. On the contrary to the expectation, the cross sections for the demethanation and the carbide formation are the largest at the sizes of 4 and 7–8, respectively, because the height of the other energy barrier toward the reactions from the chemisorption should be taken into account. The branching fraction between the demethanation and the carbide formation is related to interatomic distance in the nickel cluster ion. The carbide formation needs a chemisorbed precursor occupying two nickel atoms while the demethanation does one. Actually, the interatomic distances below the size of 6 seem to be too short to prepare the chemisorbed precursor for the carbide formation. More thorough experimental and theoretical studies are expected to elucidate the fundamental feature of the size-dependent reaction.

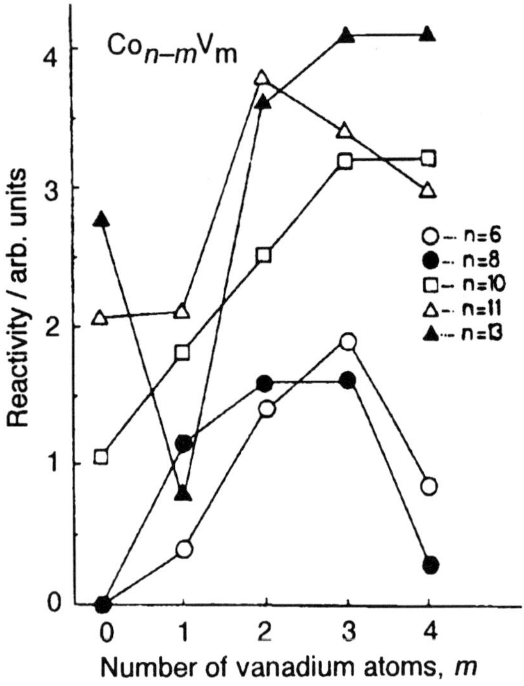

Fig. 21. Reactivity of cobalt-vanadium binary clusters, $Co_{n-m}V_m$, in hydrogen sorption. Note that $Co_{12}V$ is extremely less reactive than other clusters of $n = 13$. (Adapted from Ref. 32.)

5.3. *Effect of Vanadium Atom on the Reactivity of Cobalt Cluster*

The reactivity of metal clusters is deeply influenced by addition of foreign atoms to them. It is shown in this subsection that hydrogen sorption on cobalt clusters is influenced dramatically by addition of vanadium atoms to them.[32] Binary clusters are produced by the laser ablation on separate two metal rods. The hydrogen sorption takes place in a narrow reaction tube, in which the binary clusters react with hydrogen molecules. The relative reactivity for the hydrogen sorption obtained from the intensity ratio between cluster ions with and without a hydrogen gas in the reaction tube. Change of the reactivity of a cobalt cluster, Co_n, is examined by substituting a constituent cobalt atom with a foreign vanadium atom one by one. Figure 21 shows the reactivity change thus obtained as a function of the number of the substituted vanadium atoms, m, for primary cobalt clusters with different initial sizes, n. The reactivity generally increases with the number of the substituted vanadium atoms in the range $m \leqq 3$ except for $Co_{12}V$, whose reactivity is significantly low. The result is explained in such a manner that the vanadium atom is located in the center of $Co_{12}V$ having an icosahedral structure and attracts electrons from the surrounding cobalt atoms.

6. Clusters Supported on Solid Surfaces

As described in the previous sections, we exemplified remarkable size-dependent characteristics of isolated clusters and cluster ions. Recently, many studies have been conducted on how to use these remarkable characteristics in practical applications; for instance, deposition of size-selected cluster ions on a solid surface. The deposited clusters can be used as individual functional elements and the cluster-deposited layer itself can be utilized as a functional material. In particular, the reactivity of a solid surface covered with metal clusters is given in this subsection in order to exemplify a vital utility of clusters on a solid surface.

6.1. *Silver Clusters: Latent Images in Photographic Plates*

A photographic plate is made of a glass or organic polymer film, on which fine grains of silver bromide are dispersed. Under exposure of light, electrons liberated from the grains reduce interstitial silver ions into silver atoms repeatedly. The liberated silver atoms in the grains aggregate into silver clusters, which are regarded as latent images formed in the photographic plate. If the latent images (silver clusters) are sufficiently large, they catalyze to reduce the grains of silver bromide in their vicinity into real images (much larger aggregates of silver atoms) by the development of the plate. It is obvious that the smallest number of the constituent silver atoms in a latent image limits the maximum sensitivity of the photographic plate. Then, a question arises: what is the smallest silver cluster that catalyzes the reduction of silver bromide in a reducing environment? This was a long-standing issue in the photographic chemistry, and has been solved by depositing silver clusters of a given size on a virgin photographic plate and examining whether the clusters reduce silver bromide in the vicinity of them in a reducing environment.[33,34]

In the actual experiment, size-selected silver ions are softly landed on a transparent conductive glass plate covered with fine grains of silver bromide with a diameter of about 1 μm; the cluster ions are rapidly neutralized to

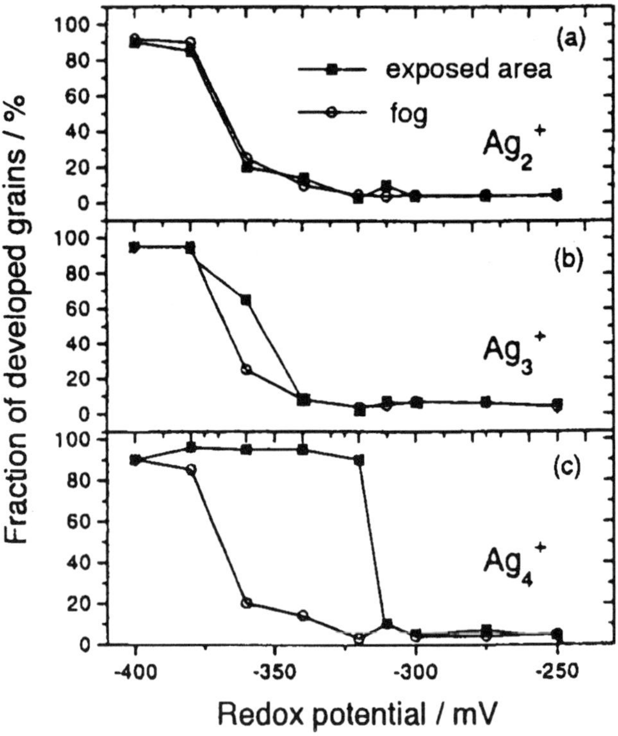

Fig. 22. Fractions of developed silver-bromide grains after deposition of size-selected silver clusters (Ag_n^+; $n = 2$–4) as a function of the redox potential. Solid squares show the data in the cluster-exposed area; open circles show those in the unexposed area (fog) as references. The silver dimer, panel (a), exhibited no influence in the catalysis of the reduction of the grains. On the other hand, the tetramer, panel (c), catalyzed the reaction in the redox potential range between −360 and −320 mV. (Adapted from Ref. 34.)

be intact after being landed. When the cluster ions are being deposited, a part of the plate is covered with a shutter so as to use the shaded part as a reference. The whole plate is reduced (developed) in an electrolyte solution with changing its redox potential. Similar sample plates are prepared with varying the size of the cluster ions and the amount of the cluster-ion deposition, and reduced by changing the redox potential. In Fig. 22, the fraction of developed grains is plotted against the redox potential. Silver dimers do not give any influence on the reduction of the grains of silver bromide and trimers slightly in a narrow region of the redox potential, whereas tetramers do in a wide region between −360 and −320 mV. Silver clusters larger than the tetramer are found to catalyze the reduction of silver bromide. The dependence on the cluster dose shows that, in the case of the tetramer and larger clusters, a coverage of 20 clusters per grain is sufficient to achieve a full development of the sample. In contrast, the monomers and the dimers do not act any on the reduction even with a 1000-times higher coverage. A similar measurement by use of silver-gold binary clusters has revealed that $AuAg_3$ is the smallest cluster that acts as a catalyst on the reduction of silver bromide; the gold atom just enhances the reactivity of Ag_3.

6.2. *Palladium Clusters: Benzene Synthesis through Cyclotrimerization of Acetylene*

Trimerization of acetylene into benzene is known to proceed on a single crystal of palladium and on fine particles of palladium dispersed on a substrate. Among them, Pd (111) surface is the most active for the trimerization because the surface has a site with three fold symmetry at which three acetylene molecules are adequately adsorbed for the trimerization into benzene; geometry-controlling reaction. In the trimerization involving a palladium cluster, it is expected that the catalytic activity of the trimerization begins to appear at a critical size as the cluster size increases because a small cluster does not have such an active site with three-fold symmetry but a larger cluster should have.

The following experiment has been performed:[35] Size-selected palladium cluster ions are softly landed on a thin film of MgO(100) prepared on a Mo(100) surface at a translational energy below 0.2 eV. The landed cluster ions are neutralized very rapidly without being suffered from any decomposition. The clusters do not migrate over the surface at the temperature of 90 K and do not interact directly with each other at a low-density with a coverage less than 1% of a monolayer. Acetylene molecules are introduced over the surface by an amount that each palladium atom shares about five acetylene molecules. Products from the surface are detected mass-spectroscopically with raising the surface temperature (temperature-programmed-reaction: TPR). Figure 23 shows the intensity of the benzene signal as a function of the surface temperature. No product is desorbed from a bare MgO (100) surface (bottom trace), while benzene is produced from palladium deposited surfaces. Several features are discernible: (1) a peak (Pd_{1-30}) at 300 K is attributed to a palladium atom supported on the surface, (2) a broad structure (Pd_{5-6}) in the range of 400–700 K to cluster migration on the surface, (3) a peak (Pd_{7-30}) at 430 K to a site similar to

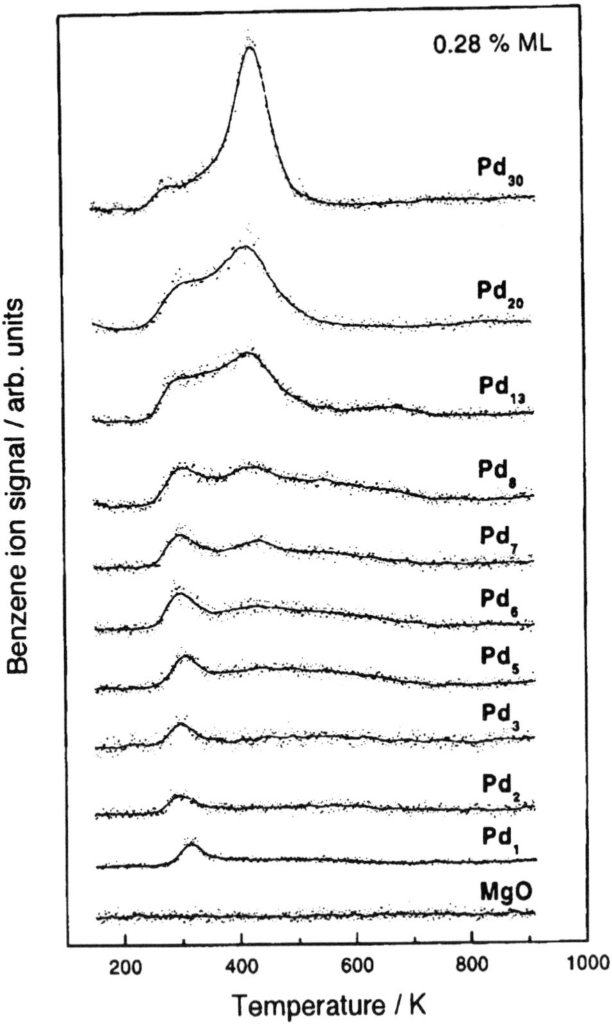

Fig. 23. Catalytic formation of benzene molecules on different palladium cluster sizes by the temperature-programmed-reaction (TPR) experiment. The benzene was produced through the cyclo trimerization of acetylene catalyzed by size-selected palladium clusters, Pd_n, supported on a MgO film. (Adapted from Ref. 35.)

the three-fold symmetry site on Pd(111). The catalytic activity of palladium atoms on the surface which has not been known so far is explained by a density-functional analysis as follows: two acetylene molecules attach to

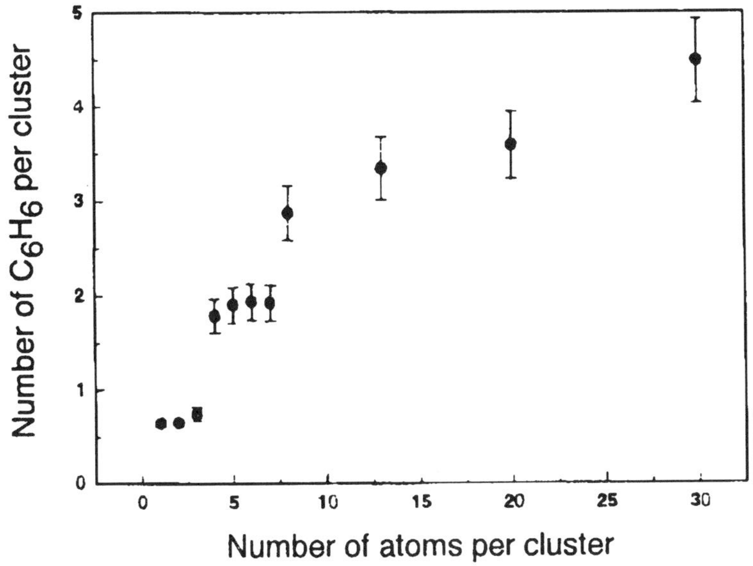

Fig. 24. Reactivity of palladium clusters supported on a MgO film as a function of the cluster size. The reactivity is expressed in the number of benzene molecules produced per cluster in the cluster-catalyzed cyclotrimerization of acetylene. (Adapted from Ref. 35.)

a neutral palladium atom into $Pd(C_4H_4)$, but the third acetylene molecule weakly attaches to it with no further reaction. Once the palladium atom is negatively charged, the structure of the palladium acetylene adduct changes so that the three acetylene molecules in the adduct react into a benzene molecule. In the MgO(100) surface, palladium atoms in defect sites are negatively charged and as a result trimerization proceeds on them. As shown in Fig. 24, the number of benzene molecules per palladium atom increases with cluster-size in a stepwise manner in accordance with the features of the trimerization described above.

6.3. *Platinum Clusters: Oxidation of Carbon Monoxide*

Monodispersed platinum clusters are prepared on a MgO(100) film by deposition of size-selected platinum cluster ions followed by neutralization.[36] Subsequently, the deposited clusters are exposed to an isotopic oxygen gas until about twenty $^{18}O_2$ molecules per platinum atom are adsorbed on the platinum clusters. Temperature-programmed-reaction proves that $^{12}C^{16}O^{18}O$ (carbon dioxide) molecules are produced, when $^{12}C^{16}O$

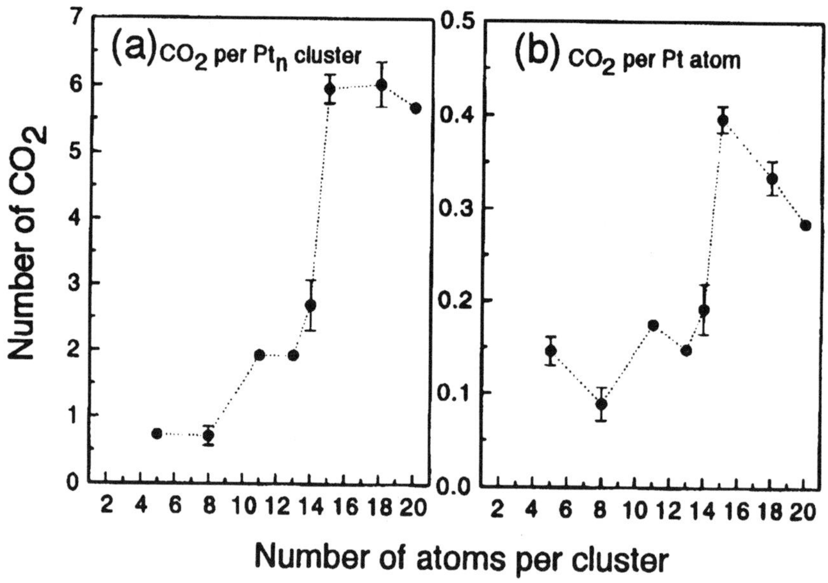

Fig. 25. Reactivity of platinum clusters supported on a MgO film as a function of the cluster-size. The reactivity is expressed in the number of carbon dioxide molecules produced either per Pt_n cluster, panel (a), or per Pt atom, panel (b), in the cluster-catalyzed oxidation of carbon monoxide. (Adapted from Ref. 36.)

(carbon monoxide) molecules are admitted to the surface. The production of the $^{12}C^{16}O^{18}O$ molecules verifies no involvement of the oxygen of MgO in the oxidation of CO into CO_2. Figure 25 shows the number of carbon dioxide molecules produced per cluster [panel (a)] and per atom [panel (b)] as a function of the cluster-size. The reactivity per cluster increases with the size and tends to level off above the size of 15, while that per atom is the highest at the size of 15.

In the oxidation process of carbon monoxide into carbon dioxide, the platinum clusters work for dissociating adsorbed O_2 into O through electron transfer from the platinum clusters to the antibonding π^* orbital of O_2. Its orbital energy is 6.16 eV. On the other hand, the orbital energy of a d electron of the platinum clusters varies between the ionization energy (9.00 eV) of a platinum atom and the work function (5.32 eV) of a solid platinum. The efficiency of the electron transfer is considered to be the highest at the size of 15, because the two orbital energies coincide at this size.

7. Conclusion

In this chapter, we have described the size- and temperature-dependent characteristics of isolated metal clusters and those supported on various substrates. A long and difficult road lies extending before us toward the complete understanding of the clusters, although much endeavor of recent several years gradually unveils the real feature of the clusters. There are still many experimental difficulties to be solved; production of size- and geometry-selected neutral clusters in a large quantity, accumulation of clusters in a trap, on a solid surface, in a liquid media, etc. ultra sensitive detection and characterization of clusters, and so on. On the other hands, the cluster, which is regarded as a few-body system, needs to be characterized by specific observables with new methodologies. Even in a technological point of view, the clusters are useful in many ways; complexity and diversity of the cluster properties facilitate our efficient search and development of novel materials with desired functions.

Acknowledgements

The author is grateful to Professor T. Kondow for his support and encouragement and to Ms M. Yamaguchi for her assistance in the manuscript preparation. A part of the author's research was supported by the Special Cluster Research Project of Genesis Research Institute, Inc.

References

1. For example, *Microcluster Physics*, edited by S. Sugano and H. Koizumi (Springer-Verlag, Berlin, 1998); *Metal Clusters at Surfaces – Structure, Quantum Properties, Physical Chemistry–*, edited by K.-H. Meiwes-Broer (Springer-Verlag, Berlin and Heidelberg, 2000); *Cluster, Selected Papers in Physics X*, edited by T. Kondow, T. Yamaguchi and A. Terasaki (Physical Society of Japan, Tokyo, 1997); *Development in Microcluster Science, Kikan Kagaku Sosetsu 38*, edited by K. Kaya, T. Kondow, O. Kajimoto and N. Nishi (Chemical Society of Japan, Tokyo, 1998).
2. For example, *Clusters of Atoms and Molecules I. Theory, Experiment, and Clusters of Atoms*, edited by H. Haberland (Springer-Verlag, Berlin, 1994); *Clusters of Atoms and Molecules II. Solvation and Chemistry of Free Clusters, and Embedded, Supported and Compressed Clusters*, edited by H. Haberland (Springer-Verlag, Berlin, 1994).
3. M. B. Knickelbein and W. J. C. Menezes, *Phys. Rev. Lett.* **69**, 1046 (1992).
4. B. A. Collings, K. Athanassenas, D. M. Rayner and P. A. Hackett, *Chem. Phys. Lett.* **227**, 490 (1994).
5. S. Minemoto, A. Terasaki and T. Kondow, *J. Chem. Phys.* **104**, 5770 (1996).

6. A. Terasaki, S. Minemoto, M. Iseda and T. Kondow, *Eur. Phys. J.* **D9**, 163 (1999).
7. V. Bonacic-Koutecky, J. Pittner, M. Boiron and P. Fantucci, *J. Chem. Phys.* **110**, 3876 (1999).
8. J. Tiggesbäumker, L. Köller, H. O. Lutz and K. H. Meiwes-Broer, *Chem. Phys. Lett.* **190**, 42 (1992).
9. C. Ellert, M. Schmidt, C. Schmitt, T. Reiners and H. Haberland, *Phys. Rev. Lett.* **75**, 1731 (1995).
10. M. Schmidt, C. Ellert, W. Kronmüller and H. Haberland, *Phys. Rev.* **B59**, 10970 (1999).
11. S. Minemoto, K. Takahashi, J. Matsumoto, M. Iseda, A. Terasaki, H. Imoto and T. Kondow, *Z. Phys.* **D40**, 13 (1997).
12. S. Minemoto, A. Terasaki, H. Imoto and T. Kondow, *J. Chem. Phys.* **109**, 9737 (1998).
13. S. Minemoto, A. Terasaki and T. Kondow, *J. Electron Spectros. Relat. Phenom.* **106**, 171 (2000).
14. A. Terasaki, A. Matsushita, K. Tono, R. T. Yadav, T. M. Briere and T. Kondow, *J. Chem. Phys.* **114**, 9367 (2001).
15. M. Iseda, T. Nishio, S. Y. Han, H. Yoshida, A. Terasaki and T. Kondow, *J. Chem. Phys.* **106**, 2182 (1997).
16. H. Yoshida, A. Terasaki, K. Kobayashi, M. Tsukada and T. Kondow, *J. Chem. Phys.* **102**, 5960 (1995).
17. K. Lee and J. Callaway, *Phys. Rev.* **B48**, 15358 (1993).
18. R. J. Van Zee and W. Weltner Jr., *J. Chem. Phys.* **89**, 4444 (1988).
19. M. F. Jarrold, A. J. Illies and M. T. Bowers, *J. Am. Chem. Soc.* **107**, 7339 (1985).
20. S. K. Nayak and P. Jena, *Chem. Phys. Lett.* **289**, 473 (1998).
21. N. Desmarais, F. A. Reuse and S. N. Khanna, *J. Chem. Phys.* **112**, 5576 (2000).
22. For example, D. M. Cox, D. J. Trevor, R. L. Whetten, E. A. Rohlfing and A. Kaldor, *Phys. Rev.* **B32**, 7290 (1985); D. C. Douglass, A. J. Cox, J. P. Bucher and L. A. Bloomfield, *Phys Rev.* **B47**, 12874 (1993); I. M. L. Billas, A. Châtelain and W. A. de Heer, *Science* **265**, 1682 (1994).
23. O. Cheshnovsky, K. J. Taylor, J. Conceicao and R. E. Smalley, *Phys. Rev. Lett.* **64**, 1785 (1990).
24. R. Busani, M. Folkers and O. Cheshnovsky, *Phys. Rev. Lett.* **81**, 3836 (1998).
25. A. Hirt, D. Gerion, I. M. L. Billas, A. Châtelain and W. A. de Heer, *Z. Phys.* **D40**, 160 (1997).
26. D. Gerion, A. Hirt, I. M. L. Billas, A. Châtelain and W. A. de Heer, *Phys. Rev.* **B62**, 7491 (2000).
27. M. Schmidt, R. Kusche, W. Kronmüller, B. von Issendorff and H. Haberland, *Phys. Rev. Lett.* **79**, 99 (1997).
28. M. Schmidt, R. Kusche, B. von Issendorff and H. Haberland, *Nature* **393**, 238 (1998).
29. R. Kusche, Th. Hippler, M. Schmidt, B. von Issendorff and H. Haberland, *Eur. Phys. J.* **D9**, 1 (1999).

30. J. Conceicao, R. T. Laaksonen, L.-S. Wang, T. Guo, P. Nordlander and R. E. Smalley, *Phys. Rev.* **B51**, 4668 (1995).
31. M. Ichihashi, T. Hanmura, R. T. Yadav and T. Kondow, *J. Phys. Chem.* **A104**, 11885 (2000).
32. S. Nonose, Y. Sone, K. Onodera, S. Sudo and K. Kaya, *J. Phys. Chem.* **94**, 2744 (1990).
33. T. Leisner, Ch. Rosche, S. Wolf, F. Granzer and L. Wöste, *Surf. Rev. Lett.* **3**, 1105 (1996).
34. P. Fayet, F. Granzer, G. Hegenbart, E. Moisar, B. Pischel and L. Wöste, *Phys. Rev. Lett.* **55**, 3002 (1985).
35. S. Abbet, A. Sanchez, U. Heiz, W.-D. Schneider, A. M. Ferrari, G. Pacchioni and N. Rösch, *J. Am. Chem. Soc.* **122**, 3453 (2000).
36. U. Heiz, A. Sanchez, S. Abbet and W.-D. Schneider, *J. Am. Chem. Soc.* **121**, 3214 (1999).

CHAPTER 4

FEMTOSECOND SPECTROSCOPY ON METAL CLUSTERS

Š. Vajda, A. Bartelt, C. Lupulescu and L. Wöste
Institut für Experimentalphysik, Freie Universität Berlin,
Arnimallee 14, D-14195, Berlin, Germany
E-Mail: woeste@physik.fu-berlin.de

Metal clusters exhibit extraordinary chemical and catalytic properties, which sensitively depend upon their size. This behaviour makes them interesting candidates for the real-time analysis of ultrafast photoinduced processes — ultimately leading to coherent control scenarios. An important parameter in this regard is the complexity of the photochemical system. The palette of size-selected metal clusters provides an interesting choice of systems, among which the amount of internal degrees of freedom can be selected. We have performed, therefore, transient two- and multi-photon ionization experiments on small metal clusters of different size in order to probe their wave packet dynamics, structural reorientations, charge transfers and dissociative events in different vibrationally excited electronic states including their ground state. The observed processes were highly dependent on the irradiated pulse parameters like wavelength range, phase and amplitude; an emphasis to employ a feedback control system for generating the optimum pulse shapes. Their spectral and temporal behavoir reflects interesting properties about the investigated system and the irradiated photochemical process. In this paper we present first the vibrational dynamics of bound electronically excited states of alkaline dimers and trimers. Then the decay dynamics of different predissociated electronic states of various homo- and heteronuclear trimers and pentamers are treated, as well as results about the photo-induced fragmentation dynamics of larger aggregates. Further schemes for observing the wave packet and reaction dynamics in the electronic ground state by means of charge reversal spectroscopy (NeNePo) or stimulated Raman pumping (STIRAP) are shown. Since the employed pulse parameters significantly influence the efficiency of the irradiated reactive pathways, like e.g. particular fragmentation channels, an active control experiment was carried out. In the experiment we excited differently branching ionization and fragmentation pathways of photo-excited Na_2K. By employing an evolutionary algoritm for optimizing the phase and amplitude of the applied laser field, the yield of the resulting parent and fragment ions could significantly be influenced and interesting features could be concluded from the obtained optimum pulse shapes.

Contents

1. Introduction

Research on metal clusters over the past twenty years has strongly been motivated by the goal to bridge the gap between the understanding of small molecules and the bulk.[1] The resulting joint theoretical and experimental efforts lead to remarkable findings like the development of the nuclear shell model,[2] the verification of the non-metal/metal transition[3] or the discovery of fullerenes.[4] Crucial elements for these accomplishments were the development of advanced computational methods and tailored experiments, specifically designed for size-selective observations. Among these the combination of supersonic molecular beam expansion techniques with photo-ionization mass spectroscopy has become a most powerful tool. It allowed, for example, the observation of specific catalytic reactions occurring on certain metal clusters of a specific size.[5] The discovery has directed much interest to an investigation of the dynamic properties of metal clusters. The development was favored by the advent of tunable femtosecond laser sources, which have opened fascinating perspectives for getting an insight into elementary intramolecular processes like bond shaking, breaking and making.[6] Such processes can now be monitored on a real-time basis by means of pump-probe spectroscopy.[7] Moreover the photo-induced dynamical sequences can even directly be influenced by using adequately shaped fs-laser pulses.[8]

In addition metal clusters are excellent model systems for time-resolved investigations in the fs-regime: they exhibit pronounced absorption bands,[9] which are well located within the tuning range of the available fs-laser sources; their detachment and photo-ionization energies[10] are also well within their reach, and the vibrational frequencies can well be resolved with the available fs-pulse lenghts. Further the influence of internal vibrational redistribution processes (IVR), which depend upon the amount of internal degrees of freedom, can simply be adapted to the experimental requirement by the right choice of cluster-size.[11] For this reason we have performed systematic pump-probe experiments first on alkaline dimers, trimers, etc. up to M_n ($n \leqq 10$) in order to study the evolution of wave packet dynamics, vibronic coupling, geometrical rearrangement, fragmentation etc. as a function of cluster-size.[12–17] We have further developed schemes for observing wave packet and reaction dynamics also in the electronic ground state by means of charge reversal spectroscopy (NeNePo)[18] or stimulated Raman pumping (STIRAP).[19] As demonstrated for Na_3, the photo-induced intramolecular dynamics of a system critically depends upon the employed pulse parameters.[20] Based on a suggestion of Rabitz *et al.*,[21] we have, therefore, carried out active control experiments, in which we excited differently branching fragmentation and ionization pathways of photo-excited Na_2K, its corresponding fragment NaK[22] and various metalorganic compounds.[23] By employing an evolutionary algorithm[24] for optimizing the phase and amplitude of the applied laser field, the yield of the resulting parent and fragment ions could significantly be influenced and interesting features about the investigated system and the irradiated photochemical process could be withdrawn from the obtained pulse shapes. Exciting perspectives arise from the experiment. Among these we will explore the possibility to perform coherent control experiments also in the electronic ground state.

2. Experimental set up of the Pump-Probe Experiment

The principle of the pump-probe observation scheme is plotted in Fig. 1. The clusters are excited from a low vibrational level of the electronic ground state to an excited state. Due to the spectral width of the employed fs-pump pulse, a coherent superposition of several vibrational states is created, which leads to the formation of an evolving wave packet. If a bound electronic state is excited [Fig. 1(a)], this wave packet will oscillate between the inner and outer turning point of the potential energy curve or surface, reflecting the vibrational motion of the excited molecule. The temporal evolution of this

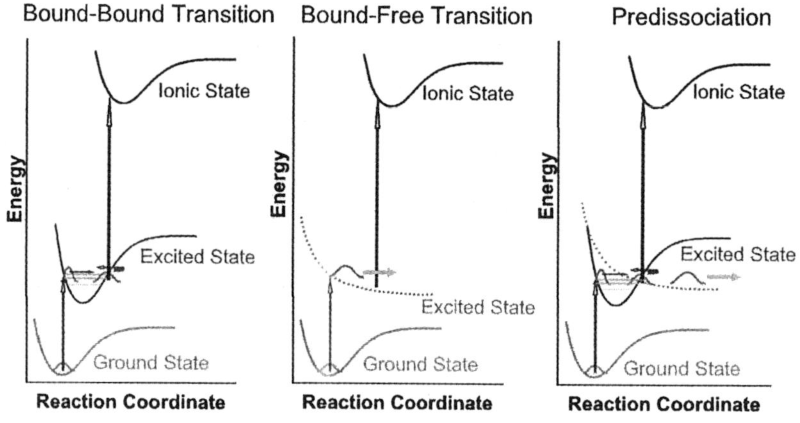

Fig. 1. The principle of pump-probe spectroscopy by means of transient two-photon ionization: A first fs-laser pulse electronically excites the particle into an ensemble of vibrational states creating a wave packet. Its temporal evolution is probed by a second probe pulse, which ionizes the excited particle as a function of the time-dependent Franck Condon-window; (a) shows the principle for a bound–bound transition, where the oscillative behaviour of the wave packet will appear; (b) shows it for a bound-free transition exhibiting the exponential decay of the fragmentizing particle, and (c) shows the process across a predissociated state, where the oscillating particle progressively leads into a fragmentation channel.

wave packet can be monitored with the probe pulse, which — at a variable delay — ionizes the particle into a size-selectively detectable state. This process may occur directly or via a higher electronic state. Since, however, the efficiency of this ionization step sensitively depends upon the position of the wave packet along the reaction coordinate, the obtained ionization efficiency (time-dependent Franck Condon factor) changes significantly as a function of time delay between excitation (pump) and ionization (probe). By tuning this time delay, the temporal evolution of the oscillating wave packet appears as an intensity modulation on the corresponding ion channel.

If, however, a bound-free transition is excited [see Fig. 1(b)], no oscillatory behaviour occurs, but the ion signal will show the temporal behaviour of the exponential decay of the photofragmenting system. The probed fragments will then appear on the corresponding ion channel as an exponential increase. If, on the other hand, a predissociated electronic state is excited [see Fig. 1(c)], the pump-probe signal will reflect both: wave packet oscillation and exponential decay.

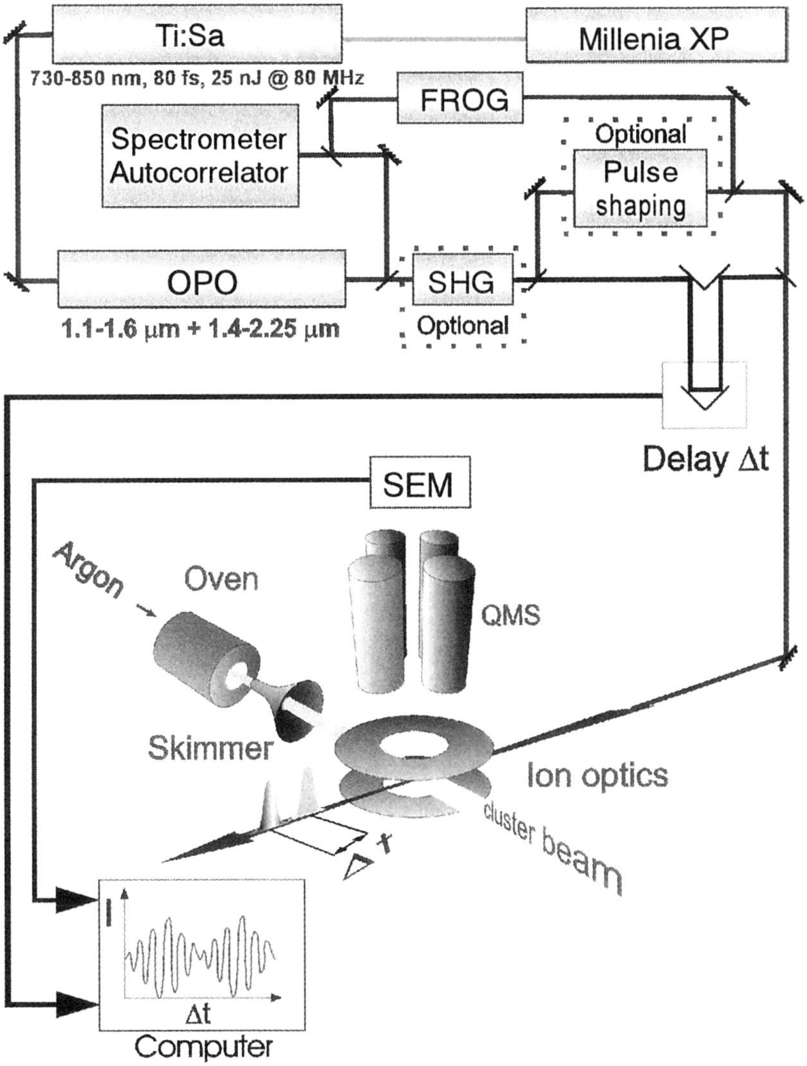

Fig. 2. Experimental setup of the pump-probe experiment showing the molecular beam irradiated by laser pulse sequences coming from a titanium sapphire fs-laser system. The resulting photoions are detected by a quadrupole mass spectrometer.

The scheme of the experiment is plotted in Fig. 2: the metal clusters are produced in an adiabatic beam expansion consisting of a radiatively heated oven cartridge, from which the alkali-metal vapor is coexpanded with three bars of argon carrier gas pressure across a nozzle of 70 μm diameter into the vacuum. For this purpose, the cartridge is surrounded with seven resistively heated hairpin-shaped tungsten filaments and protected by a threefold tantalum radiation shield. During the experiment the oven is heated to a temperature at which an alkali vapor pressure of some 10^{-1} bars is achieved with a slightly higher temperature, at the nozzle region to prevent clogging. The expansion zone is evacuated by a diffusion pump system of 10^4 l/s pumping speed, which results in a pressure of 3×10^{-4} mbar in the oven chamber during the experiment. From the expansion a molecular beam is skimmed off, using a skimmer of 1 mm diameter. It is placed approximately 7 mm behind the oven orifice. The skimmer leads into the detection chamber, which is differentially pumped by two turbomolecular pumps of 2000 and 600 l/s, which results in a pressure of about 1×10^{-5} mbar during the experiment. The mass flux of the neutral alkali cluster beam is permanently monitored with a Langmuir–Taylor detector, which is located 65 cm downstream from the skimmer. At a distance of 15 cm downstream from the skimmer a quadrupole mass spectromenter (Balzers QMG 420) is perpendicularly oriented to the cluster beam in order to extract the resulting photoions. Window ports at the detection chamber allow to irradiate the interaction zone perpendicular to both, the neutral cluster beam and the extracted ion beam.

For the experiment a commercial femtosecond laser is used (Spectra-Physics 3960 Tsunami), which is pumped by a Nd:YLF laser (Spectra-Physics Millenia X). The laser operates at a repetition rate of 80.6 MHz; it is tunable in a wavelength range between 730 and 850 nm, where it produces pulses of 80 fs duration at an average total power of 1.6 W. These conditions allow to operate the experiment at a 100% duty cycle, since each molecule of the continuous molecular beam is irradiated several times by the pulsed laser. Furthermore, the correspondingly low laser peak power prevents undesired multiphotonic transition, which would camouflage the sought information. The wavelength range of the laser can significantly be extended by using a second harmonic generation (SHG) or an optical parametric oscillator (OPO). The employed laser pulses are analyzed with a spectrometer, autocorrelator, spectrally-resolved cross correlation and SHG FROG (Femtos). In the pump-probe experiment, the light pulses were split up and recombined in a Mach–Zender interferometer system, allowing to generate pump-probe sequences of a variable delay.

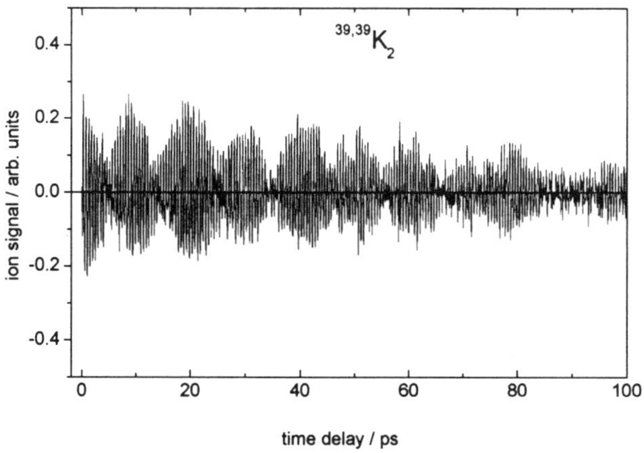

Fig. 3. Transient two-photon spectra recorded at an excitation wavelength of 834 nm (pump) for $^{39,39}K_2$. The ionization step (probe) was achieved by a delayed two-photon transition at the same wavelength.

3. Pump-Probe Spectra of Bound Electronic States

The result of a pump-probe spectrum obtained from $^{39,39}K_2$ is shown in Fig. 3. The spectrum was recorded at a pump wavelength of about 834 nm, the probe step was achieved by a delayed two-photon transition at the same wavelength. It exhibits a quite distinct oscillatory behaviour, harmonically modulated every 10 ps. More insight into the spectral features is provided by the corresponding Fourier-transform of the recorded signal, as presented in Fig. 4. The curve shows spectroresonances around 65 cm^{-1}, which is close to the known vibrational spacing of the A-state of K_2. Also the corresponding 2nd and 3rd harmonic of this frequency appear around 130 and 195 cm^{-1}. At a higher resolution (see insert of Fig. 4) the spectral signature should exhibit the anharmonic progression of those vibronic states, which were coherently excited within the bandwidth of the pump pulse. The according progression, however, shows a quite different behaviour: two largely spaced peaks at the wings of the progression dominate the spectrum. These two peaks are the cause for the 10 ps large-amplitude modulation. The progression of $^{39,39}K_2$ is obviously perturbed: at the excitation wavelength the appearing excited "light" state is strongly spin-orbit coupled to the "dark" $^{39,39}K_2$ ($b^3\Pi_u$)-state. This spectral coincidence significantly perturbs resonance positions, lifetime and Franck-Condon factors of the observed isotopomer $^{39,39}K_2$. For $^{39,41}K_2$ however, the spectral overlap

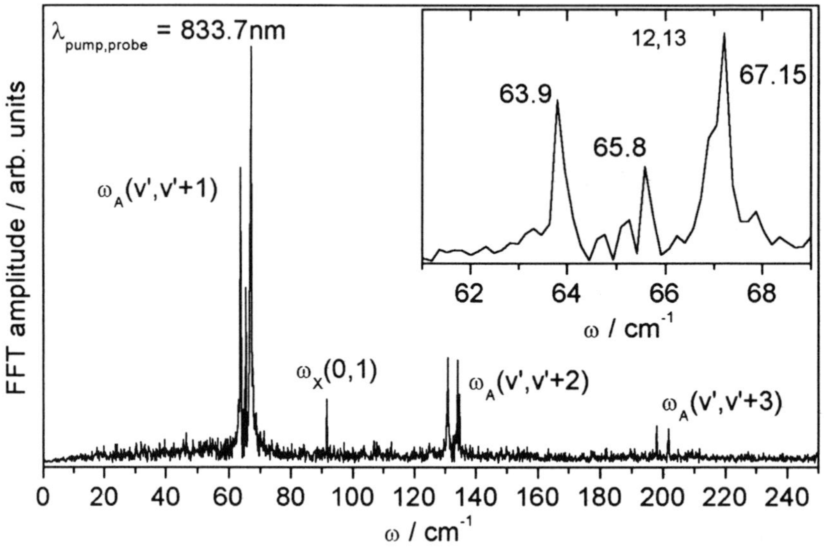

Fig. 4. Fourier-analysis of the spectrum shown in Fig. 3. The sequence exhibits around 65 cm^{-1} the fundamental frequency and at 130 cm^{-1} and 195 cm^{-1} the corresponding second and third harmonic. The insert in the figure shows the unharmonic progression at a higher resolution. The spacings indicate that the sequence is perturbed.

for such a coincidence is less pronounced, so no significant perturbations occur.[12]

Transient two-photon ionization experiments on trimer systems were motivated by a need for time-resolved verification of the pseudorotation motion, which can be considered as a superposition of the asymmetric stretch (q_x) and the bending vibration (q_y).[13] In this respect, the situation of a triatomic molecule with its three modes is quite different from an isolated oscillating dimer, which vibrates in its single mode until eventually it radiates back to the electronic ground state or predissociates. The coupling of vibrational modes in a trimer system can be considered as the onset of internal vibrational redistribution (IVR) — a crucial phenomenon in metal cluster dynamics.[18]

A typical result, which was obtained for the electronic $Na_3B \leftarrow X$-transition with transform-limited pulses of about 100 fs duration (FWHM), is shown in Fig. 5. The progression shows a pronounced molecular vibration, indicating only one vibrational mode of 320 fs duration, which corresponds

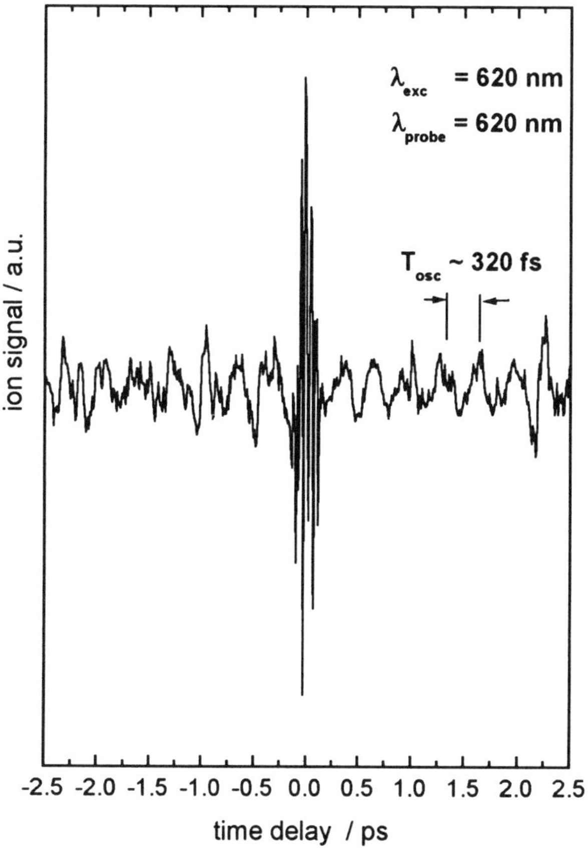

Fig. 5. Transient two-photon ionization spectrum of Na_3 recorded with transform limited 80 fs pulses at a wavelength of 620 nm. The progression exhibits the symmetric stretch mode of the electronically excited B-state.

to the symmetric stretch (q_s) mode of Na_3 in the B-state. There is, however, no indication for the asymmetric stretch, bending mode or pseudorotation.

4. Pump-Probe Spectroscopy of Dissociated and Predissociated Electronic States

Fragmentation becomes more important as the number of internal degrees of freedom inside the cluster increases. Here again fs pump-probe observation schemes provide a deep insight into the dynamics of photo-induced

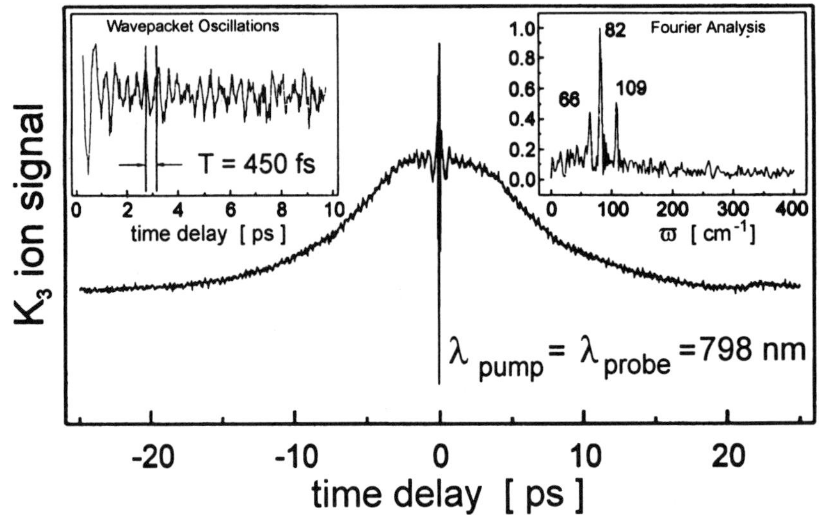

Fig. 6. Transient two-photon ionization spectrum of K_3. The progression shows a pronounced oscillation (see magnified in left insert), the corresponding Fourier-transform, indicates the three normal vibrations (see right insert). The spectrum is superposed by a fast unimolecular decay of approximately 5 ps.

cluster fragmentation. The principle of such an experiment is indicated in Fig. 1(b) and (c). The particles are electronically excited with pulses of a fs-laser (pump) into a predissociated or dissociated electronic state. There they oscillate a few times and then dissociate, or they dissociate directly. The temporal behaviour of the sequence is monitored with the probe pulse, which interrogates the system by ionizing the excited particles after a variable time delay Δt. The result of such an experiment performed on K_3 is shown in Fig. 6. At $\Delta t = 0$, the signal is at maximum. It represents the cross correlation between pump and probe pulse. For a delay time $\Delta t < 10$ ps, a pronounced oscillation occurs, which reflects the wave packet oscillation in the excited state. A magnified segment of this oscillation is shown in the left insert of Fig. 6, whereas the Fourier-transform of the result is presented in the right insert of Fig. 6. It indicates three vibrational modes, which correspond to the K_3 normal vibrations with $q_s = 109$ cm^{-1}, $q_x = 82$ cm^{-1}, and $q_y = 66$ cm^{-1}. Superposed to these oscillations there is an ultrafast unimolecular decay with a lifetime of 6 ps, which indicates that the observed state is predissociating.[14] This fast fragmentation prevented so far the observation of this excited state by means of stationary resonant multiphoton

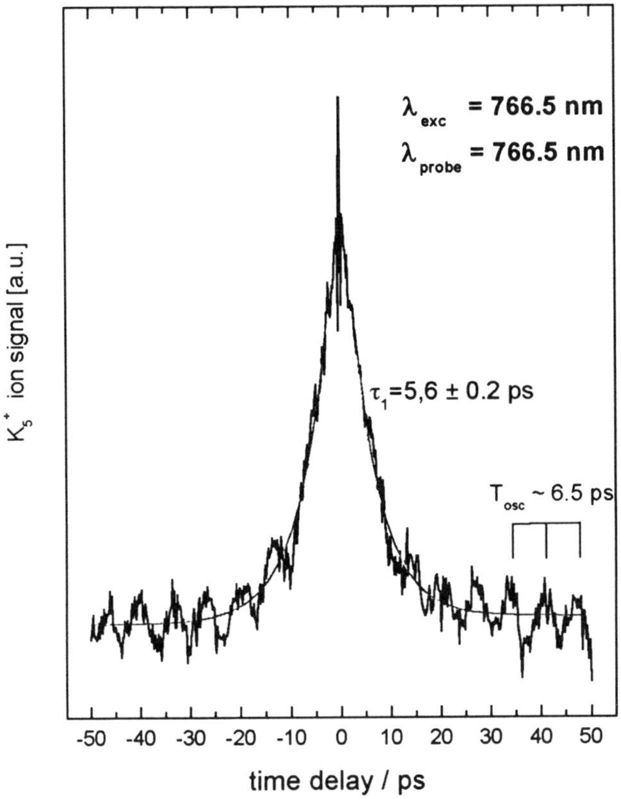

Fig. 7. Pump-Probe spectrum for K_5. The spectrum exhibits an oscillation of 6.5 ps superposed to a fast unimolecular decay of 5.6 ps.

ionization. A corresponding result obtained from the heteronuclear trimer Na_2K will be shown two chapters later.

A similar pump-probe spectrum obtained from a predissociated state of K_5 is shown in Fig. 7. The duration of an oscillation period is 6.5 ps, the exponential decay of the particle occurs in 5.6 ps; which is slightly faster than the decay of K_3.

The application of pump-probe spectroscopy to electronic transitions of larger aggregates reveals the rapidly growing number of different dissociation channels. The result of two-color pump-probe femtosecond experiments performed on sodium clusters Na_n with $3 \leqq n \leqq 10$ is shown in Fig. 8. For $\Delta t > 0$, the energy E_{pump} was 1.47 eV, whereas E_{probe}, the energy of the probe pulse, was 2.94 eV. Time decays with $\Delta t < 0$ inverted this sequence to $E_{\text{pump}} = 2.94$ eV and $E_{\text{probe}} = 1.47$ eV. The general trend shows a faster decay with growing cluster-size, which strongly depends on the excitation wavelength.[16] The size-related increase of fragmentation speed can qualitatively be explained by the growing amount of internal degrees of freedom allowing more IVR into dissociative channels. In order to describe the features which appear in Fig. 8 in more detail, several processes must besides IVR be taken into account: the direct fragmentation of the examined cluster size dissociated by the pump pulse, and those fragmentation processes that occur to particles, which have populated the observation channel temporarily with fragments of larger clusters, before they fragmentize again.[17]

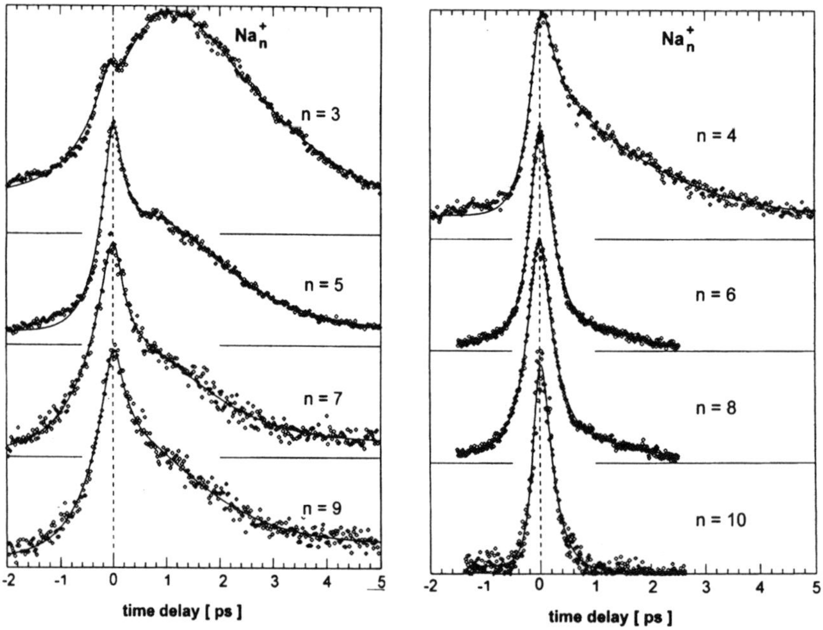

Fig. 8. Pump-probe spectra of a two-colour experiment probing the bound-free transitions in Na_n $(3 \leqq n \leqq 10)$. For $\Delta t > 0: E_{\text{pump}} = 1.473$ eV and $E_{\text{probe}} = 2.94$ eV. For $\Delta t \leqq 0: E_{\text{pump}} = 2.94$ eV and $E_{\text{probe}} = 1.47$ eV.

5. Time-Resolved Spectroscopy of Electronic Ground States

In the time-resolved experiments described so far, the pump laser simply populates the intermediate excited state. The experiment, therefore, becomes a means to study that excited state. Dynamic processes like catalysis and theoretical studies, however, are often more concerned about the electronic ground state than about excited states. For this purpose, it is useful to investigate vibrational wave packet dynamics from that ground state.

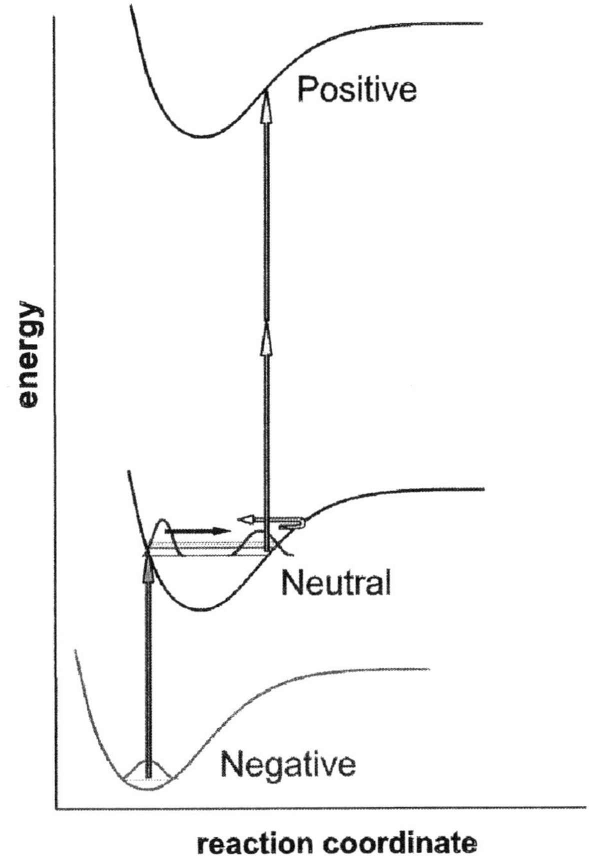

Fig. 9. Principle of NeNePo spectroscopy. The probe pulse detaches the photoelectron from the negative ion inducing a vibrational wave packet to the ground state of the neutral particle. Its propagation is interrogated by the probe pulse which ionizes it to a positive ion.

One possibility to do this is stimulated emission Raman pumping (STI-RAP). This requires to excite the species of interest to an electronically excited state and then to "down pump" from that excited state back to a higher vibrational level of the electronic ground state, which then generates a wave packet in a stationary state. This can be achieved by raising the power level of the pump laser pulses. Experiments of this kind were performed on potassium dimers.[19] Another example obtained on Na_3 will be shown in the next chapter. Besides that, we developed a new method called "Charge reversal spectroscopy" or "NeNePo" (negative-neutral-positive). It allowed us to prepare vibrational wave packets in very low electronic states or even the electronic ground state of size-selected neutral clusters.[18] The excitation scheme is shown in Fig. 9. The experiment starts with negatively charged cluster ions. This offers the advantage of mass selecting them prior to the experiment. Then, the pump laser pulse photodetaches the electrons from the negative ions to make neutrals, which generally leaves them vibrationally coherently excited, initiating wave packet dynamics. After a chosen delay, a second probe laser pulse photoionizes these neutrals to convert them to easily detectable positive ions. In the experiments, which we performed so far, this ionization was accomplished by two-photon processes. The formed positive ions then pass a quadrupole mass analyzer into an ion detector. The variation of the time-interval between detachment and ionization affects the intensity of the positive ion signal reflecting the time-dependent Franck Condon factor of the ionization process with regard to the location of the vibrational wave packet in the neutral particle.

Figure 10 shows a result obtained for Ag_4. The corresponding negative ion was produced in a sputter process, mass selected and fed into an ion-trap, where it was confined until the pump-probe sequences for initializing the charge reversal process were irradiated. The newly formed $Ag_4{}^+$ was then extracted from the trap and mass analyzed again in another quadrupole mass spectrometer. The spectrum in Fig. 10(a) shows a pronounced vibrational sequence. Theoretical calculations predict for the investigated tetramer in its electronic ground state a rhombic structure. The prediction is in good agreement with the anharmonic sequence appearing in the corresponding Fourier-transform spectrum [see Fig. 10(b)].

Theoretical calculations of Ag_3 — on the other hand — had shown that the anion should be linear, the neutral particle an obtuse triangle, and the cation an equilateral triangle.[25] From this we expected after the photodetachment of the electron from the corresponding anion an intensively vibrating neutral particle. The experimental result,[18] however, only

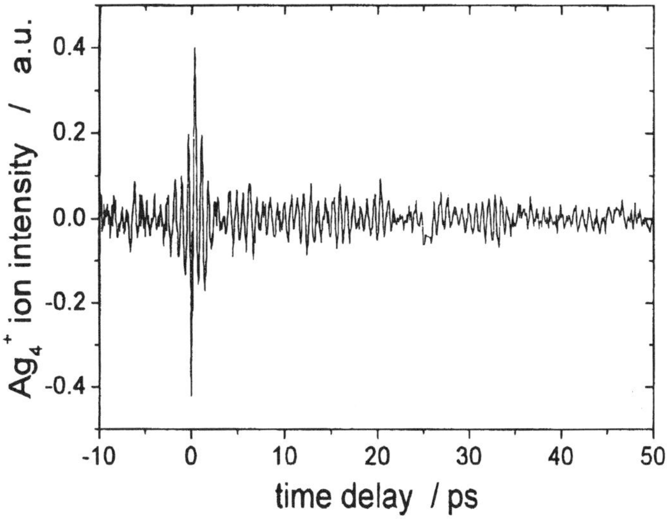

(a)

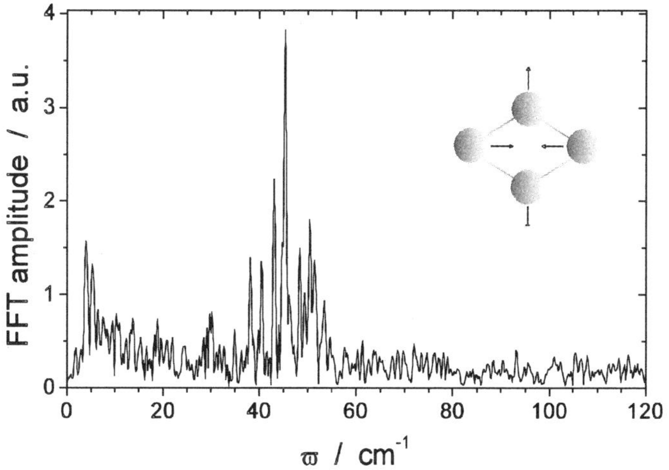

(b)

Fig. 10. NeNePo spectrum of neutral silver tetramers: (a) The progression exhibits the Ag_4 vibration in its electronic ground state; (b) shows the corresponding Fourier-transform of the unharmonic progression which corresponds to the indicated rhombus vibration.

showed one vibrational cycle, which quickly became stationary. A theoretical treatment[26] showed that fast vibrational redistribution from the bending mode into the asymmetric stretch and finally into the symmetric stretch mode was responsible for this behaviour, indicating that IVR can significantly alter the wave packet dynamics of a system. Contrary to that the pronounced oscillatory behaviour of the larger Ag_4 particle shows that the number of internal degrees of freedom of a system is not necessarily the only important parameter in this regard, and that the geometry of the system and the spacing of the relating vibrational frequencies play an important role.

6. The Influence of the Pulse Shape

In Fig. 5, a spectrum is plotted, which exhibits the oscillatory features of the symmetric stretch motion of Na_3 in its electronically excited B-state, indicating the well-known oscillation time of 320 fs. The pump-probe spectrum was obtained with transform-limited pulses of 80 fs duration at a center wavelength of 620 nm. Then the experiment was repeated by changing one experimental parameter only: the duration of the pump pulse. This was accomplished — as indicated in Fig. 11 — by passing the pump beam across a set of two parallel gratings. The assembly creates a linear frequency chirp. Its duration and spectral sequence depends only on the incidence angle

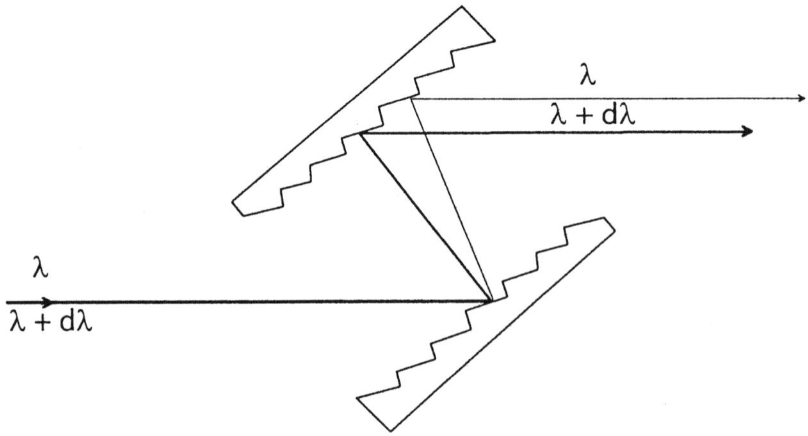

Fig. 11. Optical setup to create linearly chirped laser pulses.

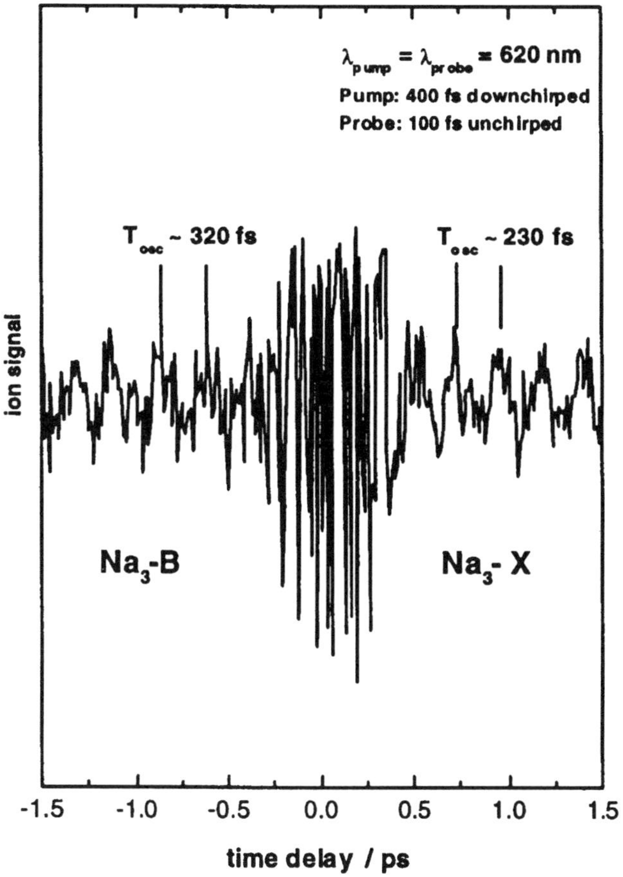

Fig. 12. Pump-Probe spectrum of Na_3 obtained under identical experimental conditions, as Fig. 5 with the only exception that the employed pump pulse was linearly downchirped to a duration of 400 fs. The resulting oscillation time of 230 fs corresponds to the symmetric stretch vibration of the electronic ground state of Na_3. At negative times (i.e. unchirped pump, chirped probe), still the vibration of the B-state is observed (see Fig. 5).

between laser beam and gratings. Figure 12 shows the corresponding Na_3 pump-probe spectrum, when a 420 fs downchirped pulse (blue preceeding red) was applied. Surprisingly, instead of the prior observed oscillation time of 320 fs (see Fig. 5) this spectrum now shows an oscillation time of 230 fs, which corresponds to the symmetric stretch mode of the electronic ground state of the sodium trimer. This observation can be explained by

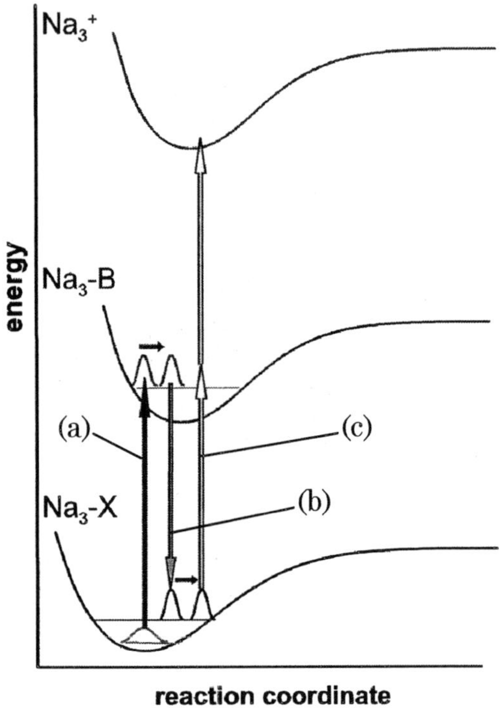

Fig. 13. Schematic view of the irradiated STIRAP process (Stimulated Raman pumping): (a) The chirped pump pulse excites the system the electronic ground state; (b) the system is stimulated back down to the electronic ground state, but a Stokes-shifted higher vibrational level. (c) The resulting wave packet is probed by the delayed probe pulse, which ionizes the particle from the oscillating ground state.

the scheme plotted in Fig. 13: the frontal frequency components (shorter wavelengths) create a propagating wave packet in the B-state of the trimer, which then is dumped down into the ground state (X-state) by the delayed longer wavelength components of the same pulse. The vibrational motion of the ground state is then monitored by the unchirped probe pulse via two-photon ionization (Stimulated Raman-pumping/STIRAP). The length of the employed pulses was determined from wavelength-resolved cross correlation measurements. The experiment clearly indicates the very sensitive dependence of the occuring intramolecular dynamics with regard to the irradiated pulse shape.

7. Reaction Control with Shaped Pulses

The advent of pulse shaping techniques, which allow to modulate the amplitude and phase of femtosecond laser pulses, has opened fascinating perspectives for driving the molecular reaction dynamics in real-time. The subject becomes most exciting, when — as proposed in 1992 by Rabitz *et al.*[21] — feedback loop algorithms are employed. This way the optimum laser pulse shapes are generated, which drive the reaction at a maximum yield on a desired path. Thermally these reactions may very well be inaccessible. An important first experimental step in this regard was performed by Gerber *et al.*,[27] who employed a genetic algorithm[24] to optimize a femtosecond laser pulse profile, which then channeled the highly fragmentative multiphoton ionization process of metal carbonyl compounds either to maximize the mother ion or a fragment ion yield. As emphasized by Rabitz, the experiment can be viewed as an analog computer, which solves the Schrödinger equation in real-time and drives the reaction into a desired direction. An important issue in this regard remains the question about the information content in the optimized laser pulse shape. The extraction of information from the acquired laser field is a most attractive aspect, since it may lead to a conceptionally new approach to the investigation of molecular dynamics. Employing evolutionary strategies — nature's concept itself — the experiment does not only learn to control the dynamics of a molecular system by optimizing a problem of high dimensionality, but also acquires valuable information about the molecular potential energy surface along the reaction path. To extract this information from the optimized laser field, a long way must still be gone and fundamental questions as, for example, the energy dissipation and the loss of coherence in a system of increasing complexity must be understood. In this regard it is most important to investigate systems, which are simple. Small clusters are good models in this regard. The solution then reflects their unique dynamical features. Based on these findings, related systems of slowly increased degrees of freedom may then become controllable and understandable as well.

The systematically performed pump-probe spectroscopy on alkali clusters provided a good indication about suited candidates for a coherent control experiment. Among these, the fragmentation dynamics of the heteronuclear trimer Na_2K appeared to us the best. The corresponding pump-probe spectrum is shown in Fig. 14(a). It clearly exhibits — superimposed on an exponential decay with a time constant of 3.28 ps — an oscillatory behaviour with a period of roughly 500 fs. The Fourier-transform of this

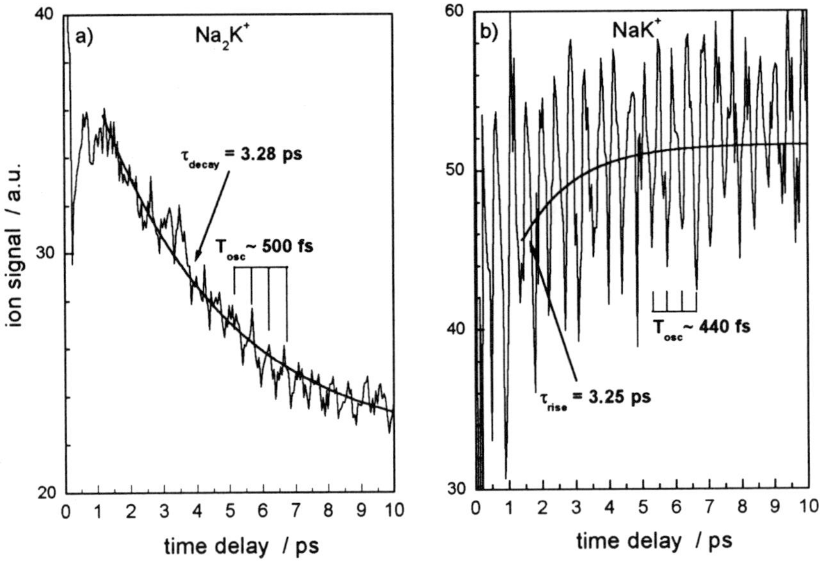

Fig. 14. Temporal evolution of a transient three-photon ionization signal of decaying Na_2K (a) and its emerging fragment NaK (b). The decay time of the excited trimer corresponds well to the rise time of its diatomic fragment. The superposed oscillations indicate wave packet oscillations in the coherently excited system.

oscillation[15] shows the two peaks at 18 cm^{-1} and 67.4 cm^{-1}, which reflect two of the three vibrational modes of the photo-excited trimer. Figure 14(b) shows the transient of the fragment NaK, where the exponential rise time of 3.25 ps indicates within the experimental error its origin from the equally fast decaying Na_2K. The observed period of oscillation 440 fs corresponds to the well-known eigenfrequency ϖ = 75 cm^{-1} of this electronically excited dimmer.[12] A schematic view of the involved transient multiphoton ionization process is shown in Fig. 15. In the pump-probe experiment the first femtosecond pump pulse excites Na_2K in a one photonic transition from its ground state to an electronically excited state, creating a coherent vibrational wave packet. According to preliminary *ab initio* calculations of Bonacic-Koutecký *et al.*,[28] the excitation wavelength corresponds to a transition into the bound excited electronic state 3 $^2A_1$. This state is crossed by a second electronic state 1 $^2A_2$ above its vibrational dissociation limit. As a result, the oscillating wave packet leaks into a fragmentation channel, forming NaK^* and Na. The temporal evolution of the oscillating mother

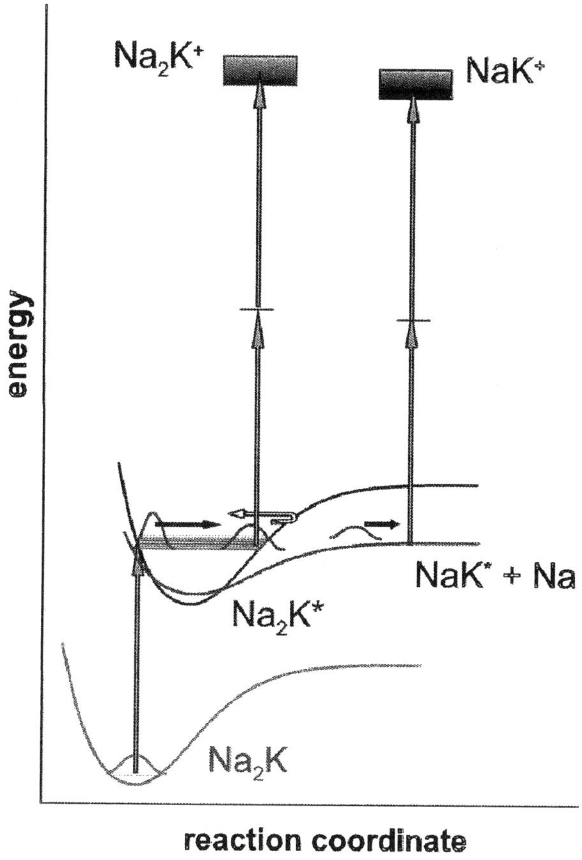

Fig. 15. Schematic view of the investigated 3-photonic fragmentation and ionization process.

molecule and its fragment are then monitored by the probe pulse, which ionizes the electronically excited species in a two-photon process, leading either to Na_2K^+ or NaK^+. It should be pointed out, that in this one-colour experiment the pump and the probe pulse originate both from the same laser source. This allows a decisive modification, in which the pump-probe scheme is replaced by tailor made pulse shapes, generated from single fs pulses.[29] In order to optimize the described three-photon ionization process to maximum yields of either mother ion or fragment signal channel, we can now establish a self-learning feedback loop[24] between signal detector and pulse modulator. With the experiment the question will be answered, if the obtained pulse shapes reflect the intrinsic properties of the product ions.

8. Setup of the Coherent Control Experiment

In the experimental setup basic components of the already described pump-probe experiment like the supersonic molecular beam apparatus, quadrupole mass spectrometer, and the femtosecond laser are reused (see Fig. 2). As shown in Fig. 16 in the optimization experiment, the laser beam is passed across a pulse shaper system, which allowes to modulate simultaneously the phase and amplitude of the laser pulses by applying voltages to a double mask liquid crystal spatial light modulator (SLM) consisting of 2×128 pixels.[29] The SLM is placed in the Fourier plane of a zero dispersion compressor, which is a linear setup of two gratings (1200 l/mm) and two plano convex lenses ($f = 200$ mm) in a $4f$-arrangement. Voltages can independently be applied on each pixel of the modulator allowing to modify the amplitude and phase of the transmitted light. By computer control, pulses of arbitrary form can thus be generated. These shaped pulses are then focussed into the detection chamber and the current of the desired mass-selected ion is taken as feedback signal.

The optimization algorithm iteratively drives the pulse shaper in order to maximize the ion signal. A schematic view of this procedure is given in Fig. 16. We apply an algorithm based on evolutionary strategies[30] to find

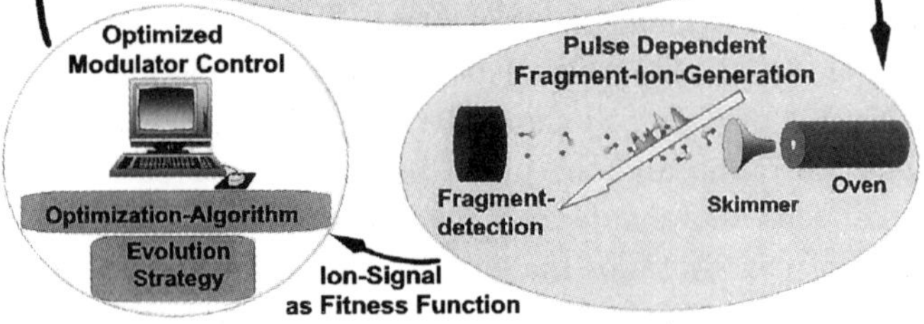

Fig. 16. Schematic view of the feedback loop for optimizing the ion yields of particular reactive channels by employing a selflearning algorithm.

the pulse shape in the 256-dimensional searching space, which yields to the highest feedback signal. Evolutionary strategies are based on the concept of evolution, applying schemes like "cross over", "mutation" and "survival of the fittest". In the beginning of the cycle 10 parent arrays of 256 random numbers are created (parent individuals). The first 128 elements specify the spectral phase shift, the others define the spectral attenuation of the modulator pixels. When sent to the modulator, a certain pulse shape is created from a bandwidth-limited pulse. In the crossover stage, 15 pairs of new individuals are created from pairs of parent individuals by randomly distributing copies of the respective parent array elements between the new pair. Thereby a population of 30 individuals is formed. Each array element is then altered by adding a random number from a Gaussian probability distribution (mutation). The whole population is subsequently tested by recording the ion yield for each modulator setting. Only those pulse shapes, which produce the highest ion yield are selected (survival of the fittest) and are taken as parents for the next generation. The optimization procedure proceeds until a convergence of the ion yield is achieved.

The thereby produced pulse forms are analyzed by SHG FROG, intensity autocorrelation and intensity cross correlation technique. When the autocorrelation trace suggests a pulse train, the analysis of the autocorrelation trace allows to retrieve the pulse form by a downhill simplex fit starting with a function consisting of up to five input Gaussian peaks. In the next step an autocorrelation trace is generated. This calculated autocorrelation trace is compared with the experimental trace. The width, height and position of the individual peaks are then searched iteratively within the downhill simplex method. When testing this procedure against the SHG FROG — as will be discussed below — data consistent for NaK were obtained. Since neither the analysis of both SHG FROG traces nor SHG autocorrelation data define the real-time ordering of the pulses contained in such pulse trains, intensity cross correlation using an unchirped fs-laser pulse as a reference pulse was applied. The intensity cross correlation transients provide an intuitive and direct information about the time ordering of the pulse elements. In the case of the Na_2K optimization experiment discussed in this paper, the full FROG trace of the emerging pulse form could not be recorded with our present setup due to the long temporal structure of the wave form consisting of a leading intense double pulse sequence followed by two low-intensity pulses occurring at times 3.7 and 8.0 ps (observed via autocorrelation) and thus exceeding the time-window of our SHG FROG configuration (part of the FROG trace was cut off). In this case the pulse from was obtained from intensity autocorrelation measurements and the time sequence of the leading pulses was determined by intensity cross correlation. The two weak pulses appearing at later times contribute to the overall intensity by $\sim 10\%$. Taking into account the fragmentation time of Na_2K^* and that under given experimental conditions no larger aggregates were present in the molecular beam, the occurrence of the two delayed weak peaks can be explained only by a failure of the optimization algorithm to diminish them.

9. Results and Discussion

Figure 17 shows a typical evolution of the ion yield during the optimization procedure for the triatomic mother ion Na_2K^+ (a) and the diatomic fragment ion NaK^+ (b). In the first generation all pulses are random and the resulting ion yield is consequently small. The dashed lines indicate the yield of transform-limited pulses. As the iteration proceeds, the ion yield increases and reaches convergence roughly after 150 and 70 generations for Na_2K^+ and NaK^+, respectively. In Fig. 18, a typical mass spectrum obtained with the pulse form, which produces the highest yield of unfragmented mother

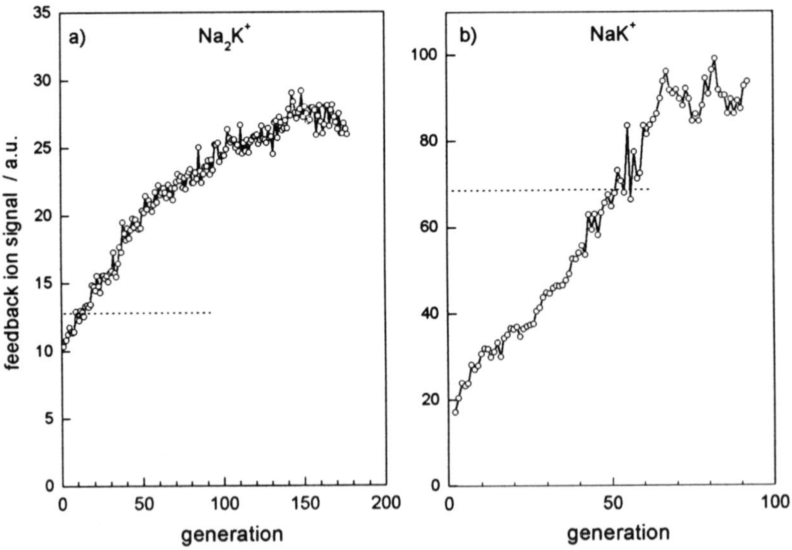

Fig. 17. Evolution of the ion yield during the optimization experiment plotted for the Na_2K-signal (a) and the NaK^+-signal (b).

ions (b), is compared with a reference spectrum produced by an unchirped pulse (a). It shows an increase of the mother ion signal by roughly 150%. A sharp decrease of the potassium mass peak can be observed as well. The optimization of NaK^+ yields to an increase of approximately 45%, when compared to short pulse ionization. We have to point out, that under the given experimental conditions the overall NaK^+ ion signal consists not only of photofragments, but there is still another contribution from directly photoionized NaK dimers, which are also present in the molecular beam. As estimated from the pump-probe spectrum of NaK, the contribution of ionic fragments to the overall signal is about 20%. It should be noted, that under certain experimental conditions also an Ar^+-mass peak emerges between the two K^+-isotopes, which is due to multiphoton ionization of the largely abundant carrier gas.

The evolution of the ion yields during the optimization procedure (as shown in Figs. 17(a) and (b) and the comparison of the mass spectra in Fig. 18 clearly indicate, that we have successfully performed a control experiment, in which the efficiency of competing reactive pathways was optimized

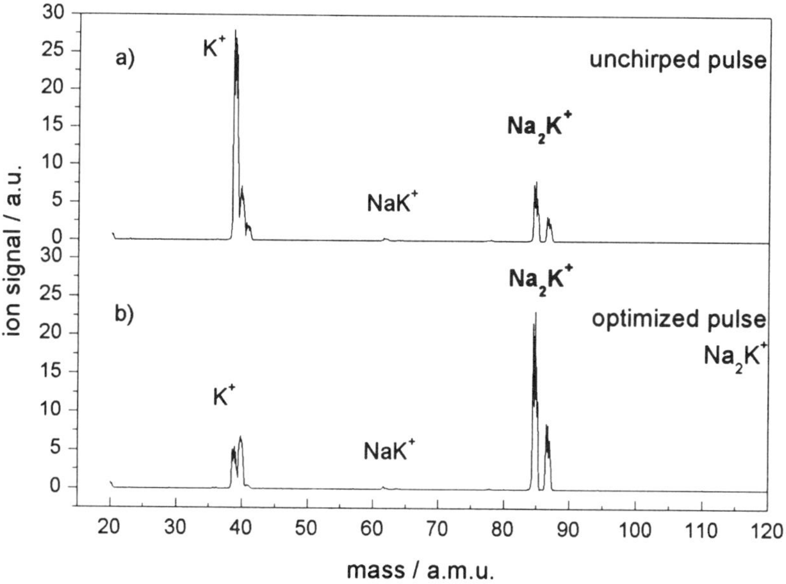

Fig. 18. Comparison of mass spectra obtained (a) with unchirped femtosecond pulses and (b) with optimized pulse shape for obtaining a maximum yield of the Na_2K^+-signal.

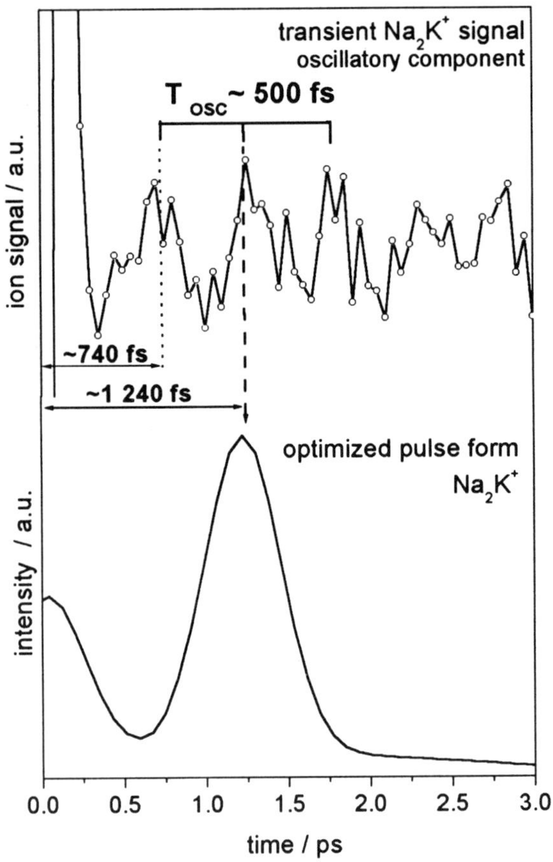

Fig. 19. Oscillatory component of the transient pump-probe Na_2K^+-signal compared with the acquired pulse shape for producing a maximum yield of Na_2K^+.

by applying feedback looped, phase-shaped femtosecond laser pulses. A most important aspect in this regard is the information content, which is acquired during the optimization procedure. In the case of the mother trimer Na_2K, a double pulse sequence (as shown in Fig. 19) was retrieved from the analysis of the autocorrelation trace. The time difference between the first and second pulse is approximately 1240 fs. This corresponds exactly to 2.5 oscillation periods of the electronically excited trimer. The intensity of the second pulse is about 80% higher than the first pulse. This can easily be understood by the fact, that the first pulse excites the system in a one-photonic

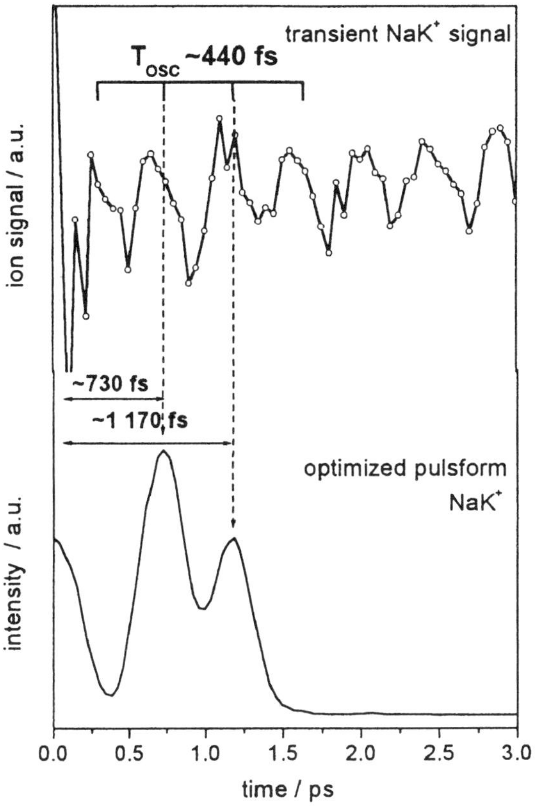

Fig. 20. Time-resolved pump-probe signal for NaK^+ compared with the acquired shape for producing a maximum yield of NaK^+.

transition, whereas the following ionization step requires two photons. It is known from pump-probe spectroscopy,[15] that the photo-ionization of the electronically excited Na_2K molecule occurs in the Franck–Condon window at the outer turning point of the propagating wave packet. The observation, that the time delay for the feedback loop optimized ion yield only occurs 2.5 oscillation periods after excitation, is less evident. It may result from the fact, that the procedure is started at random phases, which corresponds to rather long pulses, so earlier passages at the open FC-window are missed. It may also be due to complicated non-adiabatic dynamics involving several electronically-excited states, which have to be further investigated.

In Fig. 20, the pulse form obtained from the optimization process for the maximum NaK fragment ion yield is shown together with the relating pump-probe spectrum. In this case the optimum laser field could be analyzed with the SHG-FROG system. The pulse train consists of three pulses. The second pulse arrives approximately 730 fs after the first pulse and is followed by a third pulse delayed by 440 fs. The timing of the second pulse corresponds very well to 1.5 oscillation periods of the excited dimer, the timing between the second and third pulse of the pulse train matches exactly the well-known oscillation period of this dimer. The peak intensity of the second pulse is about 70% higher than the intensity of the first pulse, supporting the one-photon excitation and two-photon ionization scheme known from previous measurements.

The leading pulse is considerably longer than the other two pulses. The analysis of the temporal phase profile of the pulse train elements revealed fourth order components for the first pulse and third order components for the second and third pulse. In the time-domain, all three pulses possess a negative chirp (down chirp): a cubic and a quadratic down chirp for the leading pulse and the two after pulses, respectively. In general, strongly downchirped excitation pulses can cause a very effective focussing of the propagating wave packet in other locations of the involved unharmonic potential energy suface(s) at later times after electronic excitation. In the case of NaK, we assume that the strongly downchirped leading pulse focuses the propagating wave packet at the outer turning point of the potential energy surface of NaK*, known from the pump-probe experiments as a Franck–Condon window for ionization, and thus leading to the highest possible ion yield at the arrival of the ionising probe pulse. An intuitive assignment of the chirp contained in the second and third pulse cannot be made at this point.

Other optimization experiments performed on the Na_2K/NaK system led to similar pulse forms: mostly double pulses for the trimer Na_2K and pulse trains consisting of three pulses for the dimer NaK. In the case of the dimer, one run ended up with a double pulse and in the case of the trimer one optimization run with a triple pulse. These pulseforms were characterized mainly by intensity cross correlation only. The cross correlation signals confirmed the time sequence of the pulses mentioned above: a leading weaker pulse followed by a more intense one. In different optimization runs, the peak intensity of the second pulse was about 50–80% higher than the peak intensity of the leading pulse. It is worth to note here that two optimization runs led to pulse forms with non-intuitive time separation between pulse train elements (the separation of the pulses seemed to be independent on the oscillation period of the investigated electronic excited state). A similar feature of the pulse train has been observed recently in the potassium dimer system.[12] It is an open question, which (other) electronic states were involved in this multistep photo-ionization process. The answer may be coded in the phase profile of the pulses.

10. Summary

In conclusion we can say that we have performed systematic pump-probe measurements on small alkali clusters, analyzing the temporal behaviour of the system after bound–bound and bound-free excitations including predissociated states. The results could strongly be influenced by the shapes of the irradiated pulses. The observation prepared the way to perform a feedback control experiment on the photofragmenting system Na_2K, which shows pronounced wave packet dynamics on the signal channels of the mother ion and the NaK fragment. As a result of the achieved coherent control process it was possible to extract from the optimized laser field specific information about molecular vibration of the investigated particles. Certainly we have not yet found all information that is acquired in the optimum laser field, as for example, fingerprints of the mother molecule on the fragment channel. In its final perfection, the experiment will provide the complete information on the occurred reaction and all partners involved. Also control experiments in the electronic ground state will be another challenge.

Acknowledgements

The authors thank Prof. Thomas Leisner, Dr. Soeren Rutz, Dr. Elmar Schreiber, Dr. Porfirio Rosendo-Francisco, and Dr. Henry Hess for their participation in the earlier experiments. In addition we thank Prof. Herschel Rabitz, Prof. Vlasta Bonacic-Koutecký, Prof. Joern Manz, and Dr. Michael Hartmann for stimulating discussions concerning the interpretation of our data. Dr. Thomas Feurer and Georg Stobrawa collaborated with us in developing the evolutionary algorithm. One of the authors, Cosmin Lupulescu, acknowledges the Gottlieb Daimler and Karl Benz-Stiftung for receiving financial support for his Ph.D. studies. Porfirio Rosendo-Francisco thanks the DAAD for supporting his graduate studies at the Freie Universitaet Berlin. We also thank Prof. Tamotsu Kondow for encouraging us to write this paper. The generous support of this research project by the Deutsche

Forschungsgemeinschaft in the frame of the SFB 450 research project is gratefully acknowledged.

References

1. H. Haberland, in *Clusters of Atoms and Molecules*, Vol. I-II, edited by H. Haberland (Springer-Verlag, 1994.)
2. W. Knight, K. Clemenger, W. De Heer, W. Saunders, M. Chou and M. L. Cohen, *Phys. Rev. Lett.* **52**, 2141 (1984).
3. C. Brechignac, M. Broyer, P. Cahuzac, G. Delacrétaz, P. Labastie, J. P. Wolf and L. Wöste, *Phys. Rev. Lett.* **60**, 275 (1988).
4. H. W. Kroto, J. R. Heath, S. C. O'Brien and R. E. Smalley, *Nature* **138**, 165 (1985).
5. U. Heiz and W.-D. Schneider, in *Metal Clusters at Surfaces*, edited by K.-H. Meiwes-Broer (Springer-Verlag, Berlin, 2000).
6. J. Manz and L. Woeste, in *Femtosecond Chemistry*, Vol. I-II, edited by J. Manz and L. Woeste (VCH Publishers, Inc., New York, 1994).
7. E. D. Potter, J. L. Herek, S. Pedersen, Q. Liu and A. H. Zewail, *Nature* **355**, 66 (1992).
8. N. Dudovich, D. Barak, T. Gallagher, S. M. Faeder and Y. Silberberg, *Phys. Rev. Lett.* **86**, 47 (2001).
9. S. Vajda, T. Leisner, S. Wolf and L. Woeste, *Philos. Mag.* **B79**, 1353 (1999).
10. A. Herrmann, S. Leutwyler, E. Schumacher and L. Woeste, *Chem. Phys. Lett.* **52**, 413 (1977).
11. R. S. Berry, V. Bonacic-Koutecky, J. Gaus, T. Leisner, J. Manz, B. Reisch-Lenz, H. Ruppe, S. Rutz, E. Schreiber, St. Vajda, R. De Vivie-Riedle, S. Wolf and L. Woeste, *Adv. Chem. Phys.* **101**, 101 (1997).
12. S. Rutz, E. Schreiber, L. Woeste, in *Ultrafast Processes in Spectroscopy*, edited by O. Svelto, S. De Silvestri and G. Denardo (Plenum Publ., New York 1996).
13. G. Delacrétaz, E. Grant, R. Whetten, L. Wöste and J. Zwanziger, *Phys. Rev. Lett.* **56**, 1598 (1986).
14. H. Ruppe, S. Rutz, E. Schreiber and L. Woeste, *Chem Phys. Lett.* **257**, 365 (1996).
15. S. Vajda, S. Rutz, J. Heufelder, P. Rosendo, H. Ruppe, P. Wetzel and L. Woeste, *J. Chem. Phys.* **A102**, 4066 (1998).
16. E. Schreiber, K. Kobe, A. Ruff, S. Rutz, G. Sommerer and L. Woeste, *Chem. Phys. Lett.* **242**, 106 (1995).
17. A. Ruff, S. Rutz, E. Schreiber and L. Woeste, *Z. Phys.* **D37**, 175 (1996).
18. S. Wolf, G. Sommerer, S. Rutz, E. Schreiber, T. Leisner, L. Woeste and R. S. Berry, *Phys. Rev. Lett.* **74**, 4177 (1995).
19. R. De Vivie-Riedle, K. Kobe, J. Manz, W. Meyer, B. Reischl, S. Rutz, E. Schreiber and L. Woeste, *J. Phys. Chem.* **100**, 7789 (1996).
20. R. De Vivie-Riedle, J. Gaus, V. Bonacic-Koutecky, J. Manz, B. Reischl, S. Rutz, E. Schreiber and L. Woeste, in *Femtochemistry*, edited by M. Chergui (World Scientific, Singapore, 1995) p. 319.

21. R. S. Judson and H. Rabitz, *Phys. Rev. Lett.* **68**, 1500 (1992).
22. S. Vajda, A. Bartelt, E.-C. Kaposta, T. Leisner, C. Lupulescu, S. Minemoto, P. Rosendo-Francisco and L. Woeste, *Chem. Phys.* **267**, 231 (2001).
23. C. Daniel, J. Full, l. Gonzalez, C. Kaposta, M. Krenz, C. Lupulescu, J. Manz, S. Minemoto, M. Oppel, S. Vajda and L. Woeste, *Chem Phys.* **267**, 247 (2001).
24. I. Rechenberg, *Evolutionstrategie* (Frommann-Hozboog Stuttgart 1994).
25. V. Bonacic-Koutecky, P. Fantucci, J. Pittner and J. Koutecky, *J. Chem. Phys.* **100**, 490 (1994).
26. M. Hartmann, J. Pittner, V. Bonacic-Koutecky, A. Heidenreich and J. Jortner, *J. Chem. Phys.* **108**, 3096 (1998).
27. A. Assion, T. Baumert, M. Bergt, T. Brixtner, B. Kiefer, V. Seyfried, M. Strehle and G. Gerber, *Science* **282**, 919 (1998).
28. V. Bonacic-Koutecky, private communication.
29. A. Weiner, D. E. Leaird, J. S. Patel and J. R. Wullert II, *IEEE J. Quant. Electron.* **28**, 908 (1992).
30. H.-P. Schwefel, *Evolution and Optimum Seeking* (Wiley, New York, 1995).

CHAPTER 5

CORE LEVEL EXCITATION OF CLUSTERS

E. Rühl
Lehrstuhl für Physikalische Chemie I, Institut für Physikalische Chemie,
Universität Wörzburg Am Hubland
97074 Wörzburg, Germany
E-mail: eruehl@phys-chemie.uni-wuerzburg.de

Core excitation of clusters is reviewed as a unique approach to probe element- and site-specifically size-dependent properties of free clusters in the gas phase. The fundamental characteristics of core level spectroscopies that rely on resonant excitation are briefly reviewed. Specifically, this includes primary photoabsorption as well as subsequent processes, such as photoionization, electronic relaxation, radiative relaxation, and finally fragmentation of the singly or multiply charged clusters. Experimental techniques as well as selected results on simple model systems, such as variable size rare gas clusters, are presented.

Contents

1. Introduction

Clusters bridge the gap between the gas phase and the condensed phase of matter.[1] It is therefore of primary interest to be able to follow size dependent changes of properties of matter by specific experimental and theoretical approaches in order to understand whether changes of cluster size occur gradually, in distinct steps, or if there are size regimes with unique properties.

Resonant excitation of core levels involves the use of tunable soft- and hard-X-rays. This energy regime has widely been used since decades for a variety of investigations on atomic,[2] molecular,[3] surface[4] and condensed matter physics,[5] where synchrotron radiation is currently the primary X-ray source for experimental work in this field. Earlier work on resonant excitation of core levels has already shown that there are numerous unique properties, such as e.g. element specific excitations. Methods for determining structural properties,[6] as well as site-selective fragmentation,[7] promised important applications with respect to fundamental and applied research on size effects of matter.

Resonant excitation of core levels of clusters has become an active field of research quite lately. Size dependent properties of clusters have been investigated by various spectroscopic approaches utilizing radiation of considerably longer wavelengths, i.e. the regime from microwaves to vacuum ultraviolet (VUV) radiation.[1,8] Therefore, the first issue to be addressed in this context concerns the question why another spectroscopic technique, such as core level excitation, is needed to obtain more specific information than can be obtained from other spectroscopic approaches. This is of importance to consider, since the regime of core level excitations appeared at a first glance to be not too attractive for the following reasons: (i) The absorption cross sections are in general weaker than in the UV- and VUV-regime.[9] Therefore, the interaction between the photon beam and clusters is less efficient, so that the sensitivity of core level excitation is tightly related to the availability of high intensity, tunable X-ray sources, such as synchrotron radiation.[10] There is only a limited number of synchrotron radiation facilities in the world. However, their number has grown considerably over the years and their performance has been tremendously improved; (ii) it may be anticipated that size effects occurring with core level excitation of clusters are by far too small, so that they cannot be resolved with the spectral resolution of standard X-ray monochromators. It will be shown below that state-of-the-art facilities fully overcome this possible drawback that was evident in early work; (iii) it is well-known that high energy photons in the X-ray regime lead to massive fragmentation of molecules. This is even more true for weakly bound systems, such as clusters. Therefore, high energy photon sources are not the most obvious choice to study fragile species, such as e.g. van der Waals clusters.

One of the motivations of this review is to demonstrate the advantageous use of tunable X-rays for research on clusters. The sum of all possible drawbacks is more than compensated by the following unique advantages:

(i) *Site-and element-selective excitation*: The excitation of core levels is element specific. This property can already be deduced from the well-known Moseley's law, indicating that the square root of the X-ray transition frequencies versus the atomic number gives a linear relationship. It will be shown in the following that not only X-ray emission is element-selective, but also resonant absorption as well as the emission of photoelectrons. This can be used to obtain site- and element-specific excitations in clusters.

(ii) *Local geometric and electronic structure*: Excitations of a defined site within a cluster can also be applied to obtain structural properties of the local surroundings of the absorbing atom. This follows from single scattering of the photoelectron at neighboring atoms, which is well-known as EXAFS (*extended X-ray absorption fine structure*).[6] EXAFS is widely used for investigations of structural properties of condensed matter, especially in the following fields of fundamental and applied research: materials science, catalysis, magnetism, biology, etc. Note that structural properties of amorphous matter can be determined by X-ray absorption.

(iii) *Preparation of clusters in defined charge states*: Core level excitation leads to the formation of highly charged ions, since the core hole is stabilized by the emission of electrons from the valence shell. The charge state varies strongly near the absorption edges.

These unique properties of core level excitation spectroscopies are briefly reviewed. The fundamentals of core level spectroscopies, such as X-ray absorption, X-ray absorption fine structure, inner-shell photoionization, electronic and radiative relaxation, and fragmentation in the regime of core level excitation, are outlined along with their characteristics to size effects of matter. Some illustrative examples are taken from recent work on rare gas clusters, which are regarded as ideal model systems to demonstrate clear evidence for size effects in free clusters.

2. Inner-Shell Excitation, Photoionization, Relaxation, and Fragmentation

2.1. *X-ray Absorption*

The attenuation of an X-ray beam by matter, in the typical photon energy regime of core level excitations (typically: 100 eV $\leqq E \leqq$ 100 keV), is given by the Beer–Lambert rule:[11]

$$I(x) = I_0 e^{-\mu x} \tag{1}$$

where I_0 is the intensity of the incident radiation, $I(x)$ is the transmitted intensity, μ is the attenuation coefficient, and x is the thickness of the absorber. Note that the attenuation coefficient is given by the sum of the scattering coefficient, μ_s, and the absorption coefficient, α. We will concentrate in the following on the dominant process that governs photoabsorption in the regime of core level excitation of light elements belonging to the second or third row of the periodic table, where the photoelectric effect dominates the regime of photon energies below 100 keV. Therefore, contributions to the attenuation coefficient from Rayleigh and Compton scattering, as well as pair production are neglected. The absorption coefficient α is given by:

$$\alpha = n\sigma_{\mathrm{A}} \tag{2}$$

where n is the number density of the target and σ_A is the atomic absorption cross section. Often, the mass absorption coefficient κ_a is a useful quantity to describe the absorption strength:

$$\kappa_a = \frac{\alpha}{\rho} = \frac{\sigma_{\mathrm{A}}}{m_{\mathrm{A}}} \tag{3}$$

where ρ is the mass density and m_{A} is the atomic mass.

The absorption of an X-ray photon leads to the excitation of an inner-shell electron. The transition probability per unit time P_{if} from an initial core level $|i\rangle$ to a final state near the vacuum level $|f\rangle$, is given according to Fermi's Golden Rule:[12]

$$P_{\mathrm{if}} = \frac{2\pi}{\hbar}|\langle f|\overline{V}|i\rangle|^2 \rho_{\mathrm{f}}(E) \tag{4}$$

where $\rho_{\mathrm{f}}(E)$ is the energy density of the final state, $V(t) = \overline{V}\exp(-i\omega t)$ is a harmonic time-dependent perturbation, and $\hbar = h/2\pi$, where h is the Planck constant. $\overline{V}$ is replaced by the dipole operator $\mathbf{p}$, using the dipole approximation, so that the absorption cross section σ is expressed as follows:[12]

$$\sigma = \frac{4\pi^2\hbar^2}{m^2}\frac{e^2}{\hbar c}\frac{1}{\hbar\omega}|\langle f|\mathbf{e}\cdot\mathbf{p}|i\rangle|^2\rho_f(E)\,. \tag{5}$$

Here, $\hbar\omega$ is the photon energy, c is the speed of light, e is the elementary charge, m is the electron mass, and $\mathbf{e}$ is the unit vector. Note that the expression in brackets corresponds to the electrical dipole matrix element.

Resonant excitation of a core level leads to an abrupt increase in the atomic absorption cross section, which is commonly termed "*absorption edge*" or "*edge jump*". An absorption edge is found at energies, where resonant absorption of an X-ray photon leads to the emission of a core photoelectron, i.e. this energy corresponds to a core ionization energy. Figure 1 visualizes schematically these characteristic changes in absorption cross section. Figure 1(a) shows an atomic Coulomb potential, where occupied electronic shells (e.g. K- and L-shells) are deeply bound. Thus, resonant excitation of these core levels requires tunable soft- or hard-X-rays. In the case of heavier elements, core levels may also include M- or N-shells. Figure 1(b) shows qualitatively the shape of the atomic absorption cross section. The atomic absorption cross section is strongly energy dependent, as visualized

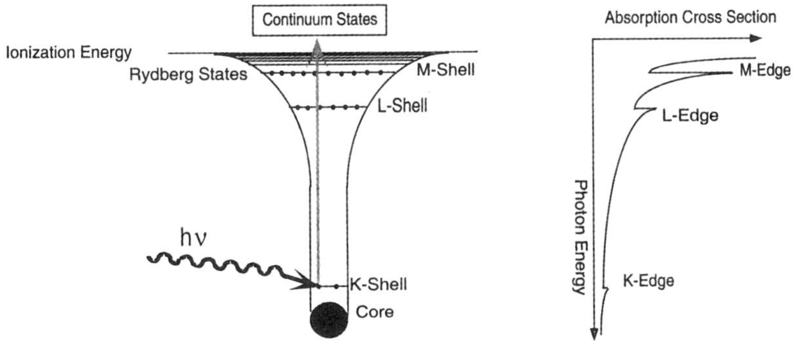

(a) (b)

Fig. 1. (a) Schematic atomic Coulomb potential, where occupied core levels (K-, L- and M-shells) are indicated. Resonant core level excitation (e.g. K-shell excitation) leads to transitions into unoccupied Rydberg states or into the corresponding core ionization continua. (b) Schematic shape of the atomic absorption cross section in the core level regime. Fine structure, such as spin-orbit splitting and resonant excitations into Rydberg states, is not shown for simplicity.

in Fig. 1(b). More quantitatively, the variation of the atomic absorption cross section is given by Ref. 13:

$$\sigma_A \cong CZ^4\lambda^3 \tag{6}$$

where C is a constant, Z is the nuclear charge, and λ is the X-ray wavelength. The shape of the total absorption cross section depends strongly on the composition of the sample, since this quantity is due to the sum of the atomic absorption cross sections.

In addition to the shape of the atomic absorption cross section shown in Fig. 1(b), there occur various discrete features below or above core absorption edges. This is often called "*near-edge X-ray absorption fine structure*" (NEXAFS).[12] Alternatively, this structure is sometimes called XANES ("*X-ray absorption near-edge structure*"). It is meant to be distinguished from EXAFS ("*extended X-ray absorption fine structure*"), a process that is discussed in detail in the following section. Nowadays, processes occurring either in the near-edge regime or in core ionization continua are called XAFS ("*X-ray absorption fine structure*").[6]

Near core level absorption edges one finds discrete XAFS-features which are due to resonant excitations of:

(i) *Rydberg states* converging to the core ionization thresholds. They follow simple series laws similar to atomic hydrogen:

$$E = E_0 - \frac{R}{(n-\Delta)^2} \tag{7}$$

where E is transition energy, E_0 is the series limit, i.e. the core ionization energy, R is the Rydberg constant, n is a quantum number, and Δ is the quantum defect. Rydberg states are known to occur in atoms, molecules, and small clusters. Transitions into Rydberg states are lifetime broadened as a result of the limited lifetime of the core hole that is created by the absorption of an X-ray photon. The line width can be estimated according to the Heisenberg uncertainty principle ($\Delta E \cdot \Delta t \geq \hbar \approx 6.6 \cdot 10^{-16}$ eV · s). This effect becomes important with increasing excitation energy. As a result, the natural line widths of resonances near the K-edge of nitrogen ($E \approx 400$ eV) is of the order of 0.1 eV, whereas a natural line widths of ca. 0.4 eV is observed near the K-edge of sulfur ($E = 2472$ eV).[12,14] Therefore, mostly the lowest members of Rydberg series can be resolved in X-ray absorption spectroscopy.

(ii) *Core excitons*: Resonances in the near-edge regime are also known to occur in the condensed phase, such as van der Waals solids (e.g. condensed rare gases).[15] These are usually termed as "*core excitons*", which are formed by optical transitions similar to excitons in the valence excitation regime, where electron hole pairs are known to be formed by photoabsorption. These are called Mott–Wannier excitons, if the separation of the electron-hole pair is greater than the lattice constant. Mott–Wannier excitons follow series laws, where the transition energy is proportional to $1/n^2$, similar to the Rydberg formula [see Eq. (7)]:[16]

$$E = E_0 - \frac{b}{n^2} \tag{8a}$$

$$b = \frac{\mu_{\text{exc}} R}{\varepsilon^2} . \tag{8b}$$

Here, μ_{exc} is the reduced effective mass of the exciton, in terms of the free electron mass, and ε is the dielectric constant. In addition, there are also Frenkel excitons, which are spatially more localized than Mott–Wannier excitons. They cannot be represented by a series formula. Therefore, Frenkel excitons have been assigned to single low-lying absorption bands in the near-edge regime of solid rare gases.[15]

Discrete transitions occurring in the pre edge regime of condensed phase species are primarily due to bulk properties, since the surface-to-bulk ratio in solids and multilayer adsorbates is in general low. In contrast, clusters consist primarily of surface sites and atoms that are located in the bulk occur in small clusters with low abundance. This is even more true, if defects in cluster structure occur, enhancing the surface-to-bulk ratio. Therefore, surface excitons are known to dominate the pre edge regime of variable size clusters, as shown in Sec. 4.1.

(iii) *Valence states*: Valence transitions are found in free and condensed molecules as well as molecular clusters, resulting from optical transitions from core levels into unoccupied or partially occupied molecular orbitals.[12] These are in most cases either of non-bonding or antibonding character, i.e. valence transitions reflect the presence of chemical bonds of the absorber, unlike weakly bound systems, such as van der Waals solids or clusters. The absorption cross section of valence transitions is in general higher than of Rydberg and exciton transitions, so that inner-shell spectra of gaseous and condensed molecular species are dominated by valence transitions.[17] These may either be located below or above the core ionization energies. The energy position relative to the absorption edge affects strongly the spectral shape of valence transitions: (i) narrow shaped resonances are found below core ionization energies, where in some cases vibronic fine structure can be resolved, depending on the Franck–Condon overlap of the states that are involved in the electronic transition;[18] (ii) broad resonances are observed in core ionization continua, since the reduced lifetime leads to a rapid decay of these excited states. Strong spatial localization of chemical bonds results in negligible spectral changes between inner-shell absorption spectra of isolated and condensed molecules. As a result, the absorption cross sections of free and condensed molecules are similar in shape.[19] However, characteristic differences occur in the Rydberg excitation regimes.[18]

(iv) *Continuum features*: Besides valence excitations occurring in core ionization continua one often finds discrete weak intensity features superimposed to the atomic absorption cross section, which result from double excitations.[20] Double excitations may involve Rydberg transitions. Therefore, one observes in these spectral regimes distinct differences between the absorption cross section of the isolated gas phase and that of the corresponding condensed species.[18] Other continuum structures are due to single and multiple scattering processes. They are discussed in the following.

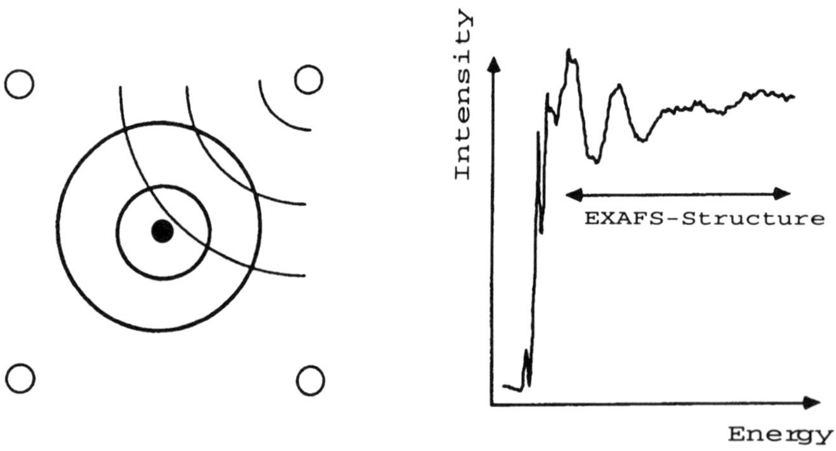

Fig. 2. Schematic diagram of the EXAFS process. The central atom (filled circle) absorbs an X-ray photon. Scattering of the outgoing wave is induced by the neighboring sites (open circles) (left); Schematic presentation of the experimental EXAFS-signal obtained from Ar $2p$-excited argon clusters (right).

2.2. *Scattering Processes*

In molecules, clusters, and the condensed phase, the emitted photoelectron may either be singly or multiply scattered at neighboring atoms. In the following we will concentrate on single scattering processes, which are commonly associated with the acronym EXAFS ("*extended X-ray absorption fine structure*"). Evidence for this process comes from characteristic oscillations of the absorption cross section, as schematically shown in Fig. 2. This makes clear that EXAFS cannot be observed in single atoms. The outgoing electron wave that is released by the absorption of the incident X-ray photon of the energy E has the wavenumber:[6]

$$k = \sqrt{\frac{2m}{\hbar^2}(E - E_0)} \tag{9}$$

where k is the photoelectron wavevector, E_0 is the core threshold energy, and m is the mass of the electron. The electron will emerge as a spherical wave of the wavelength $\lambda = 2\pi/k$ in a free atom. If there are nearest neighbors, such as in molecules, clusters, or the condensed phase, this wave releases at neighboring centers scattered waves that interfere with the primary wave. Constructive or destructive interferences may occur, depending on the distance of the neighbors and the absolute value of k. This leads to modulations of the absorption and ionization cross sections. The EXAFS-signal $\chi(k)$ is obtained according to Eq. (10):[6]

$$\chi(k) = \frac{\mu(k) - \mu_0(k)}{\mu_0(k)} \,. \tag{10}$$

Here, $\mu_0(k)$ corresponds to the atomic absorption cross section as a function of k [cf. Eq. (6)]. This quantity represents the continuous background of the EXAFS-signal (cf. Fig. 1), whereas $\mu(k)$ is due to the oscillatory part of the signal. The EXAFS-equation is given in Eq. (11)[12,21]

$$\chi(k) = -\sum_{i,j} \frac{3\cos^2\Theta_{ij}}{kR_i{}^2} F_i(k) e^{-2\sigma_i^2 k^2} e^{-2R_i/\lambda_i(k)} \sin(2kR_i + \Phi_i(k)) \,. \tag{11}$$

Equation (11) contains the polarization dependent factor $\cos^2\Theta_{ij}$, the distance R between the absorber shell j and the backscatterer shell i, $F_i(k)$ is the backscattering amplitude, which forms with both exponential terms the amplitude function; $e^{-2R_i/\lambda_i(k)}$ is the loss factor, and $e^{-2\sigma_i^2 k^2}$ is the Debye–Waller factor, which can be separated into a static and a vibrational part. The last term in Eq. (11) contains the dependence of the distance and the scattering phase shift between the primary and reflected wave. The phase

$\Theta_{ij}(k)$ of the EXAFS signal depends on the core edge of the excited atom:[6]

$$\Phi_{ij}{}^{l}(k) = \Phi_{j}{}^{l}(k) + \Phi_{i}(k) - l\pi \qquad (12)$$

where the first term is the phase shift of the absorbing atom and the second term corresponds to the phase shift of the backscatterer, l depends on the excited core-edge. K- and L_1-EXAFS, i.e. excitation of electrons from $1s$- and $2s$-orbitals, gives $l = 1$, whereas $L_{3,2}$-EXAFS, corresponding to $2p$-excitations, gives $l = 0, 2$. Note that Eq. (12) reflects a simplified approach, since contributions from the absorbing atom to the total phase shift for

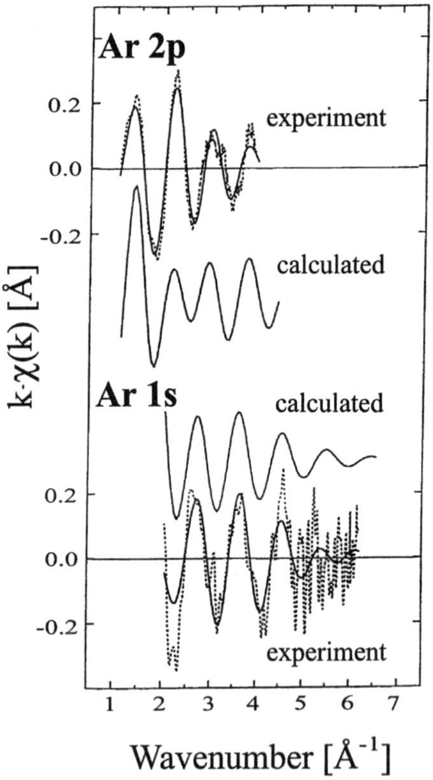

Fig. 3. Comparison of the Ar $2p$- and $1s$-EXAFS (obtained at $\langle N \rangle = 750$ (top) and $\langle N \rangle = 400$ (bottom), respectively), where both experimental and calculated EXAFS-signals are shown.[24] The dashed lines represent the experimental data while the superimposed solid lines are the first-shell component, Fourier filtered over $2.2 < R < 4.4$ Å using a Hanning apodization window (see Refs. 24 and 25 for further details).

$s(\Phi_j{}^{l=1})$- and $p(\Phi_j{}^{l=0}, \Phi_j{}^{l=2})$- ionization are not considered. This leads to deviations from the simple phase shift estimate given in Eq. (12), which is expected to be $\Phi_{ij}{}^{l}(k) = \pi$, if K- and $L_{3,2}$-EXAFS are compared. EXAFS in free atomic clusters has been investigated for the first time for variable size argon clusters.[22–25] Figure 3 shows a comparison of the Ar 2*p*- and Ar 1*s*-EXAFS-signal along with the calculated EXAFS-signal,[24] which is obtained from the spherical wave approach.[26] Both, the experimental and calculated EXAFS-signals are in good agreement, but there is a phase shift between the 1*s*- and 2*p*-EXAFS with a value between π and $\pi/2$, depending on the wavenumber. This indicates that the simple approach outlined in Eq. (12) is indeed oversimplified, but it is worthwhile to mention at this point that this result was the first consistent proof that there is experimental evidence for EXAFS in free van der Waals clusters.

The EXAFS-analysis is completed by Fourier-transforming the EXAFS signal so that the nearest-neighbor distance is obtained. Further structural properties, such as the number of nearest neighbors, the Debye–Waller factor, and shifts of the threshold energy are derived from theoretical amplitude functions and phase shifts, where plane- and spherical-waves have been considered as well as inelastic losses.[6,24,26,27] The reliability of the EXAFS-analysis increases, if a wide k-range is analyzed. The absorption and therefore the EXAFS-signal are much stronger for $L_{3,2}$-excitations than above the K-edge. In the case of argon, the absorption cross section near the K-edge is roughly by more than a factor of 50 weaker than near the L-edges.[9] However, the usable k-range is limited to ~ 4 Å^{-1} for $L_{3,2}$-EXAFS, because of the close lying L_1-edge, which disturbs with its considerable edge-jump the weak EXAFS-amplitude of the $L_{3,2}$-EXAFS. Another inherent problem with the analysis of the $L_{3,2}$-EXAFS is that this edge is spin-orbit split by 2.15 eV.[28] In contrast, the L_1-edge shows no spin-orbit splitting, but this edge is weaker in cross section and there are still blended contributions from the $L_{3,2}$-EXAFS. Thus, L-EXAFS is not attractive for a thorough EXAFS-analysis that aims to investigate size-dependent structural changes in free clusters. The EXAFS amplitude increases with Z and the K-edges shift in the same sense to higher excitation energy, which implies weaker absorption cross sections [cf. Eq. (6)]. As a result, K-edge EXAFS suffers, especially for dilute gas phase targets, such as clusters, from weak signal intensity. This inherent drawback has been compensated recently, since high intensity, high brightness X-ray sources have become available (see Sec. 3.2).

2.3. *Inner-Shell Photoionization*

Inner-shell photoionization of atoms, molecules, clusters, and the condensed phase cannot be simply described by one electron photoemission, assuming frozen orbital energies. This simple approach, which corresponds to Koopmans' theorem, is often successfully applied to describe valence-shell photoionization. However, this approach completely fails for inner-shell photoionization, where deviations of the order of 10–20 eV relative to the experimental results are found.[12,29]

Inner-shell photoionization is rather characterized by rearrangements of the outer electronic shells after the core hole is created by absorption of an X-ray photon.[30] This is a result of the reduced shielding of the nuclear attraction. It is well-known that the accompanied relaxation energy of K-shell photoionization scales with the number of valence electrons. The typical range of relaxation shifts is 2–3 eV for molecules compared to the corresponding isolated atoms. Relaxation of outer-shell electrons effectively screens the core hole potential experienced by the photoelectron. Theoretical aspects of relaxation effects in molecular photoionization are reviewed in Ref. 30. Similar results occur in free clusters, where polarization screening lowers the core ionization thresholds relative to the isolated atom. A redshift of 1.0 ± 0.1 eV has been found for the Ar $2p$-ionization energies of heavy argon clusters relative to the atomic value.[31,32] Considerably larger relaxation shifts in core ionization energies are observed, if rare gases are adsorbed on metal surfaces.[33] However, the binding energy shift depends on the number of adsorbed layers, so that for multilayers the relaxation shift is comparable to that of large homogeneous clusters. The characteristics of inner-shell photoionization is, that the surface-to-bulk ratio can systematically be varied so that, provided sufficient spectral resolution, surface and bulk core ionization energies are obtained as a function of cluster-size and composition. These values can be used for extrapolating the condensed phase value. This is of interest to insulators, where reliable core ionization energies are difficult to measure for macroscopic samples because of charging problems. From the experimental viewpoint there are two major approaches to measure core ionization energies: (i) non-resonant excitation and (ii) resonant excitation. Experimental aspects of both techniques are briefly discussed in Sec. 3.3. The former approach is equivalent to classical X-ray photoelectron spectroscopy (XPS or ESCA),[34] where besides line sources (e.g. Al $K_{\alpha 1}$-radiation: $E = 1486.70$ eV) synchrotron radiation can be used.[32] The photon energy is fixed well above the core ionization energy and the kinetic energy of the photoelectrons is measured. The latter method relies on tunable X-rays, where exclusively electrons with known kinetic energy are detected. Most favorably, zero kinetic energy (ZEKE) photoelectrons are measured as a function of photon energy.[31,35] Resonant excitation near core ionization energies have the implicit advantage to explore effects, that can preferably be detected by this detection mode. This is primarily post collision interaction (PCI),[36] which concerns the energy transfer between the low kinetic energy photoelectron and a fast Auger electron. As a result, the Auger electron can overtake the photoelectron, especially near threshold. The photoelectron may then be retained in the ionic core, so that single ionization occurs, even though the photon energy exceeds the core ionization energy. This situation changes deeper in core ionization continua, where the photoelectron has sufficient kinetic energy to be emitted, so that doubly charged final products are formed, according to the normal Auger decay (cf. Secs. 2.4, 2.5 and 4.3).

2.4. *Electronic and Radiative Relaxation*

Inner-shell photoionization leaves a core hole that is stabilized via relaxation processes (cf. Sec. 2.3). The core hole is filled by an outer-shell electron leading either to the emission of an Auger electron or an X-ray photon (cf. Fig. 4). The branching ratio between both relaxation processes depends on the nuclear charge Z.[12]

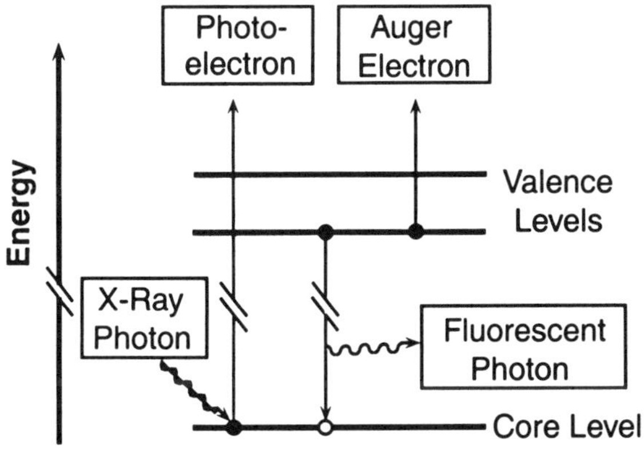

Fig. 4. Schematic diagram of inner-shell ionization and relaxation processes (adapted from Ref. 12).

Light elements undergo primarily electronic relaxation via the Auger decay, whereas core holes in heavy elements are stabilized by X-ray fluorescence. Auger electrons are easily distinguished from photoelectrons since their kinetic energy is constant as a function of excitation energy, if near-edge effects, such as post collision interaction, are neglected.[36]

There are numerous Auger channels of different kinetic energies, since electrons from various outer electronic shells can be ejected. The major part of their kinetic energy is due to the energy difference between the outer and inner electronic shells, which are involved in the relaxation process. As a result, the Auger electron spectrum covers a wide range of kinetic energies, leaving after the normal Auger decay two holes in the final products. However, there are also double and multiple Auger processes as well as Auger cascades leading to the formation of multiply charged cations.

In contrast, the resonant Auger decays occurring in the preedge regime lead to the formation of singly charged products, if the primarily excited core electron remains as a spectator in an excited electronic state, or it is further excited (shake up processes). The variety of processes, involving resonant excitations below core-edges, and direct photoemission in core ionization continua lead to strong variations in charge states of the cations that are formed. This is equivalent to variations in ionization yield, i.e. the number of charges that are produced per absorbed photon (see Fig. 5).[37] Figure 5 shows clearly for atomic argon that the thresholds of direct

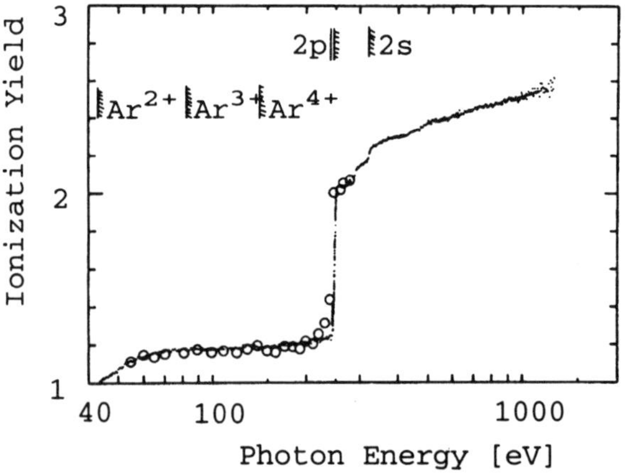

Fig. 5. Ionization yield of atomic argon (adapted from Ref. 37). See text for further details.

multiple ionization do not change significantly the ionization yield, whereas the LMM-Auger process that occurs upon Ar $2p$-excitation ($E > 248$ eV) increases steeply the ionization yield. The nomenclature of Auger processes indicates the electronic shells that are involved in electronic relaxation: (i) the core hole shell, (ii) the shell that fills the core hole, and (iii) the shell of the emitted Auger electron. In the case of van der Waals clusters, the same behavior is expected to occur as in atoms, since only one atom per cluster absorbs an X-ray photon. Unlike atoms, the number of charges that are produced in a cluster are relevant to subsequent fragmentation via charge separation (fission). These processes occur with high efficiency in core ionization continua (see Sec. 4.3).

The kinetic energy distribution of Auger electrons is known to be almost identical in free rare gas atoms and the corresponding condensed phase.[38] Therefore, no significant shifts in kinetic energy of Auger electrons are expected to occur in rare gas clusters relative to the bare atom. In contrast, adsorbed rare gases on metal surfaces show significant distance dependent shifts in photoemission energies as well as Auger electron energies.[33] These are primarily due to strong relaxation shifts, which are well-described by an image potential.

Auger yields of variable size argon clusters indicate that Auger electrons are emitted primarily from the surface of clusters.[39] This result is not too surprising, since Auger electron spectroscopy is known from numerous studies on macroscopic surfaces to be surface sensitive. However, clusters are the ideal test case to give clear evidence for this behavior, since their surface-to-bulk ratio is high and can be widely varied by changing the average cluster-size. In the case of clusters and adsorbates, the Auger electron can also be inelastically scattered at neighboring atoms, which may be ionized.[40] In the case of condensed argon the probability of this process is reported to be of the order of 18%.[40] This intracluster electron impact ionization yields one positive charge on the primarily excited site via photoemission and the other one on a neighboring site, so that the final product is doubly charged. In contrast, more likely is the normal Auger decay, which is confined to the primarily excited atom. As a result, a doubly charged atom is formed near the surface of the cluster. Subsequently, the doubly charged atom may be stabilized by neutral neighbors forming a charge transfer excimer.[41,42] These are known to be unstable with respect to radiative relaxation. In the case of argon clusters one observes a broad fluorescence band in the VUV/UV-regime ($160 \leqq \lambda \leqq 300$ nm).[41] This relaxation channel is known since decades as the "*third continuum*" fluorescence of argon.[43] Evidence for this assignment, implying that this fluorescence channel is due to double ionization, comes also from the fact that "*third continuum*" fluorescence is only observed in the Ar 2*p*-continuum. Double ionization is the dominant ionization mechanism in this spectral regime (cf. Fig. 5). "*Third continuum*" fluorescence cannot be observed in atomic argon, since it requires neighbors to the absorbing site. In contrast, atomic argon is known to undergo different fluorescence processes in the Ar 2*p*-regime. These are found to occur upon resonant excitation of Rydberg states, resulting in the emission of fluorescence light in the UV/VIS-regime.[44] These relaxation processes are of different origin than radiative relaxation of core-excited argon clusters: resonant excitation into low-lying Rydberg states (e.g. Ar $2p_{3/2} \rightarrow 3d$, $E = 247$ eV) is followed by electronic relaxation (resonant Auger decay).[45] This process creates excited, singly charged atoms, where the electron is excited into the 4*d*-Rydberg state via shake up processes. Subsequent radiative relaxation, that occurs primarily in the UV/VIS-regime, results in deexcitation of Ar^+ in the nanosecond time regime.[44] Consequently, these processes in atomic argon can only be observed in a narrow energy regime near the Ar 2*p*-edge, where single ionization dominates as a result of resonant Auger processes. In contrast, radiative relaxation in argon clusters occurs in the Ar 2*p*-continuum, since this process relies on double ionization.

2.5. *Fragmentation*

Core level excitation followed by inner-shell photoionization and subsequent relaxation leads to the formation of multiply charged species. The ionization yield, i.e. the number of charges produced per absorbed X-ray photon, increases significantly as soon as the photon energy reaches core ionization continua (cf. Fig. 5). In the case of the Ar 2*p*-excitation, double ionization is the dominant process. In contrast, upon Ar 1*s*-excitation multiple ionization is found as a result of Auger cascades.[46] Thus, highly charged cations are formed as a result of the KLL- and KLM-Auger decays, so that most abundantly Ar^{n+} is formed with $3 \leqq n \leqq 6$.

Stable doubly or multiply charged clusters have been investigated using mass spectrometry, where the occurrence of stable multiply charged clusters is unequivocally assigned according to their mass-to-charge ratio (m/z).[47] The appearance size of stable doubly charged clusters is described by the following scaling law:

$$n_2 T_c \nu^{1/3} = \text{constant} \tag{13}$$

where n_2 is the appearance size of stable doubly charged clusters, T_c is the critical temperature, and ν is the molar volume. In the case of rare gas clusters the following critical cluster sizes are reported: $n_2(\mathrm{Ar}) = 91$;[48] $n_2(\mathrm{Kr}) = 71$.[49] Considerably higher values are found for the appearance size of stable triply charged clusters, such as $n_3(\mathrm{Ar}) = 226$.[50]

Multiply charged clusters decay most likely into singly charged products via fission (charge separation).[47] This process is also known as "*Coulomb explosion*",[51] reflecting the dominant electrostatic repulsion of the charges. Molecular dynamics simulations on doubly charged xenon clusters indicate that singly charged cation pairs are formed in the picosecond time regime, if their size is too small to accommodate both charges.[52] As a result of electrostatic repulsion of the charges, singly charged fragments are formed with considerable amounts of kinetic energy. The total kinetic energy can be estimated according to Coulomb's law. This assumption appears to be reasonable for atomic clusters, whereas for molecular species, properties of highly charged intermediates need to be considered.[53]

The detection of cation pairs is accomplished by various coincidence techniques (see Sec. 3.3 for experimental details). The most simple approach involves the measurement of time-of-flight differences of correlated cation pairs using a time-of-flight mass spectrometer. This technique is called Photoion–photoion-coincidence spectroscopy (PIPICO). Firstly, we assume that fission produces two correlated cations that come from the same doubly charged precursor, according to the process: $\mathrm{M}^{++} \to \mathrm{A}^+ + \mathrm{B}^+$. The width of the time-of-flight difference distribution w, which is obtained from PIPICO-spectroscopy, is given by the following simple formula:[54]

$$w = \frac{5766}{E}\sqrt{\frac{U_0 m_{\mathrm{A}^+} m_{\mathrm{B}^+}}{m_{\mathrm{M}^{++}}}} \tag{14}$$

where E is the electric field strength (in V/cm) that is applied for acceleration of the cations in the ionization region of the time-of-flight mass spectrometer, $m_{\mathrm{M}^{++}}$, m_{A^+}, and m_{B^+} are the relative masses (in g/mol) of the doubly charged parent and the singly charged fragments, respectively, and U_0 is the kinetic energy release (in eV). Similarly, fission of a doubly charged cation, which is often called "*dication*", into two singly charged fragments and one neutral n, corresponding to the fission $\mathrm{M}^{++} \to \mathrm{A}^+ + \mathrm{B}^+ + n$, is described by:

$$w = \frac{2822}{E}\sqrt{\frac{U_0 (2m_n + m_{\mathrm{B}^+})^2 m_{\mathrm{A}^+}}{m_{\mathrm{M}^{++}}(m_{\mathrm{B}^+} + m_n)}}\,. \tag{15}$$

It has been shown that the simple approach of PIPICO measurements works satisfactory only for small molecules, where the number of possible fission processes is limited. In the case of polyatomic molecules and clusters, this approach leads to ambiguities.[55] Therefore, the use of multicoincidence experiments, where the individual flight times of the correlated cations, $t_A{}^+$ and $t_B{}^+$ are measured separately, gives more detailed information on the fission mechanisms. One important approach, that is discussed in detail below, is Photoelectron-photoion–photoion-coincidence (PEPIPICO) spectroscopy (cf. Secs. 3.3 and 4.3). The results that are obtained from PEPIPICO spectra are three-dimensional shapes of correlated cation pairs (cf. Sec. 4.3), where the relative intensity of these processes is plotted over the flight times of the correlated cations. Alternatively, the intensity of the coincidence signals can be represented in contour plots, as shown in Fig. 6. Note that only one half of the array is shown because of the mirror symmetry with respect to the main diagonal.

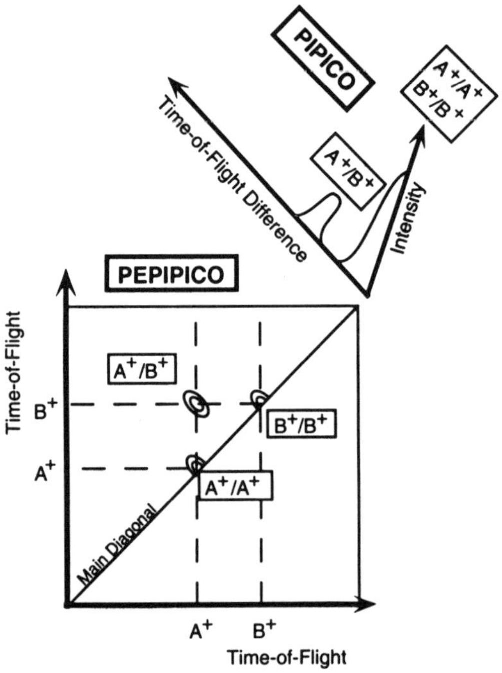

Fig. 6. Schematic comparison of PEPIPICO and PIPICO-spectra leading to the formation of correlated cation pairs. Further details are described in the text.

Figure 6 also shows that PIPICO represents just a projection perpendicular to the main diagonal of the $t_A{}^+$ and $t_B{}^+$ -array, leading often to blended signals and therefore to ambiguous results. Useful information on the fission mechanisms can be extracted from the PEPIPICO signal shapes and slopes.[56] PEPIPICO signals are often long-shaped, where the maximum signal length is found perpendicular to the main diagonal, corresponding to a considerable kinetic energy release (KER) of the fission process, typically of the order of several eV (cf. Fig. 6). The width parallel to the main diagonal corresponds to the KER of secondary processes. It is often considerably smaller, especially if secondary losses of neutrals follow primary fission.

The signal slope contains further information on the fission mechanisms.[56] A two-body decay leads to a long-shaped coincidence signal with a slope of -1. This is a result of conservation of linear momentum in the fission process, where

$$(t - t_0)_A{}^+ + (t - t_0)_B{}^+ = 0\,. \tag{16}$$

Here, t is the flight time of the cation and t_0 is the flight time of the cation which is initially at rest. However, simple two body decays are only observed for small molecular dications. The majority of fission processes of clusters, that have been investigated in the past, give rise to considerably more complicated coincidence signal shapes and slopes that are different from -1. These are rationalized by mechanisms, involving secondary evaporation of neutrals (cf. Sec. 4.3).[56] The variety of possible fission mechanisms have been discussed earlier for free molecules and clusters.[56–59] The main slope of the coincidence signals reflects primarily the mass ratio of the fragments, if secondary losses of neutrals occur: the model decay of a dication $M^{++} \to AB^+ + C^+$, $AB^+ \to A^+ + B$ gives a main slope m:

$$m = -\frac{m_A{}^+}{m_{AB}{}^+} \tag{17}$$

if $m_A{}^+ > m_C{}^+$, i.e. $m > -1$, whereas the reciprocal value with $m < -1$ is expected if $m_C{}^+ > m_A{}^+$. Often, both singly charge fragments loose neutral moieties after fission. Then, the main slope is given by the ratio of the corresponding mass ratios, similar to secondary losses of neutrals from monocations. Further details on shape analyses of multicoincidence signals can be found in Refs. 56, 58 and 59.

3. Experimental Techniques

3.1. *Cluster Production*

Variable size clusters are efficiently produced by adiabatic expansion of neat gases or gas mixtures into the vacuum. This method is well-known since decades.[1] It makes use of a small nozzle of typically 10–100 μm diameter and a skimmer, which separates the expansion chamber from the detection chamber, where the clusters are analyzed.

One important property of adiabatic expansions is, that they give size distributions, so that the average cluster-size, often denoted as $\overline{N}$ or $\langle N \rangle$, characterizes the maximum of the size distribution of neutrals that are produced in jet expansions. The average cluster-size can be easily obtained from scaling parameters, which have been introduced by Hagena.[60] The advantage of scaling parameters is, that they reduce various experimental quantities that affect the gas expansion, such as the temperature and pressure in the high pressure reservoir (stagnation temperature T_0 and stagnation pressure p_0), the nozzle diameter D, and properties of the expanded gas. The reduced scaling parameter Γ^* is defined as follows:[60]

$$\Gamma^* = \frac{\Gamma}{\Gamma_{\mathrm{ch}}} = \Gamma {R_{\mathrm{ch}}}^{3-q} {T_{\mathrm{ch}}}^{(1.5-0.25q)} \tag{18}$$

where q is an experimental parameter, which is for argon 0.85, T_{ch} is the characteristic temperature: $T_{\mathrm{ch}} = \Delta {h_0}^0/K$, where $\Delta {h_0}^0$ is the sublimation enthalpy at $T = 0$ K, k is the Boltzmann constant; R_{ch} is the characteristic distance: $R_{\mathrm{ch}} = (m/\rho)^{1/3}$, where m is the atomic mass and ρ is the density of the solid; Γ is the scaling parameter, with $\Gamma = n_0 \Delta {T_0}^{(0.25q-1.5)}$, where n_0 is the number density in the high pressure volume. Typical values for argon are: $R_{\mathrm{ch}} = 3.39$ Å and $T_{\mathrm{ch}} = 927$ K. This yields for argon the simple formula:[61]

$$\Gamma^* = \frac{1646 p_0[\mathrm{mbar}] d^{0.85}[\mu m]}{T^{2.29}[\mathrm{K}]} \,. \tag{19}$$

The reduced scaling parameter Γ^* can be used in combination with experimental observations in order to obtain a first qualitative estimate on the cluster-size in the jet:[60]

$\Gamma^* < 200$:	no clusters
$200 < \Gamma^* < 1000$:	small clusters
$\Gamma^* > 1000$:	large clusters.

Table 1. Average cluster-size $\langle N \rangle$ of krypton clusters for typical experimental expansion conditions using a nozzle diameter of 50 μm. The correlation of Γ^* with $\langle N \rangle$ is obtained from Ref. 63.

p_0	T_0 [K]	Γ^*	$\langle N \rangle$
1	298	180	2
2	298	370	4
3	298	550	12
4	298	730	18
5	298	920	30
5	273	1120	40
5	243	1460	90
5	213	1970	140
5	183	2800	350
5	163	3640	1000
5	158	3910	1250

Various experimental approaches have been used in the past to correlate Γ^* with $\langle N \rangle$. These include electron diffraction,[62] mass spectrometry,[63] and molecular beam scattering experiments.[64] The latter approach is especially suitable for the low $\langle N \rangle$-regime. The typical value range that is available for core level excitation on krypton clusters in our experiment is $\langle N \rangle <$ 1700,[65] if the correlation of Farges *et al.* is used.[62] Somewhat lower values are obtained, if the more recent, but likely more realistic calibration of Karnbach *et al.* is applied (cf. Table 1).[63]

The experiments that have been carried out in the field of core level excitation made use of portable jet expansions. These are flexible in use and easy to mount at storage ring facilities.[23–25,66] Compact experimental setups are required for these experiments, since there is only limited access to synchrotron radiation sources. As a result, table top expansions, which make use of high throughput turbomolecular pumps and relatively small nozzle diameters $D < 100$ μm, stagnation pressures $p_0 < 10$ bar, and stagnation temperatures ranging between room temperature and 70 K were preferentially used in the past. Moreover, it turns out that the requirements that make these experiments extremely flexible have no severe drawbacks, since the $\langle N \rangle$-regime that is accessible reaches from purely atomic conditions up to several thousands of atoms per clusters. This is sufficient for most investigations on size effects of clusters.

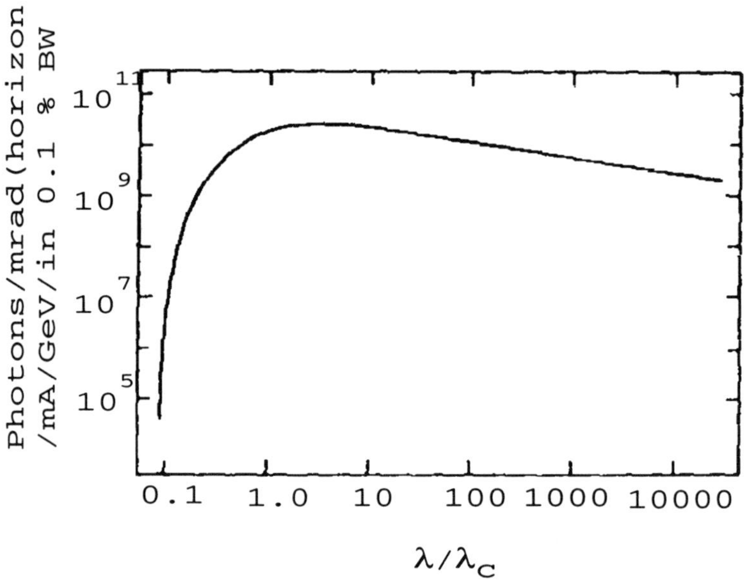

Fig. 7. Shape of the radiation distribution of an electron on a curved trajectory per GeV (adapted from Ref. 67).

3.2. *X-ray Sources*

Suitable X-ray sources for investigations of resonant core level excitations are intense continuum sources that cover wide energy regimes in the soft- and hard-X-ray regime (cf. Fig. 7). This is primarily synchrotron radiation, which is provided by storage ring facilities. Currently, only these X-ray sources allow resonant photon excitations of dilute gas phase targets, such as clusters, in spectral regimes where the absorption cross sections are low. The advantage of storage ring facilities is that they deliver high intensity, high brilliance, and widely tunable radiation. On the other hand, other classical light sources, such as bremsstrahlung-continua or X-ray line sources are not suitable for the experiments described below (see Sec. 4).

Synchrotron radiation emerges as a broad continuum from the infrared to the X-ray regime (see Fig. 7), as a result of transversally accelerated relativistic particles, such as electrons or positrons.[67–69]

Most properties of synchrotron radiation may be derived starting from classical electrodynamics, where an oscillating dipole is subjected to a Lorentz transformation. Assuming that relativistic electrons move on curved trajectories in a bending magnet of the radius R, the radiated power

P is given by:[68]

$$P = \frac{e^2 c}{6\pi\varepsilon_0} \frac{1}{(m_0c^2)^4} \frac{E^4}{R^2} . \tag{20}$$

Here, e is the elementary charge, ε_0 is the dielectric constant of the vacuum, m_0c^2 is the electron rest mass energy ($m_0c^2 = 0.511$ MeV), c is the speed of light, and E is the energy of the charged particle, which is typically of the order of 1–5 GeV. From Eq. (20), it is evident, that synchrotron radiation is preferentially emitted from electrons or positrons, since e.g. protons require considerably higher acceleration energies ($m_0c^2 = 938.19$ MeV). The energy loss per electron and turn is:

$$\Delta E = \frac{e^2}{3\varepsilon_0(m_0c^2)^4} \frac{E^4}{R} . \tag{21}$$

If typical dimensions are used for the quantities in Eq. (21), one obtains:

$$\Delta E\ [keV] = 88.5 \frac{E^4\ [\mathrm{GeV}^4]}{R\ [m]} . \tag{22}$$

Similarly, the critical wavelength λ_c of the spectral distribution is given by (cf. Fig. 7):[67]

$$\lambda_c\ [\text{Å}] = 5.59 \frac{R\ [m]}{E^3\ [\mathrm{GeV}]} = \frac{18.64}{B\ [\mathrm{T}]\ E^3\ [\mathrm{GeV}]} \tag{23}$$

where B is the magnetic field in the bending magnet. Another important property of synchrotron radiation is its strong forward collimated emission, with the small emission angle Θ:

$$\Theta \approx \tan\Theta \approx \frac{1}{\gamma} \tag{24}$$

where $\gamma = E/m_0c^2 = 1957\ E\ [\mathrm{GeV}]$. Then, assuming that $E = 1$ GeV, one obtains $\Theta = 0.5$ mrad. This allows to guide synchrotron radiation in narrow beam lines to the X-ray monochromators and to the experimental end stations. In addition, synchrotron radiation has a defined time structure, with pulses of typically 100 ps length and a high degree of polarization. It is fully polarized with the electric vector parallel to the orbital plane of the circulating electrons or positrons. Elliptic polarization is found above and below this plane.

Modern storage ring facilities use insertion devices, such as wigglers and undulators, in order to increase the intensity of the emitted radiation. Insertion devices consist of periodic structures of permanent magnets. These can be regarded as a series of small bending magnets. The dimensionless parameter K is used to distinguish wigglers from undulators:[68]

$$K = \frac{\lambda_u e \widetilde{B}}{2\pi m_e c} \tag{25}$$

where λ_u is the period length of the magnetic structures, $\widetilde{B} = B_0/(\cosh\ (\pi g/\lambda_u))$, B_0 is the magnetic field in the middle between the magnetic structures, and g is the width of the gap between the magnetic structures. Undulator conditions are obtained, if $K \leqq 1$, indicating that deflection of the electron beam is weak. The radiation that is coherently emitted with high intensity from an undulator is monochromatic. It also contains higher harmonics and there is broadening of the emitted lines, which is due to the relativistic Doppler effect. The spontaneously emitted broad band synchrotron radiation is in this case by several orders of magnitudes weaker in intensity. The amplitude of the emitted lines of an undulator increases with the number of periods. Scanning the emitted photon energy is accomplished by changing the gap width.

Synchrotron radiation or undulator radiation emerging from a storage ring is monochromatized by suitable monochromators before it can be used for core excitation experiments. First experiments on core-excitation of clusters made use of toroidal grating monochromators,[22,70] which can be used in the 150–700 eV photon energy regime. These photon energies are sufficient to excite K- and L-edges of light (second and third row) elements. Toroidal mirror monochromators have a high photon throughput, which is essential for studies on dilute gas phase targets, such as clusters. Their inherent disadvantage is the broad focal point and the broad band width of the monochromatic X-ray beam. As a consequence, spectral details that are essential to investigate changes in electronic structure of variable size clusters are difficult to investigate and represent an inherent limit of the feasibility and success of such experiments. Recent progress has been achieved by using insertion devices, such as undulators, which enhance the photon flux and brilliance of the emitted radiation by several orders of magnitude.[32,66,71–74] This is essential for investigating subtle size effects in clusters, which give rise to small spectral shifts of the order of several meV in the soft X-ray regime, corresponding to $E/\Delta E > 10^4$.[72,74]

Photon energies above $\sim$ 1500 eV are monochromatized using double crystal monochromators, which make use of the Bragg reflection condition:

$$n \cdot \lambda = 2d \sin\Theta\,. \tag{26}$$

Here, $n = 1, 2, \ldots, \lambda$ is the wavelength of the X-ray photon, d is the lattice spacing of the crystal, and Θ is the glancing angle.

3.3. *Detection Methods*

Core level excitation leads after photoabsorption of an X-ray photon to photoemission of a core electron, if the photon energy exceeds the core electron binding energy. The core hole is stabilized by electronic or radiative relaxation (cf. Secs. 2.3 and 2.4). As a result, double or multiple ionization processes govern core ionization continua. Therefore, the detection of charged products, such as electrons or cations, is the most straightforward way to investigate size effects in variable size clusters by core level spectroscopies. Figure 8 shows schematically a typical experimental setup, where all vacuum equipment is not shown for clarity. Jet experiments require high vacuum, whereas in beam lines and soft X-ray monochromators ultrahigh vacuum is needed. Therefore, an efficient differential pumping stage is essential between the jet experiment and the beam line. The jet experiment consists of (i) a continuous jet expansion for cluster production, (ii) the beam of monochromatic X-rays provided by the storage ring facility, (iii) a time-of-flight mass spectrometer for cation detection, and (iv) an electron detector. Radiative processes, leading to the emission of fluorescence light have been investigated as well.[41] However, direct photoabsorption experiments, according to Eq. (1), have not been performed to date, since the cluster beam is too dilute, the path length is too short, and the absorption cross sections are in general too low in order to observe an attenuation of the X-ray beam.

Electrons are most efficiently detected using total electron yield detectors. These consist either of a channeltron multiplier or a stack of micro channelplates, that are mounted close to the ionization region, where the photon beam crosses the cluster jet. The amplified electron current of *"all"* electrons is measured as a function of the incident photon energy. Total electron yield spectra may be regarded as pseudo-absorption spectra, but there are distinct differences in physical processes compared to photoabsorption, especially since one X-ray photon may lead to the emission of several electrons (cf. Fig. 5). Partial electron yields are also useful for investigations on clusters. Techniques that have been applied in the past were: (i) energy selected electron yields[32,39,66] and (ii) zero kinetic energy (ZEKE) photoelectron spectroscopy.[31] Energy selected electrons are obtained by using e.g. electrostatic energy analyzers, such as cylindrical mirror analyzers

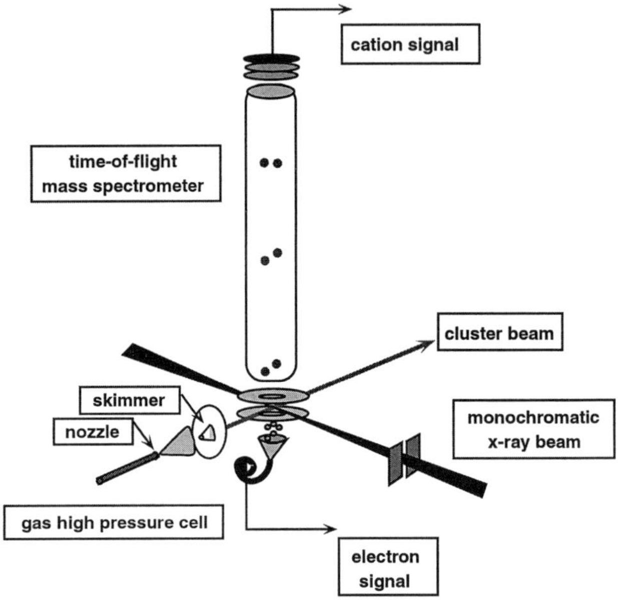

Fig. 8. Schematic view of the experimental setup for core level excitation of clusters. Further details are described in the text.

(CMA)[75,76] or electron time-of-flight analyzers.[32,66] These devices allow the detection of both photoelectrons and Auger electrons. Zero kinetic energy (ZEKE) photoelectrons are those electrons that do not carry any or very little kinetic energy.[35] These electrons are selected by using the time structure of the storage ring, mostly when a single electron bunch is stored in a storage ring.[31] The general idea of selecting ZEKE photoelectrons is straightforward: if the photoionization process occurs in a field free region, then the ZEKE photoelectrons will be at rest, whereas other energetic electrons leave the ionization region as a result of their considerable kinetic energy. A low drawout voltage pulse, corresponding to 5–10 V/cm, extracts the ZEKE photoelectrons into the ZEKE photoelectron spectrometer. Discrimination of energetic electrons is fourfold: (i) The device consists of a narrow flight tube with a length-to-diameter ratio of $\geqq 10$. Therefore, energetic electrons cannot efficiently penetrate to the detector because of angular discrimination;[77,78] (ii) the electron detector is mounted off the line-of-sight, so that straight into the detector flying electrons are not detected; (iii) the voltage pulse for extracting ZEKE photoelectrons is properly delayed after

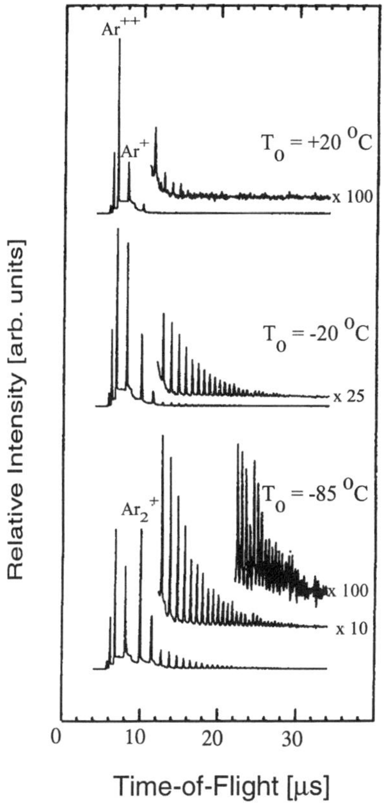

Fig. 9. Series of photoionization mass spectra of argon clusters recorded at 415 eV photon energy at various expansion conditions corresponding to $\langle N \rangle = 10$ (top), $\langle N \rangle = 50$ (middle), $\langle N \rangle = 300$ (bottom).

photoionization, it is applied after the energetic electrons have disappeared; (iv) the electron flight time is used as an additional way to suppress the detection of energetic electrons.[31]

The most common detection technique in the field of core level excitation of clusters is mass spectrometry, where the positively charged ions are mass-selected. Excitation in the core level regime leads to massive fragmentation of the singly and multiply charged clusters (cf. Sec. 2.5). As a result, the mass distribution observed in mass spectra is by far smaller than the neutral mass distribution that is characterized by the average cluster-size $\langle N \rangle$ (cf. Sec. 3.1 and Table 1). Figure 9 visualizes this discrepancy for 415 eV excitation energy. The calculated average cluster-size of neutral clusters $\langle N \rangle$ ranges between 10 (Fig. 9, top) and 300 (Fig. 9, bottom), whereas the heaviest cluster mass observed in the time-of-flight mass spectra is $Ar_7{}^+$ (Fig. 9, top) and $Ar_{37}{}^+$ (Fig. 9, bottom), respectively. In the case of small clusters, the jet is dominated by atomic argon, which gives rise to the dominant Ar^{2+} mass line at $E = 415$ eV excitation energy, corresponding to excitations of the Ar $2p$- and $2s$-continua, where double ionization dominates (cf. Fig. 5). The ionic fragmentation pattern changes considerably as $\langle N \rangle$ is increased. Especially, the $Ar_2{}^+$ mass line becomes intense, since this cation is a stable product of ionic cluster fragmentation. Photoion yields of mass-selected dimer cations, such as Ar_2^+, serve to obtain cluster-specific information on the size-dependent electronic structure. They also serve to suppress the dominant contribution of atomic species, especially if the cluster density is weak, e.g. near the threshold of cluster formation. This is illustrated in Sec. 4.1.

Time-of-flight mass spectrometry is among various methods of mass spectrometry the most suitable approach in the regime of core level excitation of free clusters. This technique has the advantage that the transmission function of the mass spectrometer is high, typically of the order of 30%.[79] The cation extraction conditions can be optimized so that even high kinetic energy fragments, that are formed by Coulomb explosion (cf. Secs. 2.5 and 4.3) are efficiently detected. Another advantage is that all cations of interest are measured simultaneously. Individual masses are selected by applying proper time gates. This allows one to measure simultaneously electronic properties of isolated and clustered species. Small energy shifts of resonant excitations, which can be as small as a few meV in the regime of core level excitation have been reliably identified in this way.[72,74] However, time-of-flight mass spectrometry requires a defined time structure. This is either accomplished by pulsing the cation drawout field (typically 100–2000 V/cm, 10–100 kHz repetition rate),[55] or by using directly the time structure of the storage ring.[73] Especially large circumference storage rings, which are characterized by relatively low repetition rates (typically 1 MHz) in the single bunch mode, can be used in combination with continuous extraction voltages for direct cation flight time measurements. This method turns out to be extremely efficient for cation detection of mass-selected cluster ions, since dead times are minimized.[73] It is noted that this approach requires the use of a short time-of-flight mass spectrometer, so that the cation flight times are kept as short as possible.[32] Then, high electric field strengths for cation extraction are required to obtain proper space focusing of the cation masses at the detector.[80]

The decay mechanisms of singly and multiply charged clusters are commonly measured by various coincidence techniques (see also Sec. 2.5), involving the simultaneous detection of electrons and ions.[56,81,82] These approaches are straightforward, since they make use of the total yield signals. In Photoelectron-photoion–photoion coincidence (PEPIPICO) spectroscopy cation flight time measurements are started by the arriving electron. Subsequently, the cations stop the flight time measurements either in a single stop or in a multistop mode. Note that total electron yield detectors, consisting just of a channeltron, detect less efficiently high kinetic energy Auger electrons. Therefore, there is a higher detection efficiency of low kinetic energy photoelectrons. Single cation stops lead to Photoelectron-photoion-coincidence (PEPICO) spectra. Their general shape is somewhat different from time-of-flight mass spectra, obtained from pulsed cation extraction. This behavior can be rationalized in terms of state-selective photo-ionization and fragmentation, which is obtained from PEPICO spectroscopy. PEPICO spectra are often recorded by using simply a time-to-amplitude converter (TAC) for time measurements. The output of the TAC is fed into a pulse height analyzer, so that the cation intensity is recorded as a function of the cation arrival time at the detector of the time-of-flight mass spectrometer. If the start-signal from the electron detector is replaced by a cation signal, while keeping the stop cation-signal, then time-of-flight differences between correlated cation pairs are measured (PIPICO spectroscopy, cf. Sec. 2.5).[55] This approach utilizes, as PEPICO, in most cases continuous extraction voltages. It is noted that the number of false coincidences needs to be kept as low as possible. Therefore, the primary cation count rate is limited to a couple of kHz. Additionally, the raw PIPICO spectra are corrected with respect to false coincidences, by using the same number of start events from a pulse generator instead of the cation start-signal. PEPIPICO spectroscopy requires a total electron detector that provides the start-signal and two independent cation stop signals.[56,57] Instead of a TAC one uses for multicoincidence spectroscopies multihit time-to-digital converters (TDC). Typical experimental results on variable size clusters are discussed in Sec. 4.3.

4. Selected Examples

4.1. *Size Effects in Electronic Structure*

Near-edge spectroscopy serves to investigate size-dependent changes in electronic structure. Early work in this field made use of low spectral resolution.

Therefore, only major effects were investigated, such as the size-dependent occurrence of surface and bulk excitons in rare gas clusters.[22–24] Similar work on molecular van der Waals clusters, containing diatomic and triatomic molecules, indicated that the spectral resolution was insufficient to resolve any discrete size dependent features in the near-edge regime.[83,84] It was rather found that fragmentation of large clusters into the mass channels of small cluster fragment ions leads to enhanced intensity in core ionization continua. More recently, high spectral resolution beamlines became available, especially since insertion devices, such as undulators, provide sufficient photon flux for high resolution work, so that nowadays size dependent changes in electronic structure can be explored in considerably greater detail.[74] The spectral profiles that are resolved with current high resolution beam lines are primarily limited by the core hole lifetimes, yielding line shapes that are dominated by Lorentzians.[72] Table 2 gives an overview on free variable size clusters that have been studied to date in the field of core level excitation.

Table 2 indicates that rare gas clusters were the first test systems, but more recently the field has developed into the direction of relevant species

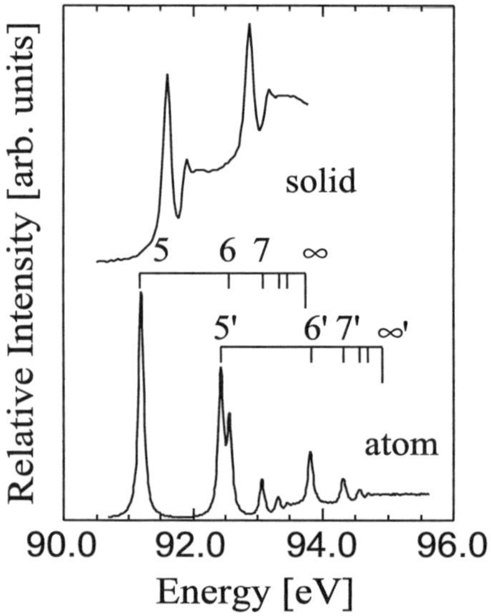

Fig. 10. Comparison of the Kr 3*d*-near-edge-structure of solid and atomic krypton. The numbers corrspond to n (cf. Eq. (7); see text for further details).

Table 2. Compilation of core level excitation spectra, where size effects in electronic structure are reported. The quoted energies correspond to core level binding energies. They are used to indicate the energy regime of the core edge (cf. Ref. 85).

Cluster	Core Edge	References
Ne	Ne 2*p* (48.5 eV)	86
	Ne 1*s* (870.2 eV)	71
Ar	Ar 2*p* (248.4 eV)	22, 23, 66, 87
	Ar 2*s* (326.3 eV)	25
	Ar 1*s* (3205.9 eV)	24, 73
Kr	Kr 3*d* (93.8 eV)	65, 72
	Kr 2*p* (1678.4 eV)	88
Xe	Xe 4*d* (67.5 eV)	89
S	S 2*p* (162.5 eV)	59, 90
	S 1*s* (2472 eV)	91, 92
CH_4	C 1*s* (284.2 eV)	93
N_2	N 1*s* (409.9 eV)	74, 83, 84
N_2Ar	Ar 2*p*, N 1*s* (248.8 eV, 409.9 eV)	84
CO	C 1*s*, O 1*s* (284.2 eV, 543.1 eV)	94
O_2	O 1*s* (543.1 eV)	95
CO_2	C 1*s* (284.2 eV)	95
N_2O	N 1*s* (409.9 eV)	83
H_2O	O 1*s* (543.1 eV)	96
C_2H_4	C 1*s* (284.2 eV)	97
CH_3OH	C 1*s* (284.2 eV)	98
C_6H_6	C 1*s* (284.2 eV)	98
NaCl	Cl 2*p*, (200 eV), Na 1*s* (1070.8 eV)	99, 100
CsF	Cs 4*d* (89.9 eV)	101
CdS	S 2*p* (162.5 eV)	102
FeS	S 2*p* (162.5 eV), Fe 2*p* (706.8 eV)	103

in materials science and biochemistry. We outline in the following, as a typical example, the spectral evolution near the Kr 3*d*-edge in variable size krypton clustersindexkrypton clusters.[72] Figure 10 shows a comparison of the Kr 3*d*-edge of gaseous and solid krypton. The gas phase spectrum shows two series of *np*-Rydberg excitations that arise from spin-orbit splitting, converging to the Kr $3d_{5/2}$- and Kr $3d_{3/2}$-series limits at 93.80 eV and 95.00 eV, respectively.[104]

The convergence limits are denoted in Fig. 10 as ∞ and ∞'. In contrast, the solid shows less discrete features.[15] The condensed phase features are broader in width and shifted to higher photon energy. Both dominant features are due to spin-orbit splitting of the lowest core excitons, reflecting considerable changes in electronic structure from the gas to the condensed phase. This size dependent evolution of spectral features is shown

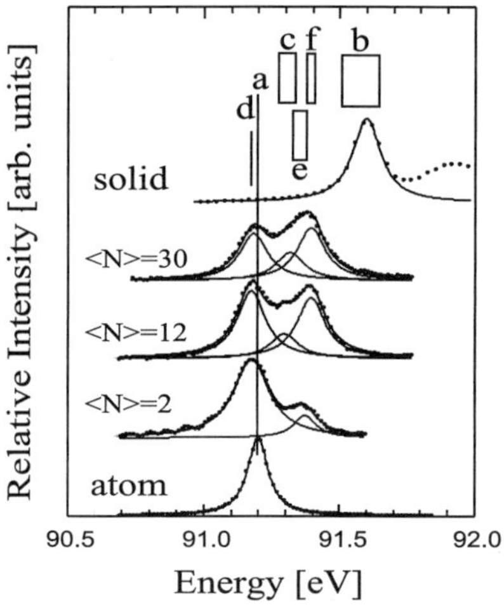

Fig. 11. Photoion yields of Kr^+ (atom); $Kr_2{}^+$ at various average cluster-sizes $\langle N \rangle$, and the total electron yield of solid Kr (top) in the Kr $3d_{5/2} \rightarrow 5p$ regime (cf. Ref. 72). Further details are described in the text.

in greater detail for the lowest-energy resonance in the Kr $3d$-regime (Kr $3d_{5/2} \rightarrow 5p$) (cf. Figs. 11 and 12). The atomic transition is represented by a single Lorentzian line with a width of 95 meV. This line width corresponds to the lifetime of the Kr $3d$-core hole. Changes in electronic structure of small clusters are investigated by photoion yields of mass-selected cluster cations, such as $Kr_2{}^+$ (see Fig. 11). This cation channel is, similar to argon clusters, one of the most preferred stable products of ionic cluster fragmentation (cf. Sec. 3.3). Near the threshold of cluster formation, corresponding to $\langle N \rangle = 2$, it is expected that $Kr_2{}^+$ is mostly formed either by: (i) direct photoionization of Kr_2 or (ii) by photoionization and subsequent ionic fragmentation of somewhat larger clusters $Kr_n (n > 2)$, which are also present in the jet at $\langle N \rangle = 2$. The photoion yield of $Kr_2{}^+$ shows that the lowest-energy transition (Kr $3d_{5/2} \rightarrow 5p$) is slightly redshifted, i.e. shifted to lower energy, by ~ 25 meV with respect to the atomic transition (cf. Fig. 11).[72] The line width is somewhat increased, which indicates that there is another line broadening mechanism besides the lifetime of the core hole, which is likely due to short-lived final states in krypton clusters. There is also a weak blueshifted shoulder in the $Kr_2{}^+$-yield ($\langle N \rangle = 2$, cf. Fig. 11), which is a result of fragmentation of larger cluster cations into the selected $Kr_2{}^+$-channel. This assignment comes primarily from the fact,

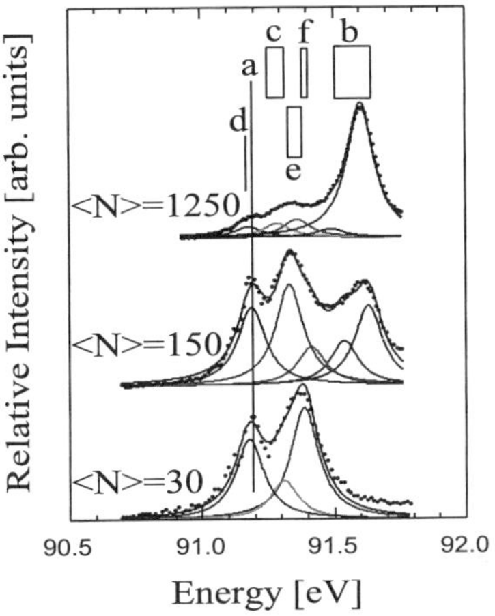

Fig. 12. Total electron yields (TEY) of Kr clusters recorded at various average cluster-sizes $\langle N\rangle$ in the Kr $3d_{5/2} \rightarrow 5p$-regime (cf. Ref. 72). Further details are described in the text.

that this blueshifted feature grows in intensity with increasing $\langle N\rangle$, where fragmentation of larger clusters contributes more strongly to the Kr_2^+-yield via photofragmentation. Therefore, photoion yields of mass-selected cluster cations, such as Kr_2^+, are suitable to detect changes in electronic structure of core excited neutrals. The Kr_2^+-signal is deconvoluted into three surface components in order to obtain a reasonable fit with the experimental spectra (see Fig. 11). The surface components are assigned to different geometric sites, where dimer- (d), edge- (e), and corner- (c) sites absorb at slightly different energies. Face- (f) and bulk- (b) sites are not expected to be observed in small clusters, since these occur only in large clusters.

This assignment is confirmed by model calculations based on the non-structural core exciton theory of Resca *et al.*,[105] indicating that different geometric sites in clusters give rise to characteristically shifted transition energies relative to the bare atom, as indicated in Fig. 11.[72] The dimer absorption is expected to occur at lower energy than the atomic transition, corresponding to a polarization shift. Atoms that are located either on various surface- or bulk-sites are expected to give rise to blueshifted resonances, which result from screening. The competition between screening of the electron, due to repulsive scattering, and the increased electron binding, due to cluster polarization, determines the sign of the overall spectral shift relative to the atomic transition. Considerably larger blueshifts than for surface sites are expected for subsurface atoms and atoms that are located in the bulk of the clusters, leading to blueshifts that are greater than 400 meV.[72] It is noted that the model calculations consider besides perfectly shaped clusters of icosahedral geometry,[106] which represent the lowest energy structures of small rare gas clusters, also imperfectly shaped isomers. This results in the model calculations in energy regimes, where the geometrical sites are expected to absorb, instead of a well-defined transition energy (cf. Fig. 11). Imperfectly shaped clusters are expected to be dominated by edge sites. This is consistent with the experimental results shown in Fig. 11, where the edge sites are the dominant surface component with increasing $\langle N \rangle$. Consistently, these results indicate that the local surroundings of the absorbing atom is of crucial importance to the energy of the transition. The relative intensities of the Kr $3d_{5/2} \rightarrow 5p$-transitions give an estimate on the relative occurrence of the individual sites in variable size Kr clusters. From this it becomes evident that there are no bulk features, even at $\langle N \rangle = 30$. This is not anticipated, if exclusively perfect isosahedral cluster shapes are considered.[106] The second icosahedral shell that surrounds an atom is closed at Kr_{13}. This cluster is expected to have already 8% of its atoms in the bulk, whereas for the next shell, corresponding to Kr_{55}, the contribution of bulk sites increases to 24%.[106] This is unlike the experimental findings, where first evidence for bulk excitons occurs at $\langle N \rangle > 100$. This is further evidence to consider imperfect cluster structures, with a considerably enhanced surface-to-bulk ratio. However, it is not possible to assign a single cluster structure to the spectral features, since there is a distribution of various cluster sizes and shapes present in the jet. Note, that these results reflect properties of *neutral* van der Waals clusters of weak binding energy. The results are evidently different from those that are obtained e.g. from mass spectrometry, where intense mass lines occur, which reflect stable geometric structures, such as icosahedral shells. These stable mass lines are known as "*magic numbers*".[1,107] They reflect properties of ionic clusters, which are considerably stronger bound as a result of the positive charge. Therefore, it is evident that structural properties, such as the electronic and geometric structure, of neutral and charged clusters are different from each other.

It is also important to note that the Kr_2^+-yield decreases if $\langle N \rangle$ is further increased, indicating that fragmentation of large clusters does not support the formation of small fragment ions. Thus, another detection technique, that is sensitive to heavy clusters is applied, such as total electron yields (TEY). Figure 12 shows the continuation of the size evolution toward higher $\langle N \rangle$, where the atomic contribution is subtracted in order to observe exclusively size effects of clusters. The TEY corresponding to $\langle N \rangle = 30$ is almost identical to that shown in Fig. 11, indicating that both experimental approaches yield equivalent results. Clear evidence for different bulk sites is found at $\langle N \rangle = 150$, where edge sites are still the dominant component. Finally, in the large cluster limit, bulk sites are found to be dominant. This example indicates that systematic size dependent changes in electronic structure are tightly related to the geometric structure in weakly bound systems. An alternative approach to obtain structural information on free clusters is outlined in the following section, where results from EXAFS spectroscopy are discussed. Another aspect that goes beyond the scope of this review concerns solvation effects that can be investigated by heterogeneous clusters.[84,108] Core level excitations are the ideal approach to investigate heterogeneous clusters because of the element- and site-selectivity of core level spectroscopies.

4.2. *Size Effects in Geometric Structure*

The lowest energy structures of free rare gas clusters are multishell icosahedra. Clusters of increasing size are expected to complete shell-by-shell these five-fold symmetry structures (see Fig. 13). Closed shells are expected to

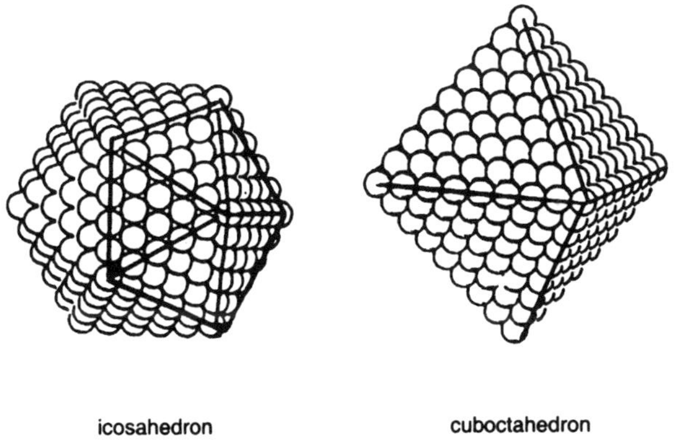

Fig. 13. Comparison of a hypothetical Ar_{561}-icosahedron (left) and a Ar_{489}-cuboctahedron (right) (adapted from Ref. 109).

occur at Ref. 106:

$$N = \frac{1}{3}(10s^3 - 15s^2 + 11s - 3) \tag{27}$$

where N is the total number of atoms in a cluster and s is the number of shells in a multishell icosahedron. The number of surface bound atoms N_s is[106]

$$N_s = 10s^2 - 20s + 12\,. \tag{28}$$

Closed shells are expected for $13, 55, 147, 309, 561, 923, \ldots$ atoms. One may expect that clusters grow further in five-fold symmetry (cf. Fig. 13), but it is known that the heavy condensed rare gases, such as Ne, Ar, Kr, and Xe, form a face-centered-cubic-(fcc) lattice. Theoretical work predicts that in small clusters the surface effect dominates the cluster stability, so that more than 10^4 atoms per cluster are needed for the formation of stable cuboctahedra (cf. Fig. 13), which have a higher binding energy than the corresponding icosahedra.[110] In contrast, earlier experimental work suggests that at least 750 or 3500 atoms per cluster are required to show fcc-structures.[62,111] One of the first issues that has been addressed in the field of core excitation of free clusters was to investigate structural changes of argon clusters as a function of $\langle N \rangle$.[22–24] EXAFS is primarily sensitive to the nearest-neighbor shell of the absorbing atom (see Sec. 2.2), but other shells that are collinear to the nearest neighbors are also observed with high-intensity in the EXAFS-signal. This is a result of a strong forward scattering component of the nearest-neighbor shell in the EXAFS-process. In the case of the fcc-lattice of solid argon, the fourth shell is collinear with the nearest-neighbor shell, which gives rise to two strong EXAFS-signals, whereas icosahedra only show the nearest-neighbor shell.[73] Thus, EXAFS is sensitive to the occurrence of fcc-moieties in variable size rare gas clusters. This is demonstrated by simulations of the EXAFS-signal on 1s-excited argon clusters using the spherical wave approach (see Fig. 14).[112] Two different geometries of variable size are considered: (i) fcc-structures and (ii) multishell icosahedra. Figure 14 shows that the EXAFS-signal of variable size multishell icosahedra reflects exclusively the shell of nearest-neighbors, since there are no other close lying shells collinear with the nearest-neighbor shell.

In the case of argon clusters one finds from the calculated EXAFS-signal that the nearest-neighbor distance is 3.76 Å. This quantity is independent on the cluster-size and shape. The only quantity that is expected to change is the number of nearest neighbors, which increases with cluster-size, so that

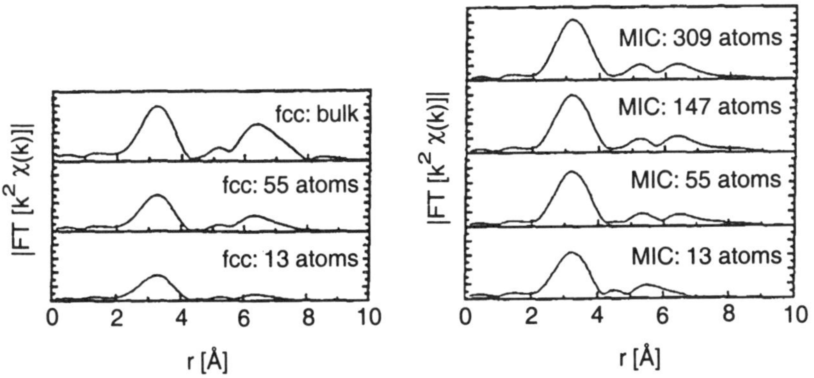

Fig. 14. Simulated EXAFS signals of variable size argon clusters: fcc-structures (left), multishell icosahedra (MIC) (right). Further details are described in the text and in Ref. 73.

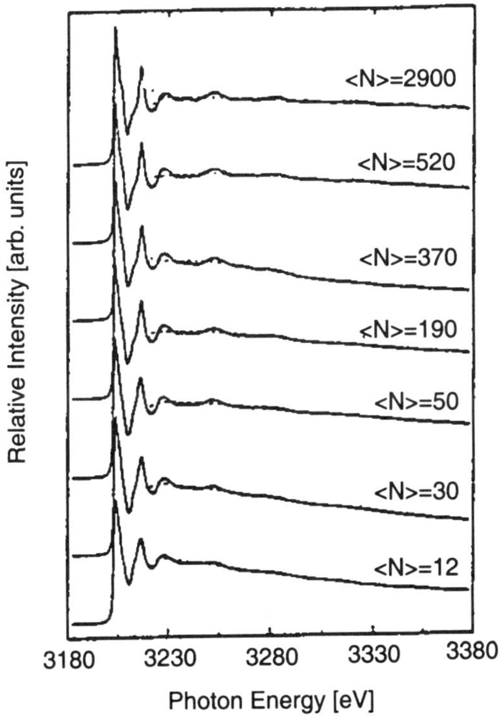

Fig. 15. Experimental $Ar_2{}^+$-yields recorded at various average cluster sizes $\langle N \rangle$ in the Ar 1*s*-regime (cf. Ref. 73).

finally the bulk value of 12 is reached. The simulations shown in Fig. 14 indicate that there is a second maximum in the EXAFS-signal near ~ 6.5 Å in the fcc-lattice as well as polycuboctahedra, which is clear evidence for scattering from the fourth-shell.[73] The intensity of this maximum grows with increasing the number of fcc-sites near the absorbing atom. These findings are in agreement with the experimental EXAFS-results.

The raw experimental spectra are shown in Fig. 15. Unlike predictions that are based on the most stable icosahedral cluster structures,[110] the fcc-sensitive second maximum in the EXAFS-signal occurs already with appreciable intensity at an average cluster-size $\langle N \rangle \sim 200$, as shown in Fig. 16. The question arises why such small clusters form fcc-units, which

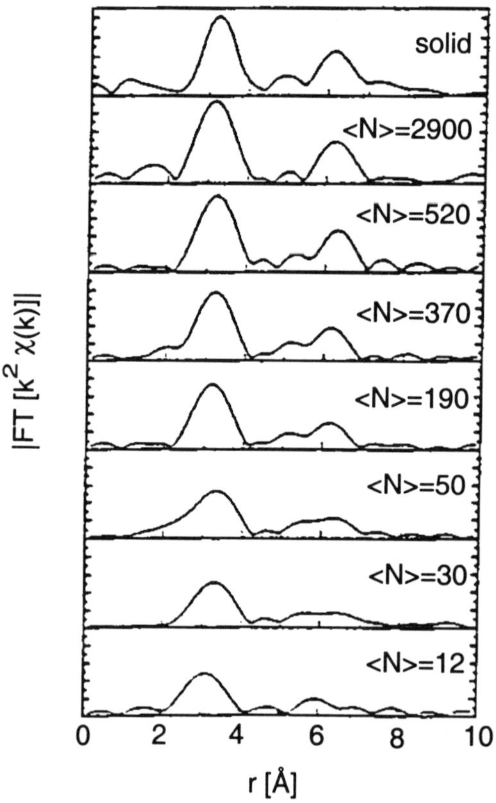

Fig. 16. Experimental EXAFS signals obtained from the raw spectra shown in Fig. 15 (cf. Ref. 73).

are expected to occur at considerably larger cluster sizes, according to theoretical work[110] and previous electron diffraction experiments,[62] where a critical size for the occurrence of fcc-structures of at least 750 atoms per cluster was deduced. The results obtained from EXAFS are rationalized as follows: the clusters are formed in the adiabatic nozzle expansion, where the supersaturated gas undergoes spontaneous nucleation by fast cluster growth. Perfectly shaped clusters with the lowest free energy are expected to grow slower than those containing defects, such as stacking faults.[109,113] More recent theoretical work indicates that there is no evidence for a structural transition from icosahedra to fcc,[114] rather than a persisting multiply twinned core of five-fold symmetry that is surrounded by a faulted shell with defects that stimulate the fcc-crystal growth. This growth mechanism also serves to prevent hcp-growth. These results are in-line with previous electron diffraction work.[62]

We conclude that one of the important advantages of EXAFS-spectroscopy is, that it is sensitive to the occurrence of fcc-moieties in clusters, so that these are found at considerably lower $\langle N \rangle$ than in other experimental approaches.

4.3. *Size Effects in Fragmentation of Multiply Charged Clusters*

Fragmentation of core excited clusters becomes already evident from mass spectra, where the mass distribution is concentrated at much smaller masses than estimated from the average cluster-size $\langle N \rangle$ (cf. Fig. 9). In addition, it is shown in Fig. 11 that photoion yields of mass-selected cluster fragments reflect changes in electronic structure as a function of $\langle N \rangle$. However, these results give no specific information on the mechanisms that lead to stable product ions.

Reliable fragmentation mechanisms can be deduced from coincidence experiments, as outlined in Secs. 2.5 and 3.3. We concentrate in the following only on results from photoelectron-photoion–photoion coincidence (PEPIPICO) spectroscopy on argon clusters in order to demonstrate how size-dependent fragmentation mechanisms of doubly charged argon clusters are derived.[57] Note that considerably more work has been performed in the field of multicoincidence spectroscopies of doubly and triply charged atomic, molecular, and heterogeneous clusters.[39,59,84,90,98] Figure 17 shows a typical PEPIPICO spectrum of argon clusters.

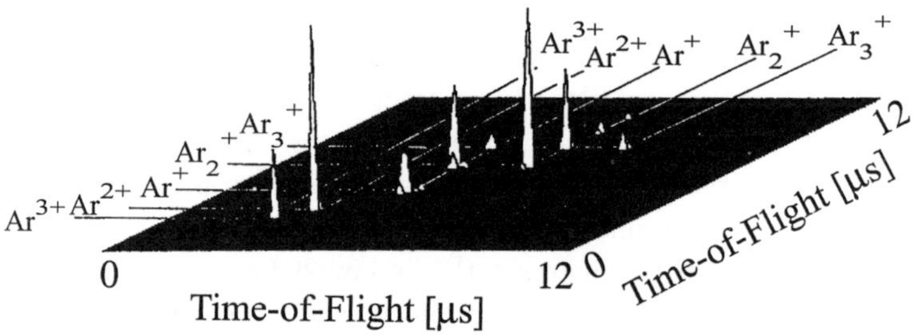

Fig. 17. Photoelectron–photoion–photoion-coincidence (PEPIPICO) spectrum of argon clusters ($\langle N \rangle = 60$) recorded at 260 eV photon energy (cf. Ref. 57).

The three-dimensional spectrum shows the flight times of correlated cation pairs and their corresponding intensity for Ar 2p-excited (E = 260 eV) argon clusters at $\langle N \rangle = 60$. These cation pairs result in the Ar 2p-continuum primarily from fission of doubly charged clusters (cf. Fig. 5). Two different fission processes are observed: (i) symmetric fission and (ii) asymmetric fission. Along the main diagonal, where $t_{A^+} = t_{B^+}$ (cf. Fig. 6), one observes symmetric fission, where both correlated cations are of the same mass. The following symmetric products are observed: Ar^+/Ar^+, Ar_2^+/Ar_2^+, and Ar_3^+/Ar_3^+. There are no larger symmetric fission products, even though $\langle N \rangle = 60$ produces considerably larger neutral clusters. Starting from a purely atomic jet, one observes no PEPIPICO signal at all, since the atoms are not correlated to each other. An increase in $\langle N \rangle$ shows that primarily symmetric fission occurs. However, the relative intensity of the symmetric channels decreases with increasing $\langle N \rangle$,[57] whereas increasing intensity of asymmetric charge separations is found off the main diagonal. These are cation pair series consisting of a light cation that is correlated with a heavier one: Ar^+/Ar_n^+, Ar_2^+/Ar_n^+, and Ar_3^+/Ar_n^+, with $n \leqq 12$.

Further insight into the fission mechanisms is obtained from the peak shape and slope analysis of the coincidence signals (cf. Sec. 2.5). The PEPIPICO-signals are found to change their shape and slope with increasing $\langle N \rangle$. This is clear evidence for size-dependent fission mechanisms, which occur for each PEPIPICO-signal.[57] Figure 18 shows, as a typical example, the evolution of PEPIPICO-shape with increasing $\langle N \rangle$ for the Ar^+/Ar_2^+-coincidence: firstly, the occurrence of a Ar^+/Ar_2^+-coincidence originates at least from Ar_3 or an even larger neutral cluster (Ar_n, $n > 3$). Photoionization and subsequent relaxation of the core hole forms likely an intermediate Ar_n^{++}, with $n > 3$. This dication decays into the observed cation

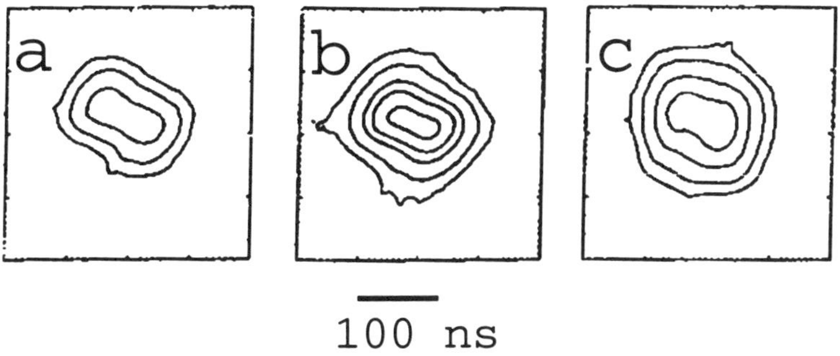

Fig. 18. Shapes of the experimental $Ar^+/Ar_2{}^+$-coincidence signal at various average cluster-sizes: (a) $\langle N \rangle = 4$; (b) $\langle N \rangle = 10$; (c) $\langle N \rangle = 60$.

pair $Ar^+ + Ar_2{}^+$, evaporating an unknown quantity of neutral atoms. The minimum dication size ($Ar_3{}^{++}$) involves a two-body decay, implying an experimental slope of -1, according to the considerations given in Sec. 2.5 [cf. Eq. (17)]. Figure 18(a) shows that the experimental slope is significantly smaller than -1 at the threshold of cluster formation. It is rather found that the slope corresponds to -0.67 at $\langle N \rangle = 4$. Therefore, this cation pair cannot be formed by the simple two-body decay at the threshold of cluster formation rather than by a more complicated mechanism that involves the loss of one or more neutrals from the heavier moiety *after* the fission step. The experimental PEPIPICO-signal slope is rationalized by the following fission mechanism, where the mass ratio of the second step [cf. Eq. (17)] and Eq. (29) determines the slope of $-2/3$:

$$Ar_4{}^{++} \rightarrow Ar^+ + Ar_3{}^+ \tag{29a}$$

$$Ar_3{}^+ \rightarrow Ar_2{}^+ + Ar\,. \tag{29b}$$

At this point it is not clear, whether the cluster dications are already smaller in size than the neutral precursors. This cannot be investigated by PEPIPICO-spectroscopy. However, the size-dependent shape of the $Ar^+/Ar_2{}^+$-PEPIPICO-signal, shown in Fig. 18, is a clear indication that losses of neutrals occur preferably after the fission process. An increase of the average cluster-size to $\langle N \rangle = 10$ yields a twisted signal, where the main slope is decreased to ~ -0.35 [cf. Fig. 18(b)]. This slope is rationalized using the same mechanism, as outlined in Eqs. (30a) and (30b) but starting from a larger neutral aggregate, where more neutrals are evaporated in the second step that follows primary fission:

$$Ar_7^{++} \rightarrow Ar^+ + Ar_6^+ \tag{30a}$$

$$Ar_6^+ \rightarrow Ar_2^+ + 4\ Ar\,. \tag{30b}$$

A further increase to $\langle N \rangle = 60$ gives a round-shaped coincidence signal [cf. Fig. 18(c)], where it is impossible to derive an experimental slope. This signal is thought to be due to fission, where most likely the momenta of both singly charged fragments are not directed into opposite direction, i.e. the simple picture of Coulomb explosion is not any more applicable to the size regime that is close to the critical stability, where two charges can be accommodated in a stable cluster ($N_{crit}(Ar) = 91$).[48] Thus, the singly charged fragments may be ejected at arbitrary angles which is reflected by the round-shaped of the Ar^+/Ar_2^+-coincidence signal.

5. Summary and Outlook

Tunable X-rays are shown to be useful to investigate properties of variable size clusters. The examples that are presented in this chapter represent intentionally simple model systems that are selected to demonstrate the potential of this experimental approach.

In summary, Fig. 19 shows a common scheme that has been developed from results on 2*p*-excited Ar clusters. It includes all processes that follow primary photoabsorption of a soft-X-ray photon by an atom that is located near the surface of a cluster. This includes: core hole formation, photoionization, relaxation, and finally fragmentation. In the case of molecular clusters additional fragmentation processes are known to occur: besides van der Waals fragmentation, yielding intact molecular moieties that are either neutral or singly charged, chemical fragmentations occur.[84] In the case of nitrogen clusters one obtains e.g. product ions that contain odd numbers of nitrogen atoms, such as $N(N_2)_n^+$.

Recent progress in this field has been made by using high spectral resolution from light sources that rely on insertion devices, such as undulators. Small spectral shifts are found in the soft X-ray regime of molecular clusters that include vibrational resolution of excited valence states.[74]

Focus of present work are investigations of technologically relevant species that are bound in homogeneous and heterogeneous clusters (cf. Table 2). One of the challenges of future work is certainly related to the development of stable and intense cluster sources that are compatible with

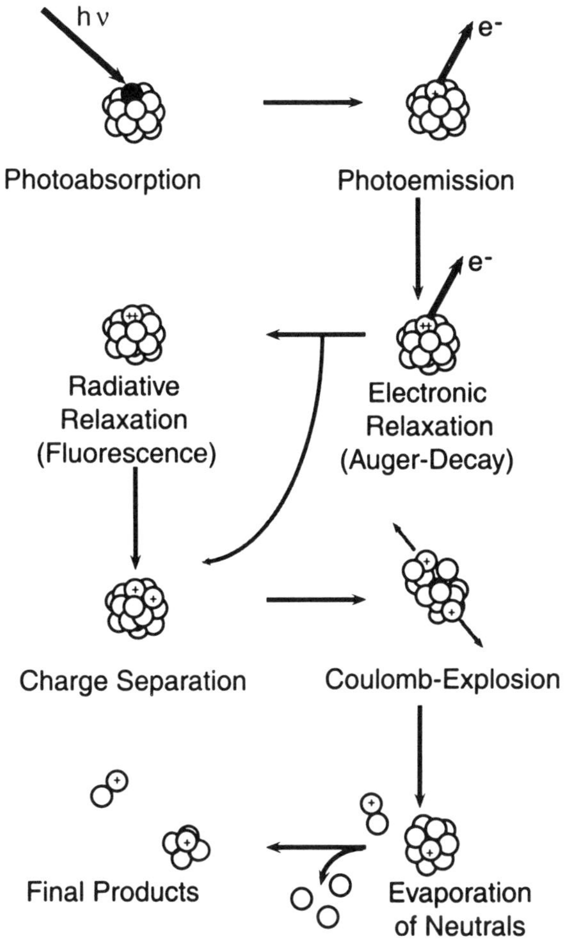

Fig. 19. Schematic diagram of the sequence of processes occurring upon core excitation of clusters.

the environment of synchrotron radiation research. The control of structural properties, especially, if the clusters are size-selected and deposited on proper substrates are expected to be of outstanding interest to future developments of nanostructured matter. Nanoparticles with defined properties are also expected to be of relevance to catalytic chemical processes. Present work is strongly stimulated by new experimental opportunities that arise from state-of-the-art light sources, such as high brightness synchrotron radiation sources from storage ring facilities including combinations with laser radiation. Future opportunities will also include free electron lasers that are expected to reach the soft X-ray regime.[115]

Acknowledgments

It is a pleasure to thank all co-workers and collaborators, who have participated in this research during the last years: C. Heinzel, B. Wassermann, H. W. Jochims, H. Baumgärtel, J. Geiger, A. Knop, C. M. Teodorescu, R. Flesch, D. Gravel, D. Schwarz, M. Kuhlmann, J. Blumberg, C. Sonntag, A. P. Hitchcock, A. A. Pavlychev, T. Möller, W. Drube, R. Frahm, P. A. Dowben, D. N. McIlroy, J. J. Neville, C. Reynaud, H. Schmelz, P. Tzekeris, K. Kosmidis, M. Simon, P. Morin and M. Lavollée.

Financial support by the Deutsche Forschungsgemeinschaft (DFG), the Bundesministerium für Forschung und Bildung (BMBF), and the Fonds der Chemischen Industrie are gratefully acknowledged.

References

1. *Clusters of Atoms and Molecules* Vols. I and II, *Springer Ser. Chem. Phys.* Vols. 52 and 56, edited by H. Haberland (Springer, Berlin, 1994, 1995).
2. U. D. Becker and A. Shirley, *VUV and Soft X-Ray Photoionization* (Plenum, New York, 1996) V. Schmidt, *Electron Spectroscopy of Atoms using Synchrotron Radiation* (Cambridge University Press, Cambridge, 1997).
3. D. M. Hanson, Adv. Chem. Phys. **77**, 1 (1990) I. Nenner and P. Morin, in *VUV and Soft X-Ray Photoionization*, edited by U. Becker and D. A. Shirley (Plenum, New York, 1996), pp. 291.
4. J. Stöhr, in *Chemistry and Physics of Solid Surfaces V*, edited by R. Vanselow and R. Howe (Springer, Berlin, 1984), pp. 231.
5. G. S. Brown and D. E. Moncton, *Handbook on Synchrotron Radiation*, Vol. 3, (North Holland, Amsterdam, 1991).
6. B. K. Teo and D. C. Joy, *EXAFS Spectroscopy* (Plenum, New York, 1981); K. Baberschke and D. Arvanitis, *XAFS VIII* (North Holland, Amsterdam, 1995).
7. K. Müller-Dethlefs, M. Sander, L. A. Chewter and E. W. Schlag, *J. Phys. Chem.* **88**, 6098 (1984).
8. *Structure and Dynamics of Weakly Bound Molecular Complexes, NATO-ASI Series C 212*, edited by A. Weber (Reidel, Dordrecht, 1987); *Linking the Gaseous and Condensed Phases of Matter, NATO-ASI Series B 326*, edited by L. G. Christophorou, E. Illenberger, W. F. Schmidt (Plenum, New York, 1994).
9. M. Thomas, *Optical Grapher Program* (Center of X-Ray Optics, U.C. Berkeley, 1992).

10. C. Kunz, *Synchrotron Radiation, Topics in Current Physics*, Vol. 10 (Springer, Berlin, 1979).
11. F. Yang and J. H. Hamilton, *Modern Atomic and Nuclear Physics* (McGraw-Hill, New York, 1996).
12. J. Stöhr, NEXAFS *Spectroscopy, Springer Ser. Surf. Sci.* Vol. 25 (Springer, Berlin, 1992).
13. B. K. Agarwal, X-ray *Spectroscopy, Springer Ser. Opt. Sci.* Vol. 15 (Springer, Berlin, 1979).
14. F. C. Brown, *Solid State Phys.* **29**, 1 (1974).
15. R. Haensel, N. Kosuch, U. Nielsen, U. Rössler and B. Sonntag, *Phys. Rev.* **B7**, 1577 (1973).
16. I. T. Steinberger, C. M. Teodorescu, D. Gravel, R. Flesch, B. Wassermann, C. W. Hutchings, A. P. Hitchcock and E. Rühl, *Phys. Rev.* **B60**, 3995 (1999).
17. A. P. Hitchcock and D. C. Mancini, *J. Electron. Spectrosc. Relat. Phenom.* **67**, 1 (1994).
18. B. Kassühlke, P. Averkamp, S. Frigo, P. Feulner and W. Berthold, *Phys. Rev.* **B55**, 10854 (1997). C. T. Chem, Y. Ma and F. Sette, *Phys. Rev.* **A40**, 6737 (1989).
19. D. Blechschmidt, R. Haensel, E. E. Koch, U. Nielsen and T. Sagawa, *Chem. Phys. Lett.* **14**, 33 (1972).
20. U. Kuetgens and F. Hormes, *Phys. Rev.* **A44**, 264 (1991).
21. W. Malzfeldt, W. Niemann, P. Rabe and R. Haensel, *Springer Proc. Phys.* **2**, 445 (1984) W. Malzfeldt, Ph.D. Thesis (University of Kiel, 1985).
22. E. Rühl, H. W. Jochims, C. Schmale, E. Biller, A. P. Hitchcock and H. Baumgärtel, *Chem. Phys. Lett.* **178**, 558 (1991).
23. E. Rühl, C. Heinzel, A. P. Hitchcock and H. Baumgärtel, *J. Chem. Phys.* **98**, 2653 (1993).
24. E. Rühl, C. Heinzel, A. P. Hitchcock, H. Schmelz, C. Reynaud, H. Baumgärtel, W. Drube and R. Frahm, *J. Chem. Phys.* **98**, 6820 (1993).
25. E. Rühl, C. Heinzel, H. Baumgärtel, W. Drube, A. P. Hitchcock, *Jpn. J. Appl. Phys.* **32**, Suppl. 32-2, 791 (1993).
26. J. Mustre de Leon, Y. Yacoby, E. A. Stern and J. J. Rehr, *Phys. Rev.* **B42**, 10843 (1990).
27. P. A. Lee, P. H. Citrin, P. Eisenberger and B. M. Kincaid, *Rev. Mod. Phys.* **53**, 769 (1981).
28. L. Petersson, J. Nordgren, L. Selander, C. Nordling and K. Siegbahn, *J. Electron. Spectrosc. Relat. Phenom.* **27**, 29 (1982) W. L. Jolly, K. D. Bomben and C. J. Eyermann, *At. Data Nucl. Data Tables* **31**, 433 (1984).
29. U. Gelius, *Phys. Scr.* **9**, 133 (1974) D. A. Shirley, *Adv. Chem. Phys.* **23**, 85 (1973) R. L. Martin and D. A. Shirley , In *Electron Spectroscopy: Theory, Techniques and Applications*, Vol. 1, edited by C. R. Brundle and A. D. Baker (Academic Press, New York, 1978).
30. J. Schirmer, M. Braunstein, M.-T. Lee and V. McKoy, in *VUV and Soft-X-Ray Photoionization*, edited by U. Becker and D. A. Shirley (Plenum, New York, 1996) p. 105.

31. A. Knop, H. W. Jochims, A. L. D. Kilcoyne, A. P. Hitchcock and E. Rühl, *Chem. Phys. Lett.* **223**, 553 (1994).
32. O. Björneholm, F. Federmann, F. Fössing and T. Möller, *Phys. Rev. Lett.* **74**, 3017 (1995).
33. G. Kaindl, T.-C. Chiang, D. E. Eastman and F. J. Himpsel, *Phys. Rev. Lett.* **45**, 1808 (1980).
34. K. Siegbahn, C. Nordling, G. Johansson, J. Hedman, P. F. Héden, K. Hamrin, U. Gelius, T. Bergmark, L. O. Werme, R. Manne and Y. Baer, *ESCA Applied to Free Molecules* (North Holland, Amsterdam, 1969).
35. K. Müller-Dethlefs and E. W. Schlag, *Ann. Rev. Phys. Chem.* **42**, 109 (1991).
36. A. Niehaus, *J. Phys.* **B10**, 1845 (1977) P. A. Heimann, D. W. Lindle, T. A. Ferrett, S. H. Liu, L. J. Medhurst, M. N. Piancastelli, D. A. Shirley, U. Becker, H. G. Kerkhoff, B. Langer, D. Szostak and R. Wehlitz, *J. Phys.* **B20**, 5005 (1987).
37. N. Saito and I. H. Suzuki, *Int. J. Mass Spectrom. Ion Proc.* **115**, 157 (1992).
38. J. D. Nuttall and T. E. Gallon, *Phys. Stat. Sol.* **B71**, 259 (1975) J. D. Nuttall and T. E. Gallon, *Surf. Sci.* **53**, 698 (1975).
39. E. Rühl, A. Knop, A. P. Hitchcock, P. A. Dowben and D. N. McIlroy, *Surf. Rev. Lett.* **3**, 557 (1996).
40. R. Scheuerer, P. Feulner, G. Rocker, Z. Lin and D. Menzel, *DIET IV, Springer Ser. Surf. Sci.* Vol. 19 edited by G. Betz and P. Varga (Springer, Berlin, 1990) 235; K. B. McAfee Jr. and R. S. Hozack, *Phys. Rev.* **A32**, 810 (1985).
41. E. Rühl, C. Heinzel and H.-W. Jochims, *Chem. Phys. Lett.* **211**, 403 (1993).
42. H. Langhoff, *Opt. Comm.* **68**, 31 (1988).
43. C. Cachoncille, J. M. Pouvesle, F. Davanloo, J. J. Coogan and C. B. Collins, *Opt. Commun* **79**, 41 (1990). C. Cachoncille, J. M. Pouvesle, F. Davanloo, J. J. Coogan and C. B. Collins, *Z. Phys.* **D23**, 984 (1990). T. D. Strickler and E. T. Arakawa, *J. Chem. Phys.* **41**, 1783 (1964). G. Klein and M. J. Carvalho, *J. Phys.* **B14**, 1283 (1981). A. Ulrich, H. J. Körner, W. Krötz, G. Ribitzi, D. E. Murnik, E. Matthias, P. Kienle and D.H. H. Hoffmann, *J. Appl. Phys.* **62**, 357 (1987).
44. R. Flesch, H.-W. Jochims, J. Plenge and E. Rühl, *Phys. Rev.* **A62**, 62504 (2000).
45. M. Mayer, E. V. Raven, B. Sonntag and J. E. Hanse, *Phys. Rev.* **A43**, 177 (1991). J. A. de Gouw, J. van Eck, J. van der Weg and H. G. M. Heideman, *J. Phys.* **B25**, 2007 (1992).
46. T. A. Carlson, M. O. Krause, *Phys. Rev.* **137**, 1655 (1965). T. A. Calson, W. E. Hunt and M. O. Krause, *Phys. Rev.* **151**, 41 (1966). M. O. Krause and T. A. Carlson, *Phys. Rev.* **158**, 18 (1967). R. D. Deslattes, *Aust. J. Phys.* **39**, 845 (1986).
47. O. Echt and T. D. Märk, In *Clusters of Atoms and Molecules, Springer Ser. Chem. Phys.* Vol. 56, edited by H. Haberland (Springer, Berlin, 1994) p. 183.

48. P. Scheier and T. D. Märk, *J. Chem. Phys.* **86**, 3056 (1987).
49. O. Echt, D. Kreisle, E. Recknagel, J. J. Saenz, R. Casero and J. M. Soler, *Phys. Rev.* **A38**, 3236 (1988).
50. P. Scheier and T. D. Märk, *Chem. Phys. Lett.* **136**, 423 (1987).
51. T. A. Carlson and R. M. White, *J. Chem. Phys.* **48**, 5191 (1968).
52. J. G. Gay and B. J. Berne, *Phys. Rev. Lett.* **49**, 194 (1982).
53. U. Werner, J. Becker, T. Farr and H. O. Lutz, *Nucl. Instr. Meth.* **B124**, 298 (1997).
54. D. M. Curtis and J. H. D. Eland, *Int. J. Mass Spectrom. Ion Proc.* **63** 241 (1985).
55. E. Rühl, C. Schmale, H. W. Jochims, E. Biller, M. Simon and H. Baumgärtel, *J. Chem. Phys.* **95**, 6544 (1991).
56. J. H. D. Eland, *Mol. Phys.* **61**, 725 (1987). J. H. D. Eland, *Acc. Chem. Res.* **22**, 381 (1989).
57. E. Rühl, C. Heinzel, H. Baumgärtel, M. Lavollée and P. Morin, *Z. Phys.* **D31**, 245 (1994).
58. M. Simon, Ph. D. Thesis (Université de Paris-Sud, Orsay, 1991).
59. C. M. Teodorescu, D. Gravel and E. Rühl, *J. Chem. Phys.* **109**, 9280 (1998).
60. O. F. Hagena, *Z. Phys.* **D4**, 291 (1987).
61. J. Wörmer, V. Guzielski, J. Stapelfeldt, G. Zimmerer and T. Möller, *Phys. Scr.* **41**, 490 (1990).
62. J. Farges, M. F. de Feraudy, B. Raoult and G. Torchet, *J. Chem. Phys.* **84**, 3491 (1986).
63. R. Karnbach, M. Joppien, J. Stapelfeldt, J. Wörmer and T. Möller, *Rev. Sci. Instr.* **64**, 2838 (1993).
64. U. Buck and H. Meyer, *Phys. Rev. Lett.* **52**, 109 (1984).
65. A. Knop and E. Rühl, in *Structure and Dynamic of Clusters*, edited by T. Kondow, K. Kaya and A. Terasaki (Universal Academic Press, Tokyo, 1996) p. 235.
66. O. Björneholm, F. Federmann, F. Fössing, T. Möller and P. Stampfli, *J. Chem. Phys.* **104**, 1846 (1996).
67. *Handbook on Synchrotron Radiation*, Vols. 1a and 1b, edited by E. E. Koch (North Holland, Amsterdam, 1983).
68. K. Wille, *Physik der Teilchenbeschleuniger und Synchrotronstrahlungsquellen* (Teubner, Stuttgart, 1992).
69. H. Winick and S. Doniach, *Synchrotron Radiation Research* (Plenum, New York, 1980).
70. S. Bernstorff, W. Braun, M. Mast, W. Peatman and T. Schroeter, *Rev. Sci. Instr.* **60**, 2097 (1989).
71. F. Federmann, O. Björneholm and T. Möller, *Phys. Rev. Lett.* **73**, 1549 (1994).
72. A. Knop, B. Wassermann and E. Rühl, *Phys. Rev. Lett.* **80**, 2302 (1998).
73. S. Kakar, O. Björneholm, J. Weigelt, A. R. B. de Castro, L. Tröger, R. Frahm, T. Möller, A. Knop and E. Rühl, *Phys. Rev. Lett.* **78**, 1675 (1997).

74. R. Flesch, A. A. Pavlychev, J. J. Neville, J. Blumberg, M. Kuhlmann, W. Tappe, F. Senf, O. Schwarzkopf, A. P. Hitchcock and E. Rühl, *Phys. Rev. Lett.* **86**, 3767 (2001).
75. D. N. McIlroy, P. A. Dowben, A. Knop and E. Rühl, *J. Vac. Sci. Technol.* **B13**, 2142 (1995).
76. C.-M. Teodorescu, D. Gravel, E. Rühl, T. J. McAvoy, J. Choi, D. Pugmire, P. Pribil, J. Coos and P. A. Dowben, *Rev. Sci. Instr.* **69**, 3805 (1998).
77. T. Baer, W. B. Peatman and E. W. Schlag, *Chem. Phys. Lett.* **4**, 243 (1969).
78. H. W. Jochims, Ph.D. Thesis (University of Freiburg, 1976). H. W. Jochims, E. Rühl and H. Baumgärtel, *Z. Naturforsch.* **B44**, 13 (1989).
79. M. Simon, T. LeBrun, P. Morin, M. Lavollée and J. L. Maréchal, *Nucl. Instr. Meth.* **B62**, 167 (1991).
80. W. C. Wiley and I. H. McLaren, *Rev. Sci. Instr.* **26**, 1150 (1955).
81. C. Y. Ng, *Vacuum Ultraviolet Photoionization and Photodissociation of Molecules and Clusters* (World Scientific, Singapore, 1991).
82. L. J. Frasinski, M. Stankiewicz, K. J. Randall, P. A. Hatherly and K. Codling, *J. Phys.* **B19**, L819 (1986).
83. E. Rühl, *Ber. Bunsenges. Phys. Chem.* **96**, 1172 (1992).
84. E. Rühl, A. P. Hitchcock, P. Morin and M. Lavollée, *J. Chim. Phys. Paris* **92**, 521 (1995).
85. G. P. Williams, in *X-Ray Data Booklet* (Lawrence Berkeley Laboratory, University of California, 1986) pp. 2–5.
86. A. Knop and E. Rühl, *BESSY Annual Report*, Berlin 189 (1995).
87. A. A. Pavlychev, E. V. Semenova, A. P. Hitchcock and E. Rühl, *Physica* **B208/209**, 187 (1995).
88. H. C. Schmelz, M. A. Gaveau, C. Reynaud, C. Heinzel, H. Baumgärtel and E. Rühl, *Phyisca* **B208/209**, 519 (1995).
89. J. O. Löfken, A. Kolmakov, C. Nowak, M. Riedler, A. Wark and T. Möller, *HASYLAB, Annual Report*, Hamburg 179 (1998).
90. C.-M. Teodorescu, D. Gravel, J. Choi, D. Pugmire, P. A. Dowben, N. Fominykh, A. A. Pavlychev and E. Rühl, *J. Electron Spectrosc. Relat. Phenom.* **101/103**, 193 (1999).
91. J. M. Durand, J. Olivier-Fourcade, J. C. Dumas, M. Womes, C. M. Teodorescu, A. Elafif, J. M. Esteva and R. C. Karnatak, *J. Phys.* **B29**, 5773 (1996).
92. E. Rühl, R. Flesch, W. Tappe, D. Novikov and N. Kosugi, *J. Chem. Phys.* **116**, 3312 (2002).
93. S. Kakar, O. Björneholm, J. O. Löfken, F. Federmann, A. V. Soldatov and T. Möller, *Z. Physik*, **D40**, 84 (1997).
94. A. A. Pavlychev and E. Rühl, *J. Electron. Spectrosc. Relat. Phenom.* **106**, 207 (2000).
95. C. Heinzel, Ph.D. Thesis (Freie Universität Berlin, 1993).
96. O. Björneholm, F. Federmann and T. Möller, *J. Chem. Phys.* **111**, 546 (1999).
97. A. P. Hitchcock and E. Rühl, in *The Physics of Electronic and Atomic Collisions, AIP Conf. Proc. 360*, edited by L. J. Dubé, J. B. A. Mitchell,

J. W. McConkey and C. E. Brion (American Institute of Physics, Woodbury, New York, 1995) p. 89.

98. J. Geiger, S. Rabe, C. Heinzel, H. Baumgärtel and E. Rühl, in *Linking the Gaseous and Condensed Phases of Matter*, edited by L. G. Christophorou, E. Illenberger and W. F. Schmidt (Plenum, New York, 1994) p. 217.
99. C. Nowak, C. Reinecker, A. Kolmakov, J. O. Löfken, F. Picucci, M. Riedler, A. V. Soldatov, M. Wollf and T. Möller, *J. Electron. Spectrosc. Relat. Phenom.* **101**, 199 (1999).
100. M. Riedler, A. R. B. de Castro, A. Kolmakov, J. O. Löfken, C. Nowak, A. V. Soldatov, A. Wark, G. Yalovega and T. Möller, *J. Chem. Phys.* **115**, 1319 (2001).
101. A. Wark, A. Komakov, J. O. Löfken, C. Nowak, M. Riedler and T. Möller, *HASYLAB Annual Report*, Hamburg 205 (1998).
102. C. Nowak, H. Döllefeld, A. Eychmüller, J. Friedrich, A. Kolmakov, J. O. Löfken, M. Reidler, A. Wark, H. Weller, M. Wolff and T. Möller, *J. Chem. Phys.* **114**, 489 (2001).
103. D. Gravel, M. Kuhlmann, D. Schwarz, J. Blumberg, C. M. Teodorescu and E. Rühl, *BESSY Annual Report*, Berlin 187 (1998).
104. G. Johansson, J. Hedman, A. Berndtsson, M. Klasson and R. Nielsson, *J. Electron. Spectrosc. Relat. Phenom.* **2**, 295 (1973).
105. L. Resca, R. Resta and S. Rodriguez, *Phys. Rev.* **B18**, 696 (1978), L. Resca, R. Resta and S. Rodriguez, *Phys. Rev.* **B18**, 702 (1978).
106. M. R. Hoare, *Adv. Chem. Phys.* **40**, 49 (1979).
107. A. Ding and J. Hesslich, *Chem. Phys. Lett.* **94**, 54 (1983), P. Scheier and T. D. Märk, *Int. J. Mass Spectrom. Ion Proc.* **76**, R11 (1976).
108. E. Rühl and B. Wassermann, to be published (1999).
109. B. W. van de Waal, *J. Chem. Phys.* **98**, 4909 (1993).
110. B. W. van de Waal, *J. Chem. Phys.* **90**, 3407 (1989).
111. J. W. Lee and G. D. Stein, *J. Phys. Chem.* **91**, 2450 (1987).
112. J. J. Rehr, R. C. Albers and S. I. Zabinsky, *Phys. Rev. Lett.* **69**, 3397 (1992), E. A. Stern, M. Newville, B. Ravel, Y. Yacoby and D. Haskel, *Physica.* **B208/209**, 117 (1995).
113. B. W. van de Waal, *Z. Phys.* **D20**, 349 (1991).
114. B. W. van de Waal, *Phys. Rev. Lett.* **76**, 1083 (1996).
115. J. Roßbach, *Nucl. Instr. Meth.* **A375**, 269 (1996). J. Feldhaus, E. L. Saldin, J. R. Schneider, E. A. Schneidmüller and M. V. Yurkov, *Opt. Commun.* **140**, 341 (1997).

CHAPTER 6

LABORATORY EXPERIMENTS ON SINGLE LEVITATED AEROSOL PARTICLES

T. Leisner
Institut fuer Physik, TU Ilmenaum Postfach 100565,
D-98684 Ilmenau, Germany
E-mail: Thomas.Leisner@Physik.TU-Ilmenau.de

Homogeneous and heterogeneous condensation is an ubiquitous phenomenon in the atmosphere and proceeds via nanometer sized critical nuclei. The application of levitation techniques and laser spectroscopy initially developed for cluster research allows to observe these processes in the laboratory under well controlled and realistic conditions. The application of appropriate models allows to deduce the size and surface energy of the critical nuclei as a function of temperature.

Contents

1. Clusters in Atmospheric Physics

Atomic and molecular clusters do not only possess fascinating and surprising physical and chemical properties, but do also form the first stage of homogeneous condensation from the gas phase to the liquid or solid state. They are therefore an intermediate in the initial formation of the solid state phase, for example, in the initial formation of condensed matter in cirumstellar dust shells of red giant stars, or in the formation of aerosol particles in the atmosphere of the earth.

Classical nucleation theory describes homogeneous phase transitions like the condensation of the liquid phase from vapor or the freezing of liquids. The following discussion of homogeneous freezing of liquids can be applied similarly to condensation from the gas phase. In both cases the homogeneous nucleation is kinetically hindered so that strong supersaturation of the vapor or supercooling of the liquid is usually observed. At a certain temperature, the free energy of the solid phase equals the free energy of the liquid. This is the equilibrium temperature of the phase transition, i.e. the melting point of the system. If the temperature drops below this point, the free energy of the solid drops below that of the liquid. However, a phase transition will not occur immediately, as it has to proceed via small solid nuclei which have a higher free energy than the bulk material due to the interface energy between the two phases. The compound system of nucleus plus liquid therefore has a higher free energy then the liquid itself.

At a given supercooling temperature, there exists a certain size of the nucleus where the free energy of the compound has a maximum. The addition of another monomer to this critical nucleus or germ reduces its free energy and the system will freeze repaidly. The germ size decreases with falling temperature. Under many realistic conditions, the size of the critical germ will be on the order of some tens of atoms or molecules and therefore fall very well into the realm of cluster physics. The strongly size dependent properties of small clusters therefore determine the homogeneous phase transition rates of vapors and liquids.

In the case of the condensation of vapors in the atmosphere, small droplets are formed which are immersed in the surrounding air, the atmospheric aerosol. Here the droplet size ranges from about 10 nm to 1 μm, and is therefore well above the region where the properties vary abruptly with size. The size dependence of the physical and chemical characteristics is rather smooth and can be described by appropriate scaling laws.[1] Nevertheless, the large surface to volume ratio of aerosol particles accounts for a multitude of effects with great scientific, technological and environmental importance.

In our laboratory, we have applied some of the methods and experimental technologies of cluster science to the analysis of atmospheric processes and this article will summarize our findings.

2. The Atmospheric Aerosol

The airborne particles together with the surrounding gas phase constitute the atmospheric aerosol. At ground level, the particle concentration in the atmosphere ranges from about 10^3 cm^{-3} over the ocean to 10^5 cm^{-3} over populated areas. The concentration drops rapidly with height to about 10^2 cm^{-3} above 5 km. In spite of the various sources of the particles their size-distribution is surprisingly universal and ranges from about 10 nm to 10 μm. Figure 1 shows typical number size distributions as they can be found over continents and the ocean. It is usually given as $\mathrm{d}N/\mathrm{d}(\log r/r_0)$, which can be converted in a linear size distribution by $\mathrm{d}N/\mathrm{d}(\log r/r_0) = \ln(10)\, r\, \mathrm{d}N/\mathrm{d}r$. Above a size of a few ten nanometers, the aerosol number distribution is usually described by a power law, this is $\mathrm{d}N/\mathrm{d}(\log r/r_0) \sim r^{-s}$, where the exponent s ranges from about three to four. Aerosol particles are produced either by condensation in the gas phase, or by abrasion of solid or liquid particles from the surface. Homogeneous condensation usually involves high supersaturation which is not often found in the atmosphere.

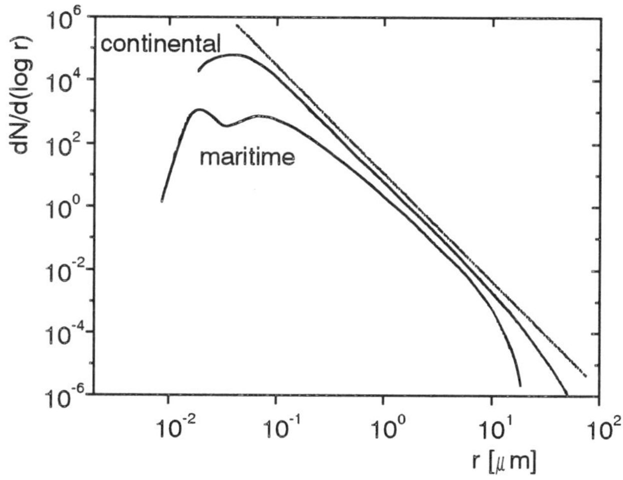

Fig. 1. Number density distribution of maritime and continental aerosol in the surface boundary layer. (Straight Line: power law distribution with exponent $s = 3.4$, see text.)

It is therefore mainly limited to the cocondensation of sulfuric acid and water. The acid in this case is produced in the atmosphere by gas phase oxidation of SO_2. A special case is the production of sea salt aerosols over the oceans, which are effectively produced by the disintegration of air bubbles on the ocean. The atmospheric aerosol is a dynamic system; the initially produced particles undergo various processes like heterogeneous condensation, coagulation, and deposition which lead to a change in their size and chemical composition and gives rise to the universality of the aerosol size-distribution.

It has become clear only recently that the atmospheric aerosol plays an important role for the climate on earth. It is common to distinguish between direct and indirect effects of the aerosols on the climate. Aerosols effect directly the radiation balance of the earth due to scattering and absorption of electromagnetic radiation (radiative forcing). On the other hand they influence the physics and chemistry of the atmosphere as condensation nuclei for cloud droplets and their chemical reactions with atmospheric trace gases. Though these indirect aerosol effects are difficult to quantify, they are at least as important as the direct radiative forcing. An especially important and complex example for the indirect influence of aerosols on the chemistry and radiation balance of the earth is the role of stratospheric aerosol particles on the polar ozone depletion, which is discussed in more detail below.

3. The Stratospheric Aerosol and Polar Ozone Depletion

Most of the atmospheric ozone is contained in a narrow layer in a height between 15 and 20 km above the earth surface. This peculiar height distribution is due to a equilibrium of photochemical ozone production in the higher stratosphere and its destruction by photolysis and chemical decomposition in lower regions. The ozone layer shields the biosphere against harmful ultraviolet radiation and therefore protected the evolution of life outside the ocean. In spite of early warnings,[2,3] the threat of anthropogenic ozone destruction by the massive release of stable halogenated hydrocarbons has been realized only after the discovery of the antarctic ozone hole.[4] This almost complete chemical destruction of the ozone layer in the antarctic spring could not be explained solely by gas phase chemistry, but was later attributed to chemical reactions on the surface and in the bulk of aerosol particles in the antarctic stratosphere.[5] Due to the extreme dryness of the stratosphere, larger aerosol particles form here only at the very low temperatures of the polar winters. Under these conditions, they grow from minute sulfuric acid or sulfate particles by condensation of gas phase water and HNO_3. These polar stratospheric clouds (PSCs) catalyze chemical reactions which transfer chlorine and bromine from relatively inert reservoir species (like $ClONO_2$, HCl) into compounds which photolyze readily (like Cl_2 and HOCl) and release halogen radicals. This process is called halogen activation. The halogen radicals can then destroy ozone in catalytic cycles. More details of the complex chemical processes involved in the halogen activation and the catalytic ozone destruction are given in a recent review article.[6]

It is of great scientific and public interest to know if a ozone hole can form over the Northern Hemisphere in the near future. The assessment of such synoptic atmospheric questions is only possible with the use of large numeric models of the properties of the atmospheric system which have to be solved on the largest computer systems available. As input, reliable data for the physical and chemical properties of the constituents of the atmosphere are needed. While these have been determined in the laboratory or *in situ* for most of the molecular constituents of the atmosphere, the aerosol particles are often not well characterized. This especially holds true for the stratospheric aerosol, which was considered merely a meteorological curiosity prior to the discovery of the antarctic ozone hole. Surprisingly, even the equilibrium properties of the bulk phase of ternary mixtures of water, sulfuric acid and nitric acid are not well-known at low stratospheric temperatures, let alone dynamic processes of the growth or of phase transitions in small aerosol particles under stratospheric conditions.

It has been pointed out recently, that the question of homogenous and heterogeneous freezing of stratospheric aerosols is one of the key questions in the understanding of the physico-chemical behavior of the PSCs.[7] According to the phase diagram of liquid sulfuric acid solutions, PSCs were initially believed to consist solely of solid particles,[8] while it now seems clear from experimental observations and theoretical considerations that most of the PSCs stay liquid above the ice frost point.[9–11] It is still not clear, however, why sometimes solid PSCs are found above the ice frost point, and how they are formed.

The phase of the aerosol particles determines the rate of chlorine activation. For liquids the reactive uptake coefficient depends strongly on the solubility and diffusivity of the chlorine containing reactants like HCl and HOCl and $ClNO_3$.

Due to the lower vapor pressure of the solids compared to supercooled liquids of the same temperature, frozen aerosol particles grow rapidly by condensation of water and nitrous oxides and are removed from the stratosphere by deposition if their size exceeds about 10 μm. Under antarctic conditions this effect can lead to a permanent removal of nitrogen oxides from the stratosphere; a process which is known as denitrification.

In the laboratory, homogenous freezing of H_2SO_4 and of ternary H_2O, H_2SO_4 and HNO_3 solutions is difficult to observe for two reasons. First impurities or imperfections of the walls of the containment can act as a nucleation center, and second the freezing process may be slowed down considerably because of the high viscosity of sulfuric acid at low temperatures. Therefore previous experiments performed with bulk solutions were only able to give upper limits of the nucleation rate.[14–17]

In our experiments, many of the above mentioned difficulties could be overcome, as we observe single levitated aerosol particles over an extended period of time. By this means we are able to perform a thorough study of the thermodynamics and kinetics of the freezing of water, sulfuric acid and PSC droplets, and perform first measurements of HCl uptake into sulfuric acid droplets under stratospheric conditions.

4. Experimental Methods

In our experiments, we are observing processes with single levitated micro particles under stratospheric conditions.The main prerequisites are a microparticle generator, a levitation device, a climate chamber, and the particle spectroscopy and analysis. We use a piezo-driven nozzle for the generation of single micrometer sized droplets, an electrodynamic balance (Paul trap) as a levitator, and laser light scattering as our main method of observation. An overview of the setup is given in Fig. 2.

The Paul trap is housed in a central climate chamber which can be cooled down to the temperature of liquid nitrogen. The pressure and composition of the background gas in this chamber can be controlled by a gas mixing and inlet system and is usually adapted to stratospheric conditions. This chamber is suspended inside an insulation chamber which is kept at high vacuum. This chamber is equipped with a quadrupole mass spectrometer to analyze the chemical composition of the atmosphere, which leaks from the central chamber through a small pinhole. This system of vacuum chambers is connected to a standard time-of-flight mass spectrometer by means of a droplet translator. This is a cold finger which can be used to

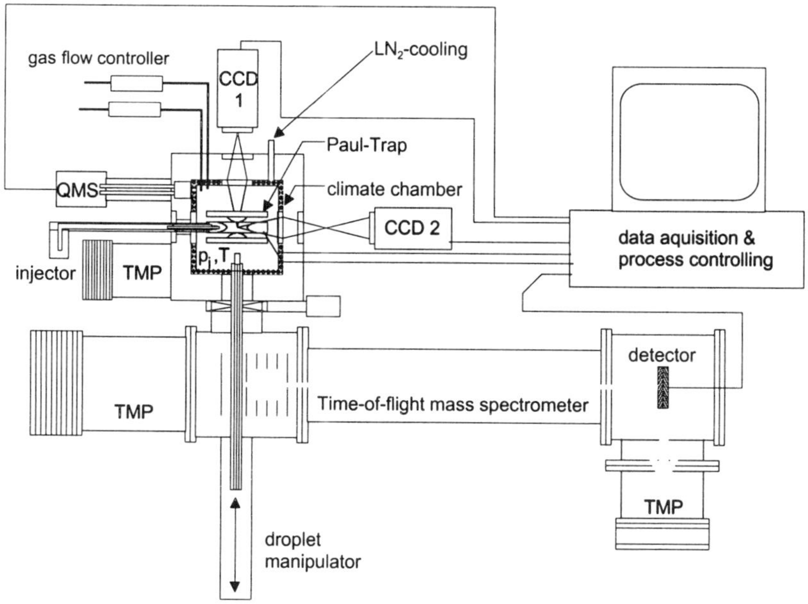

Fig. 2. Schematic setup of the aerosol experiments.

transfer the droplet from the Paul trap into the mass spectrometer. Here it can be brought into the gas phase by slow heating of the finger or by laser ablation. This allows to analyze the chemical composition of the aerosol particles after chemical processing in the artificial stratosphere.

The droplets are produced with an commercial piezo-driven injector. Upon generation, they are charged weakly by influence of an electric field at the nozzle tip. This charge is needed for the electrodynamic levitation. By varying the charge to mass ratio between $+10^{-3}$ C/kg and -10^{-3} C/kg we were able to prove that in this range the nucleation rate is not influenced by the droplet charge.

Inside the Paul trap, the particles are illuminated by means of a helium-neon laser. The scattered light is focused on a linear CCD array for the detection and automatic regulation of the particle position, and is imaged angular resolved onto a CCD camera. From the DC voltage the position control loop has to apply to the Paul trap, we obtain the charge to mass ratio of the droplet. The size and refractive index of liquid droplets can be determined to great precision by a careful analysis of the angular distribution of the light scattered from the particle. When both parallel and

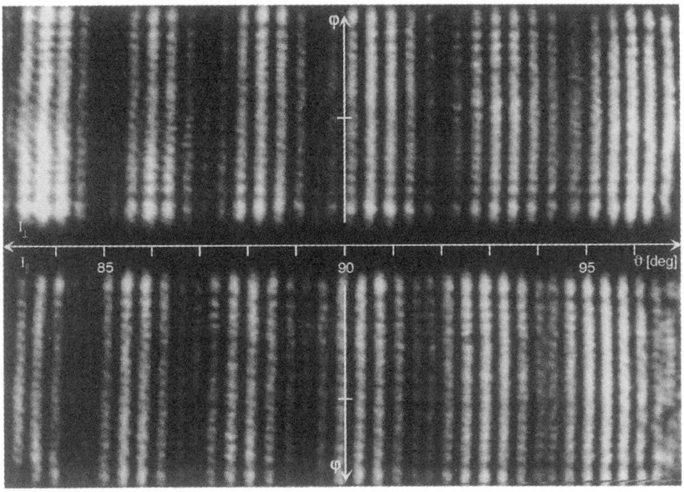

Fig. 3. Angular resolved scattering pattern from a spherical liquid droplet for incident light polarized perpendicular (upper) or parallel to the scattering plane.

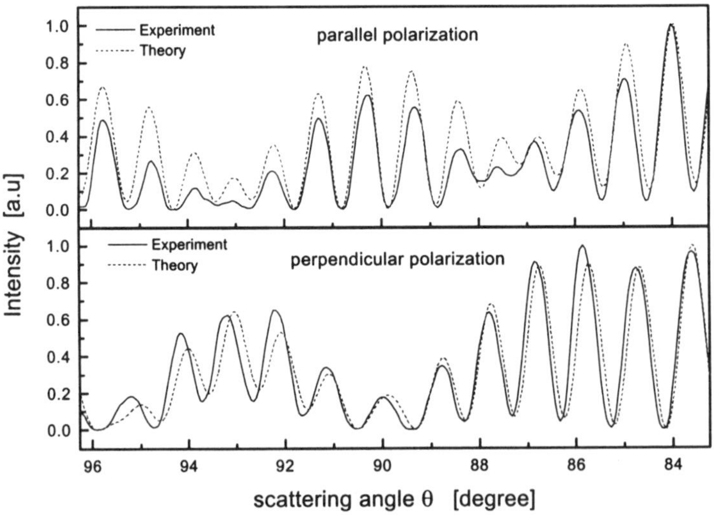

Fig. 4. Comparison between measured and calculated scattering intensities as a function of the scattering angle from a sulfuric acid droplet, concentration: 45 wt%, $T = 215$ K, $d = 39.81$ μm, $n = 1.4069$.

perpendicular polarization of the incident light are analyzed simultaneously we are able achieve a size error of less than 10^{-4} if the index of refraction is known, or an independent determination of size and refractive index with a precision of better than 10^{-3}. A typical scattering pattern from a sulfuric acid droplet is given in Fig. 3 for both polarizations of the incident light.

Its general appearance resembles a pattern of equidistant stripes with smoothly varying intensity. By comparing the patterns to calculations according to Mie theory, both the size parameter x, which is defined as the ratio between circumference of the droplet and wavelength of the light, and the index of refraction n can be determined. This is demonstrated in Fig. 4 where a comparison between measured and calculated angular dependence of the scattered light intensity is given.

The measurement of the mass to charge ratio of the droplet and the acquisition of the scattering image is performed in real-time. After analyzing the stored light scattering data, the temperature, size, charge and index of refraction of the droplet is known at every instant of the experiment. The liquid–solid phase transition was detected by an analysis of the light scattering of the droplet. Liquid droplets are spherically symmetric and the polarization of the scattered light is unchanged, if the droplet is illuminated by either parallel or perpendicular polarized light. As soon as the droplet freezes, inner surfaces appear, which break the spherical symmetry and thereby induce depolarization. This behavior is illustrated in Fig. 5.

The frozen particle depolarizes the incident parallel polarized light and the interference stripes which indicate the spherical particle are no longer

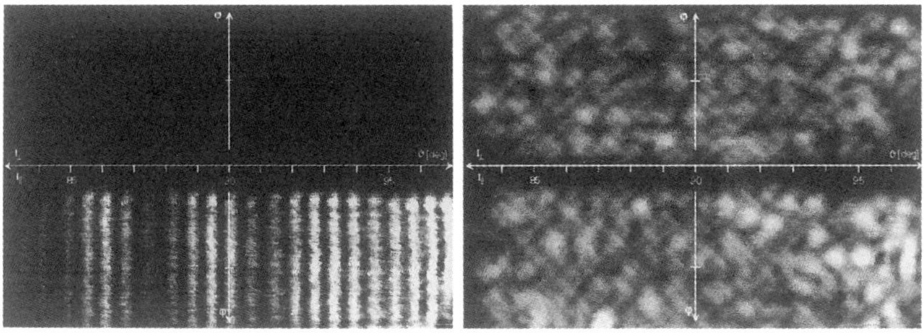

Fig. 5. Angular distribution of the light scattered from a liquid (left) and frozen (right) water droplet illuminated with parallel polarized light. The perpendicular and parallel polarized component of the scattered light is given in the upper and lower part of the image respectively.

visible. In our experiments, the appearance of scattered intensity in the depolarizing channel serves as an indicator for the freezing phase transition. Two other components of the setup received special attention and are therefore described in more detail: the piezo-driven droplet injector and the stabilization of the temperature field inside the trap.

As homogeneous freezing is a statistical process, it is necessary to investigate a large number of independent droplets in order to determine the rate constant of freezing. We therefore developed a system to inject the droplet into the cold trap at stratospheric temperatures. This allows to investigate a large number of particles without an intermediate warming/cooling cycle. In order to prevent the liquid from freezing inside the injector, the injection system had to be heated. This is achieved by two independent heating circuits, one for the tip and one for the body of the nozzle. The heating of the nozzle raises two additional problems however: first the temperature gradient from the nozzle to the trap causes convective airflow inside the trap, and second, the solution inside the nozzle can evaporate the more volatile component, thereby changing its composition. In order to overcome these problems, we took great care in insulating the nozzle from the trap and injected the droplets into the trap from a distance of about 20 mm from the trap center. Additionally, the nozzle was operated at a regular interval of about $1s$ in order to supply fresh solution to the tip. These droplets however were prevented from entering the trap by a suitable electric field. Only when a droplet had to be injected into the trap, this field was switched off. In order to insulate the trap from heat sources like the injector nozzle, it was completely shielded inside a copper cover leaving only small holes for the optical observation and the droplet entrance. By this means we were able to obtain reliable data for the nucleation rate, whereas in early stages of the experiment, where the Paul trap was operated unshielded in the climate chamber, we obtained higher and unstable nucleation rates. This indicates that the droplet was exposed to turbulent airflow and its temperature was fluctuating at a rate which was too fast for our temperature sensors to monitor.

5. Dynamics of the Liquid–Solid Phase Transition

Homogeneous freezing is the formation of a solid phase from a (supercooled) liquid. The thermodynamic freezing point is the temperature, where the free energy of the solid phase equals the free energy of the liquid. A pure liquid does not freeze homogeneously at that temperature, as freezing has to proceed via small nuclei or germs of the solid phase. From classical statistical mechanics we know that in a canonical system all possible elementary configurations are populated according to a Boltzmann distribution. Therefore the probability to find a germ of a given size inside a liquid of a given temperature can be calculated if the energies of germ formation are known for all germ sizes. Unfortunately, these energies cannot be obtained from *ab initio* calculations, due to the far ranging and non-additive intermolecular potentials found in real liquids.

The classical nucleation theory therefore tries to make reasonable assumptions on the energy of germ formation by extrapolation of the bulk properties of the solid down to very small sizes. A good overview over the progress and the problems in this field is given in Ref. 18. In order to check the various theories, reliable experimental data are needed. In the case of water, it was indeed only possible to arrive at a reasonable theory by fitting the energy of germ formation to reproduce the experimental results.[19]

6. The Homogeneous Freezing of Water

In our experiments we were able to determine the rate J of homogeneous nucleation in water droplets to a high precision by observing the freezing transition of more than 2000 individual droplets.

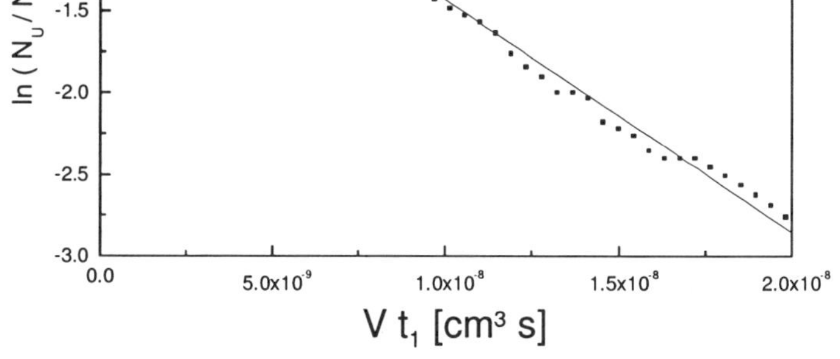

Fig. 6. Fraction of unfrozen water droplets as a function of the product of their residence time t_1 and their Volume V at $T = -236$ K.

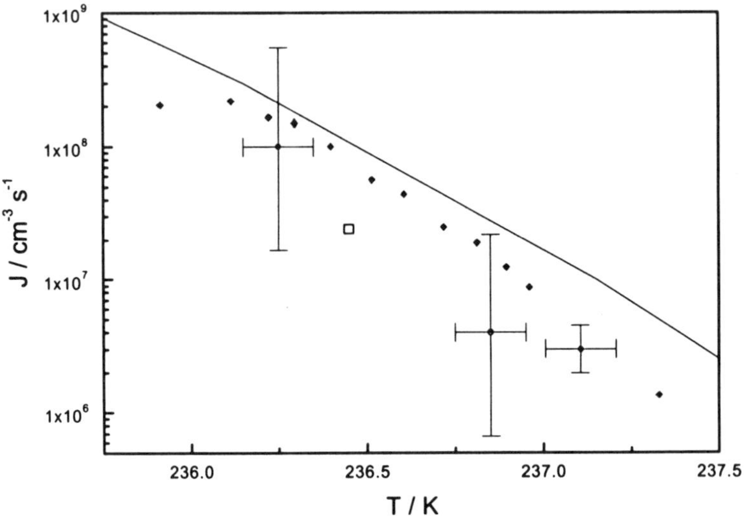

Fig. 7. Measured and calculated nucleation rates of water as a function of temperature. Diamonds: This measurement, filled squares: Ref. 20, open square: Ref. 21, solid line: theory cited from Ref. 19.

The measured individual times until the supercooled droplets freeze at a certain temperature are converted into a nucleation rate by plotting the cumulative fraction of unfrozen droplets against the product of droplet volume and time in a logarithmic diagram as shown in Fig. 6. The negative slope of the curve gives the nucleation rate J at that temperature.

After repeating this procedure at various temperatures, we arrive at a curve that gives the homogeneous nucleation rate of water as a function of temperature in an interval between 236 and 238 K (see Fig. 7). Note that the nucleation rate changes in this small interval by almost three orders of magnitude.

The measured rates lie within the large limits of uncertainty of previous measurements, and compare favorably with a recent extension of the classical theory of freezing.[19] Moreover, by comparison to that theory, we are able to extract from our data the free energy for the formation of the critical germ as a function of temperature. As shown in Fig. 8, this energy increases with temperature, as the size of the critical germ increases. It is possible to calculate from these results the critical radius of the ice germ that finally leads to freezing.

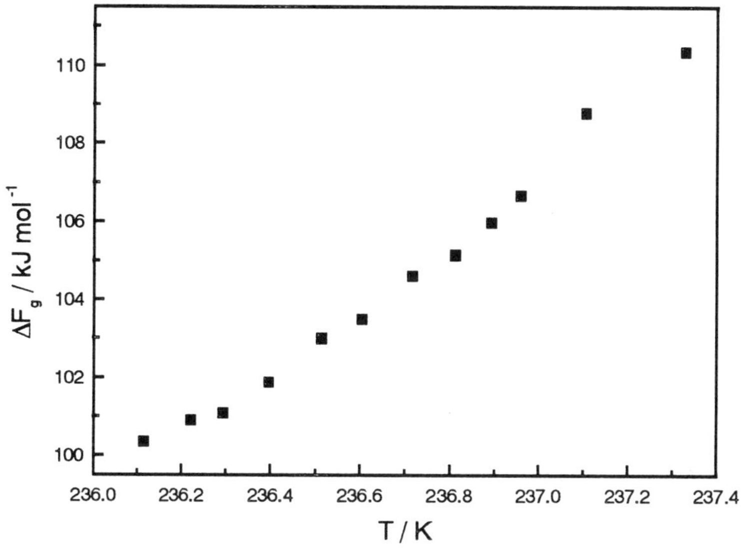

Fig. 8. Energy of germ formation for pure water as a function of temperature.

It lies in the range of 1.5 nm, which corresponds to about 400 water molecules and decreases monotonously with temperature (see Fig. 9, diamonds). To our knowledge, this is the first direct experimental determination of the size of the critical nucleus for freezing in water. The radius of the nucleus is about 30% higher than predicted by theory (compare Fig. 9, solid line), which is a good agreement given the many assumptions that go into classical nucleation theory. One of the unknown factors in theory is the liquid/solid interface energy. If this energy, which in theory is taken to be the difference between the surface tension of liquid water and of ice is larger by 70%, the two curves almost coincide.

Generally, for liquid water we find a remarkable good agreement with the extended classical nucleation theory.[19] It has to be noted however, that this theory has been obtained only by varying a key parameter, the diffusion activation energy of water as a function of temperature, in order to fit the experimental data. It indeed turns out, that at low temperatures, this energy differs significantly from the values usually extracted from viscosity measurements. Uncorrected classical nucleation theory gives nucleation rates which lie many orders of magnitude above our measured values.

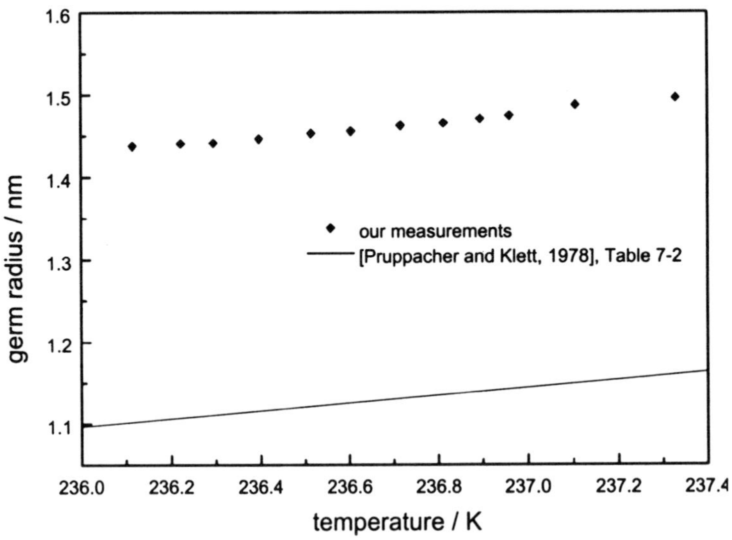

Fig. 9. Radii of the critical ice germs as a function of temperature. Diamonds: Our measurements, Solid Line: theory according to Ref. 19.

7. The Freezing Transition in Sulfuric Acid Solution

As described above, sulfuric acid solutions are the main component of the stratospheric aerosol and their state is of crucial importance for the rate of heterogeneous chemical reactions on their surface or in the bulk phase.

As compared to water, the freezing of sulfuric acid is more involved, as it is a two-component system, and the growth of a solid nucleus is not only accompanied by a transport of latent heat, but also by a transport of matter, if the composition of the solution is not by chance the composition of one of the stable hydrates of sulfuric acid. This effect, together with the high viscosity of sulfuric acid solutions makes the time for crystal growth much larger than in the case of pure water. It turns out, that in micrometer sized droplets at stratospheric temperature, the time for the complete freezing of the droplets can be much longer than the time for the formation of the initial germ. The depolarization ratio, which is our indicator for the appearance of the solid phase, is changed significantly only when a considerable fraction of the droplet is frozen. In the case of slowly freezing liquids like sulfuric acid solutions, this introduces a large error into the determination of the freezing rate. Fortunately, in binary systems the composition of the liquid phase changes as the solid germ grows. This introduces a change of the

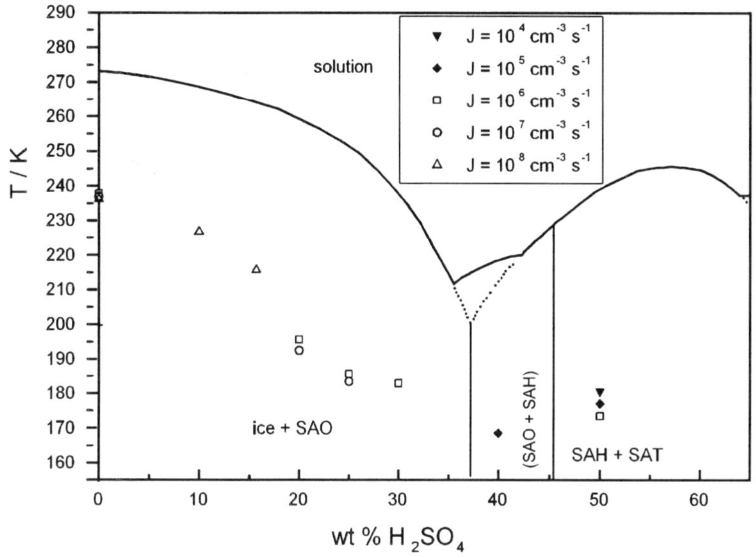

Fig. 10. Temperatures needed to reach a certain nucleation rate as a function of sulfuric acid concentration. The phase diagram according to Ref. 22 and the stable hydrates in bulk H_2SO_4 are given for comparison. (SAO: $H_2SO_4 \cdot 8H_2O$; SAH: $H_2SO_4 \cdot 6.5H_2O$; SAT: $H_2SO_4 \cdot 4H_2O$).

index of refraction of the liquid part of the solution which can be detected with high sensitivity by light scattering. We analyzed a model droplet to find a mathematical expression for the germ growth in order to correct our measured data. This allows to extract from the measurements both the time necessary to form the initial germ and the time for the complete freezing of the droplet. By this means we have investigated the freezing rate and the germ growth rate as a function of temperature and of sulfuric acid concentration.

With increasing acid concentration, the freezing rate depends less sensitively on the temperature than in the case of pure water. Our results are summarized in Fig. 10, where the temperatures which are needed to achieve a certain nucleation rate are given as a function of the sulfuric acid concentration. The respective rates are those which could easily be measured in our experiment. They decrease from low to high concentrations due to experimental limitations. At low H_2SO_4 concentrations, only fast freezing could be analyzed with high precision as the slow evaporation of water changed the composition of the droplets at longer time. At high concentrations however, the low nucleation rates are easier to determine because the time for the formation of the germ should be of the order of the time for the complete freezing of the droplet.

Nevertheless, given the strong dependence of the freezing rates on the temperature, our results (Fig. 10) indicate clearly, at which temperatures appreciable freezing of micrometer sized droplets within a second to minute time scale occurs.

These experiments are the most thorough determinations of the nucleation rate of sulfuric acid solution at stratospheric temperatures so far. For comparison the results are plotted in Fig. 11 together with previous experiments and theoretical predictions.

Figure 11 reveals big discrepancies between theory and experiment. At concentrations below 30%, the theory predicts freezing rates which are about eight orders of magnitude higher than those observed in the experiment. This might be due to the high sensitivity of the theory to the energy of germ formation, which is not well-known. It is more surprising, that the

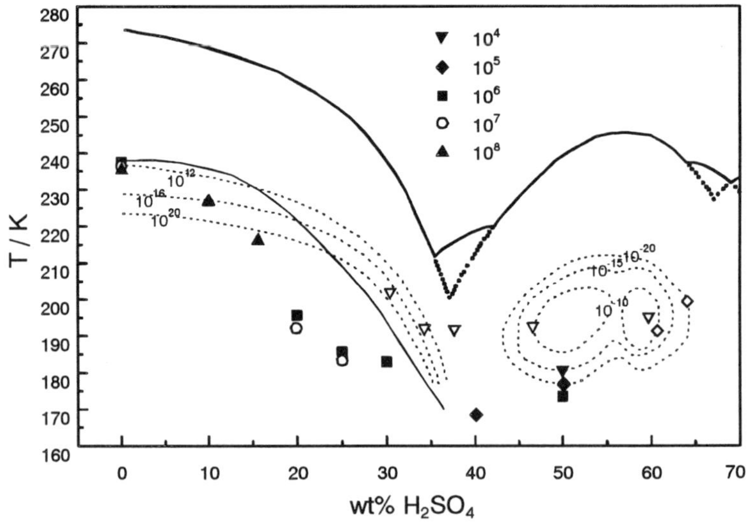

Fig. 11. Temperatures needed to reach a certain nucleation rate as a function of sulfuric acid concentration. The phase diagram according to Ref. 22 is given for comparison. solid symbols: our data; solid line: Data from Bertram *et al.* (Ref. 23); open symbols: freezing rate below 10^3 $cm^{-3}s^{-1}$ from Ref. 24. Dashed lines: Theoretical predictions according to Ref. 17; nucleation rates are given in $[cm^{-3}s^{-1}]$.

observed decrease of the slope of the log $J(T)$ curve with increasing sulfuric acid concentration is not reflected in the theory. In the case of concentrated H_2SO_4 (45 wt% – 65 wt%) we observe a even higher discrepancy in J of about 15 orders of magnitude. This might be due to the high uncertainty in the diffusion activation energy at low temperatures. As demonstrated in Ref. 19, the nucleation rates at low temperatures might be strongly underestimated, if this energy is taken from viscosity measurements.

Harder to understand are the reasons for the discrepancies between our data and those of Bertram *et al.* (Ref. 23, solid line in Fig. 11). He does not report freezing rates, but their observed onset of freezing correspondents roughly to a nucleation rate of 10^{12} $cm^{-3}s^{-1}$. Even if we take into account his uncertainty in sulfuric acid concentration of $\pm 7\%$, there remain unexplained differences around a concentration of 25 wt%. The upper limits of J given by Charleton *et al.* (Ref. 24, open symbols in Fig. 11), are in agreement with our data, but we observe much lower values of J.

The composition of the stratospheric background aerosol depends on the temperature and the height above ground. The lower the temperature and the lower the height of the aerosol, the more water is contained in the aerosol droplets. The composition of stratospheric aerosols as a function of

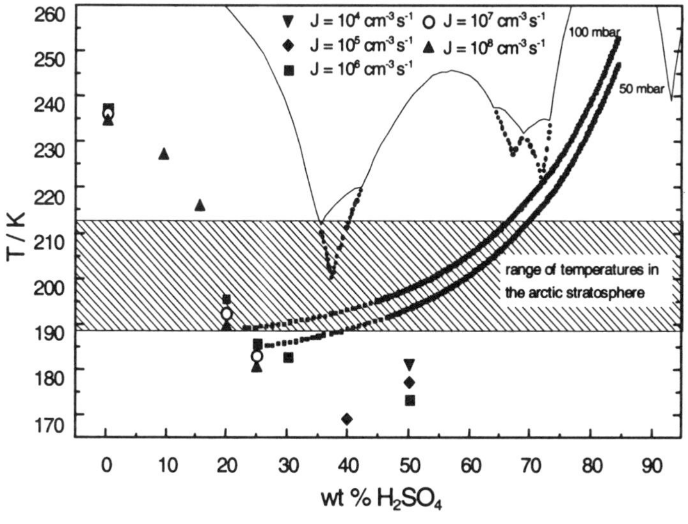

Fig. 12. Experimental data for the freezing of sulfuric acid solution compared to the composition of stratospheric background aerosol (data from Ref. 25), and the phase diagram of H_2SO_2 (Ref. 22).

temperature is plotted in Fig. 12 together with the phase diagram and our freezing data. In order to reach reasonable freezing rates, the temperature of the stratosphere has to drop to about 180 K. Under these conditions the droplets have a concentration of about 20 wt% H_2SO_4 and the freezing rate reaches 10^8 $cm^{-3}s^{-1}$. Under these circumstances, the time for freezing is between some days and some weeks depending on the size of the particle. This scenario is only realistic, if large H_2SO_4 aerosol particles are very abundant in the stratosphere. This can be the case after large volcano eruptions. Under normal conditions however, the background aerosol starts to adsorb HNO_3 from the gas phase if the temperature drops below about 195 K (dashed part of the curves in Fig. 12). The droplets are then ternary solutions of H_2SO_4, HNO_3 and H_2O and the discussion above is no longer valid. We conclude that pure sulfuric acid droplets do not freeze homogeneously under the conditions of the normal polar stratosphere. Therefore we extended our investigation of homogenous freezing to ternary systems of stratospheric relevance.

8. Homogeneous Freezing of Ternary Solutions

For the investigation of the freezing of ternary H_2SO_4 / HNO_3 / H_2O solutions, we used the experimental techniques described above for sulfuric acid solution. In this case however, we did not explore the whole composition space, but concentrated on the compositions of stratospheric relevance. From the experimentally known composition of the stratosphere, it is possible to calculate the composition of the aerosol particles as a function of temperature and height above ground. With decreasing temperature, the particles which are initially concentrated sulfuric acid take up an increasing amount of water and nitric acid. This behavior is reflected in Fig. 13, where the results of two theoretical calculations[26,27] are compared. They differ slightly as they assume a different height above ground for the particles and a different gas phase composition of the stratosphere. The good agreement between the two calculations demonstrates the narrow range of uncertainty about the composition of the stratospheric aerosol.

The symbols in Fig. 13 designate the compositions that were used in our experiments. They follow the calculated compositions with the exception of one droplet which was chosen to resemble pure nitric acid trihydrate (NAT).

We investigated the typical time for an aerosol to freeze at the temperature typical for its stratospheric appearance. Only for the droplets

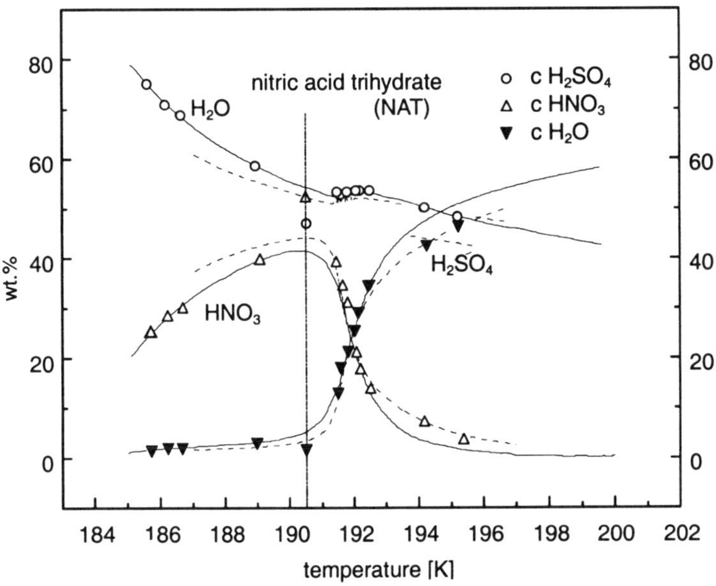

Fig. 13. Calculated composition of stratospheric aerosols as a function of temperature, and composition of the aerosol particles investigated in our experiments. Solid line: Data from Ref. 27. Dashed Line: Data from Ref. 26. Symbols: Investigated compositions.

of composition 2% H_2SO_4, 25% HNO_3 and 77% H_2O (leftmost droplet in Fig. 13) homogeneous nucleation was observed in our experimental time window, corresponding to an lower limit of the freezing rate of $J > 3 \cdot 10^5$ $cm^{-3}s^{-1}$. For all the other droplets no homogeneous freezing could be found at their respective stratospheric existence temperatures.

From these results we conclude that homogeneous freezing of stratospheric aerosols is only of importance at very low temperatures which might be reached in arctic regions only for short periods and in limited regions of space. The observed solid PSC particles are therefore most likely due to heterogeneous freezing.

9. Heterogeneous Freezing of Ternary Solutions

We performed a series of experiments to investigate the heterogeneous nucleation of ternary solutions using tiny ice crystals as a condensation germ. These germs were produced by deliberately introducing humid and warm air into the trap chamber. Upon cooling, small ice crystals form by condensation of the supersaturated water vapor. These ice germs are then adsorbed at the surface of the trapped aerosol droplets and might induce freezing. The experiments were performed using the same composition of the aerosol particle as in the homogeneous freezing experiments described above (c.f. Fig. 13). The temperature was always the stratospheric existence temperature of the particle. Heterogeneous nucleation induced by ice germs was found for all particles with an existence temperature below 193 K. For the sulfuric acid rich particles with an existence temperature above 193 K neither homogeneous nor heterogeneous nucleation could be found.

These findings are in excellent agreement with *in situ* measurements of the depolarization properties of stratospheric particles. At temperatures below 187 K, large frozen particles are frequently observed, while at higher temperatures, usually only a small fraction of the stratospheric clouds is frozen. The amount of frozen particles depends strongly on their temperature history. Frozen particles are usually observed, whenever the temperature fell below the ice frost point (typ. 187 K). In the absence of ice germs, the freezing of stratospheric aerosol particles might be induced by other germs, like meteorite dust or soot particles. We are currently preparing experiments to investigate the heterogeneous nucleation on such germs.

10. Conclusion

Using an electrodynamic trap inside a climate chamber, we were able to investigate water, sulfuric acid solutions and ternary droplets of H_2SO_4, HNO_3 and H_2O. The method of angular and polarization resolved light scattering allowed to measure the freezing kinetics of single well characterized aerosol particles. Observing a large number of single particles we were able to obtain freezing rates and freezing velocities. From these data, microphysical properties of water and sulfuric acid solutions like the free energy of germ formation and the diffusion activation energy could be obtained as a function of temperature and concentration. The experiments on ternary solutions allowed to explain the frequent observation of liquid PSC in the stratosphere. They also demonstrate, under which conditions freezing transitions can occur homogeneously or heterogeneously in the stratosphere.

11. Outlook

Currently we are investigating the uptake kinetics of stratospheric trace gases like HCl or $ClONO_2$ into PSC droplets. These experiments will yield input data for theoretical models of the chemical ozone destruction in the stratosphere. The experimental techniques demonstrated above can be applied to many other fields of physics or chemistry. Among other applications, we are observing the coulomb explosion of single highly charged droplets as they evaporate inside a electrodynamic trap and explore the possible use of microdroplets as a high quality cavity for the spectroscopic detection of trace impurities or nanoparticles inside the droplets.

Acknowledgements

Prof. Dr. Ludger Wöste at the Free University of Berlin has continuously supported this work and contributed valuable ideas. The experiments were performed by Inez Weidinger, Benedikt Kramer, Oliver Hubner and Hermann Vortisch in the course of their Ph.D. and diploma theses. The project was initiated in a close and fruitful collaboration with Prof. Helmut Baumgartel and Martin Schwell from the Department of Physical Chemistry of the Free University and Prof. Eckart Rühl from the University of Osnabruck.

Financial support from the FNK commission of the FU Berlin, the Deutsche Forschungsgemeinschaft, the Federal Ministry of Education and Research and the Deutsche Bundesstiftung Umwelt is gratefully acknowledged.

References

1. J. Joshua, *J. Z. Phys.* **D24**, 247 (1992).
2. P. J. Crutzen, *Quart. J. Roy. Met. Soc.* **96**, 320 (1970).
3. M. J. Molina and F. S. Rowland, *Nature* **249**, 810 (1974).
4. J. C. Farman, B. G. Gardiner and J. D. Shanklin, *Nature* **315**, 207 (1985).
5. S. Solomon, R. R. Garcia, F. S. Rowland and D. J. Wuebbles, *Nature* **321**, 755 (1986).
6. Th. Peter and P. J. Crutzen, *J. Aerosol Sci.* **24**, S119-S120 (1993).
7. Th. Peter, *Ann. Rev. Phys. Chem.* **48**, 785 (1997).
8. O. Toon, E. Browell, S. Kinne and J. Jordan, *Geophys. Res. Lett.* **17**, 393 (1990).
9. R. Zhang, P. Woolridge and M. Molina, *J. Phys. Chem.* **97**, 8541 (1993).
10. M. Molina, R. Zhang, P. Woolridge, J. McMahon, J. Kim, H. Chang and K. Beyer, *Science* **261**, 1418 (1993).
11. K. Stebel, R. Neuber, G. Beyerle, J. Biele, I. Benigna, P. Scheuch, H. Schütt and O. Schrems, *Proc. Quad. Ozone Symp.* edited by R. D. Bojkov and G. Visconti, (L'Aquila, Italy, 1996) p. 607.
12. B. P. Luo, K. S. Carslaw, Th. Peter and S. Clegg, *Geophys. Res. Lett.* **22**, 247 (1995).
13. T. Huthwelker, S. L. Clegg, Th. Peter, K. S. Carslaw, B. P. Luo and P. Brimblecombe, *J. Atmos. Chem.* **21**, 81 (1995).

14. K. D. Beyer, S. W. Seago, H. Y. Chang and M. J. Molina, *Geophys. Res. Lett.* **21**, 871 (1994).
15. T. Ohtake, *Tellus* **B45**, 138 (1993).
16. T. Koop, U. M. Biermann, W. Raber, B. P. Luo, P. J. Crutzen and Th. Peter, *Geophys. Res. Lett.* **22**, 917 (1995).
17. B. P. Luo, Th. Peter and P. J. Crutzen, *Geophys. Res. Lett.* **21**, 1447 (1994).
18. H. R. Pruppacher and J. D. Klett, *Microphysics of Clouds and Precipitation* (D. Reidel Publishing Co., Dordrecht, Holland, 1978).
19. H. R. Pruppacher, *J. Atmos. Sci.* **52**, 1924 (1995).
20. P. J. DeMott and D. C. Rogers, *J. Atmos. Sci.* **47**, 1056 (1990).
21. G. R. Wood and A. G. Walton, *J. Appl. Phys.* **41**, 3027 (1970).
22. C. M. Gable, H. F. Betz and S. H. Maron, *J. Am. Chem. Soc.* **72**, 1445 (1950).
23. A. K. Bertram, D. D. Patterson and J. Sloan, *J. Phys. Chem.* **100**, 2378 (1996).
24. K. L. Charleton, D. M. Sonnenfroh, W. T. Rowlins, B. E. Wyslouzil and S. Arnold, *J. Geophys. Res.* **35**, 125 (1997).
25. T. Koop, U. M. Biermann, W. Raber, B. P. Luo, P. J. Crutzen and Th. Peter, *J. Phys. Chem.* **A101**, 1117 (1997).
26. A. Tabazadeh, R. P. Turco, K. Dradla, M. Z. Jacobson and O. B. Toon, *Geophys. Res. Lett.* **21**, 1619 (1994).
27. K. S. Carslaw, B. P. Luo, S. L. Clegg, Th. Peter, P. Brimblecombe and P. J. Crutzen, *Geophys. Res. Lett.* **21**, 2479 (1994).

CHAPTER 7

CLUSTER FORMATION FROM LIQUID PHASE

F. Mafuné, N. Horimoto, J. Kohno and T. Kondow
Cluster Research Laboratory, Toyota Technological Institute and Genesis Research Institute, Inc. 717-86 Futamata, Ichikawa, Chiba 272-0001, Japan
Email: mafune@mail.cluster-unet.ocn.ne.jp

In this chapter, we describe experimental studies on formation of gas phase clusters from liquid phase. Several elaborate techniques enable us to prepare gas phase clusters which have not been produced by conventional molecular beam methods.

Contents

1. Introduction

Intrinsic properties of a molecule manifest themselves in its physical and chemical properties, when the molecule is isolated in the gas phase. In the present stage, our understandings reach a level of describing the isolated molecule involving its ultrafine states.[1] Once the molecule is dissolved in a medium, the fine structures of the molecule cannot be differentiated easily mainly because of spatial and temporal fluctuation of surrounding molecules and hence the methodologies employed in the gas phase studies are no longer applicable in a straightforward manner. Partly because of this difficulty, a molecule in a liquid medium is not well understood at a level of the corresponding isolated molecule in spite of its practical importance; almost all the practically important chemical reactions take place in a liquid medium, protein and DNA molecules function properly *in vivo* (mainly water), etc.[2,3] In this regard, it is important to develop elaborate methodologies which are well suited to the investigation of molecules in liquid media.

To begin with, a molecule in a liquid medium should be isolated into the gas phase for scrutiny by use of the methods available for the detailed investigation of a molecule isolated in the gas phase; spectroscopic and dynamic studies. Then, the molecule attached with solvent molecules (solvated cluster) is investigated similarly with an attention how the solvent molecules influence the static and dynamic properties of the solvated cluster as a function of the number of the solvent molecules.[4,5]

Clusters of volatile atoms and molecules such as Ar, N_2, O_2, H_2O, etc. are generated by a supersonic molecular beam method. The atoms and the molecules stagnant in a nozzle at a pressure of several atmospheres are introduced into vacuum through a pinhole and are cooled by an adiabatic expansion. The internally cold atoms and molecules in the expansion aggregate into clusters. On the other hand, clusters of non-volatile atoms and molecules such as metals are generated by techniques of evaporation, laser ablation, ion sputtering, plasma heating, etc. Sodium clusters, for instance, are produced by heating a solid sodium sample in an oven and appregation in a helium atmosphere.[6,7] Clusters of metals having low melting points (< 1000 K) are often prepared by a similar method. On the other hands, clusters of high melting points such as Pt (1768 K) and V (1910 K) are prepared by the laser ablation method.[8]

Difficulty arises when one is attempt to produce clusters composed of molecules vulnerable to heat. Biological molecules such as proteins and DNA molecules are in this category. The laser ablation is unlikely to prepare clusters of such biological molecules, because a considerable extent of fragmentation is inevitable concurrently with the laser ablation.[9] In order to overcome this difficulty, several techniques have been developed to isolate biological molecules and those with solvent molecules (water etc.) from liquid media; the biological molecules are known to have their bioactive forms and functionality in a buffer solution. In this chapter, we describe the development of several techniques and applications on isolation and detection of molecules from liquid media in addition to referring to the isolation mechanism of the molecules from the liquid media.

2. Experimental Methodologies

2.1. *Electrospray*

In electrospray, a solution containing a solute of interest is introduced through a metal capillary tube floated at several kV against an electrode which is located a few mm or a few cm away from the tip of the capillary tube (capillary nozzle).[10–13] The solution is continuously supplied by a syringe pump at 0.1–20 ml/min into the capillary tube. The electric field gradient between the capillary nozzle and the electrode pull the solution toward the electrode, forming a Taylor cone. If the field is sufficiently high with respect to the surface tension of the solution, the cone is further drawn to a filament that generates small charged droplets. Multiply charged ions of a solute molecule and that solvated with solvent molecules (solvated cluster) are formed by explosion of small, highly charged droplets by the Coulomb repulsion forces. In a typical experimental set-up, the capillary tube is located outside a vacuum chamber. The charged droplets are transferred from ambient atmosphere to a vacuum chamber housing a mass spectrometer. Figure 1 shows a typical mass spectrum of water clusters produced by electrospray of a 0.1 mM acetic acid in water. The capillary nozzle is floated by 4.5 kV against the electrode located 15 mm downstream from the nozzle. Peaks appearing equidistantly in the mass spectrum are assigned to the protonated water clusters, $H^+(H_2O)_n (n = 1, 2, 3, \ldots)$. Electrospray is most frequently-used to isolate molecular ions of biological interest in the gas phase.[14] Electrospray extracts ions from the solution by electrostatic forces, causing no thermal decomposition of molecules. On the other hand, multiply charged ions tend to be produced by electrospray, as expected from the formation mechanism of the molecular ions.

In attempting a chemical analysis based on mass spectrometry, molecules of interest have to be isolated in the gas phase with minimizing decomposition of the molecules. Suppose a biological molecule is heated by any kind of methods, it tends to decompose before it vaporizes. In order to prevent the molecule from being decomposed, the molecule has to be isolated swiftly in the gas phase, before the internal modes of the molecule are excited by the energy transmission. Isolation techniques of this kind

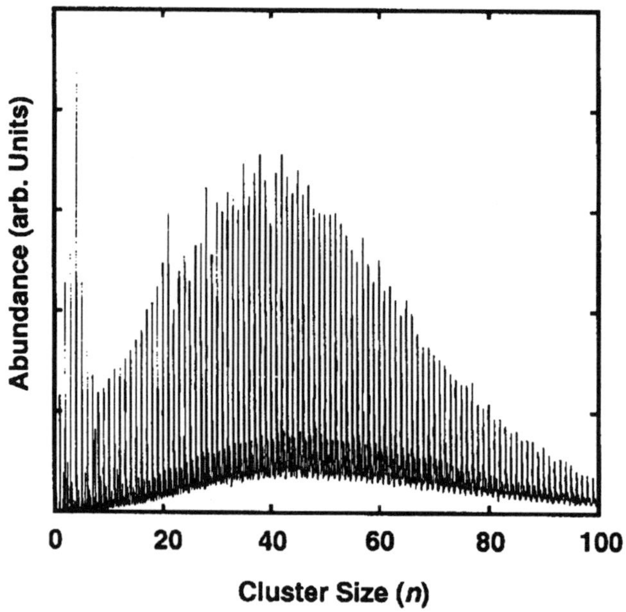

Fig. 1. Mass spectrum of water cluster ions produced by electrospray (by courtesy of T. Tsukuda and Y. Negishi, Institute for Molecular Science).

developed so far include fast atom bombardment (FAB)[15] and pulsed laser desorption,[16] which provide an impulsive force to the molecule.

2.2. *Liquid Jet*

When small liquid droplets are formed and left in a vacuum for a long time at a temperature slightly higher than a room temperature, they are subjected to a phase transition by rapid volume expansion into a mixture of gas and small droplets. Clusters are formed by further volume expansion of the small droplets. The method of preparing clusters in the gas phase by taking advantage of these phenomena is developed by Nish and his coworkers, and is named "liquid jet" method.[17–20]

As shown in Fig. 2, small droplets with roughened surface are formed at first at the exit of the capillary nozzle; the roughened surface causes a bulk liquid into small droplet. The small droplets are fragmented into clusters by rapid volume expansion, because the difference of the inner and the outer pressures increases with a decrease in the droplet diameter.

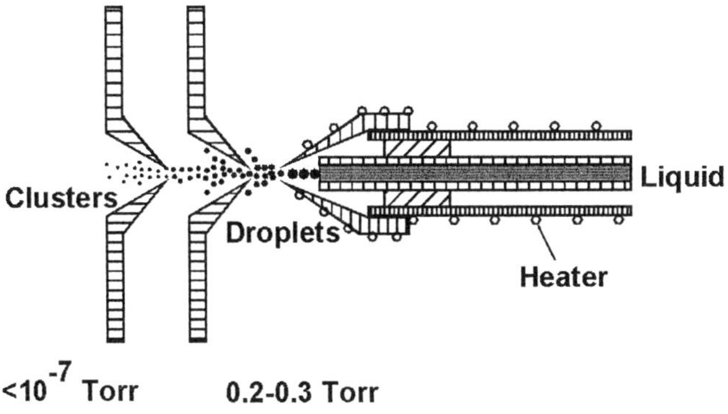

Fig. 2. Schematic diagram of liquid jet apparatus.

A sample solution is continuously supplied to the nozzle by using a pump designed for a liquid chromatograph. The small droplets are transferred from a chamber with a high ambient pressure (0.2–0.3 Torr) to a chamber differentially pumped down to a low pressure (10^{-7} Torr), which houses a mass spectrometer. Clusters are formed at the entrance of the low pressure chamber.

In some solutions, the solute and the solvent molecules form an inhomogeneous structure (solvation structure), in which the solute and solvent molecules are strongly bound each other. The liquid jet method enables to isolate clusters in the gas phase maintaining the solvation structure, because free solvent molecules (almost bound-free to the solvation structure) in the solution are likely to desert the solvation structure as a solvated cluster in the gas phase on volume expansion of the droplets. In reality, the size-distribution of a hydrated methanol cluster ion, $H^+CH_3OH(H_2O)_{n-1}$, produced from a methanol solution in water shows a linear increase in the abundance with n, whereas the abundance of $H^+CH_3OH(H_2O)_{n-1}$ produced by attachment of a methanol molecule onto water clusters in the gas phase is almost constant regardless of n. This finding indicates that a methanol molecule is mainly solvated by the water molecules in a solution. Namely, the size-distribution of the gas phase clusters thus produced provides structural information of the solutions.

2.3. *Liquid Beam*

When a liquid is introduced into a vacuum through a pinhole as a continuous stream, one can form a flowing liquid filament (liquid beam) having the same diameter as the diameter of the pinhole.[21,22] In contrast to the liquid jet described above, no liquid droplets are formed on ejection from the pinhole, or no rapid volume expansion of the liquid takes place.

The Reynolds number of the flow, R_e, is in a suitable range ($100 < R_e < 1000$) for satisfying the laminar flow condition, where the Reynolds number is defined as:

$$R_e = rd\nu/\eta \tag{1}$$

where r is the density of the liquid, d is the diameter of the nozzle aperture ($d = 20 \times 10^{-6}$ m), ν is the velocity of the liquid flow, and η is the viscosity of the liquid. Therefore, a suitable flow rate varies with the properties of the liquid employed. The flow rate of ethanol ($r = 0.79 \times 10^3$ kg m^{-3}, $\eta = 1.2 \times 10^{-3}$ kg m^{-1}) has to be in the range between 0.15 mL min^{-1} and 1.5 mL min^{-1}, while that of n-hexane ($r = 0.68 \times 10^3$ kg m^{-3}, $\eta = 0.32 \times 10^{-3}$ kg m^{-1}) ought to be higher than 0.56 mL min^{-1}. The liquid flow becomes unstable into droplets at a lower R_e, while at a higher R_e it disintegrates immediately on its leaving from the nozzle.

A temperature on the liquid beam surface decreases by the evaporative cooling with the distance from the nozzle exit. The temperature is found to drop by $\sim$ 30 K at 4 mm downstream from the nozzle exit, while the liquid beam remains to be liquid within a distance of several millimeters to centimeters from the nozzle exit. This phenomenon is proved by no substantial scattering of a laser beam either due to Rayleigh instabilities or a liquid–solid phase transition when it travels inside the liquid beam.[22]

Figure 3 shows a schematic view of the experimental apparatus, which consists of a liquid beam, IR and UV lasers, and a reflectron time-of-flight (TOF) mass spectrometer. Liquid beam is introduced through a nozzle aperture with 20 μm in diameter into a vacuum chamber evacuated down to 10^{-5}–10^{-6} Torr. The aperture plate is diverted from a commercial platinum aperture used in an electron microscope. A pump designed for a liquid chromatograph supplies the continuos liquid flow at a flow rate of 0.2–0.4 mL min^{-1} The liquid beam is trapped on a surface cooled at 77 K at 10 cm downstream from the aperture. Traveling a distance of 1 mm from the aperture, the liquid beam is crossed with the pulsed IR laser beam having a pulse width of $\sim$ 10 ns at the first acceleration region of the TOF mass spectrometer. The IR laser is resonant at the wavelength of 2.9 μm with the OH vibration of solvent water molecules so that only the solvent molecules in the liquid beam are excited vibrationally by IR-laser irradiation. The fourth harmonic of a pulsed Nd:YAG laser (UV laser of 266 nm), is focused

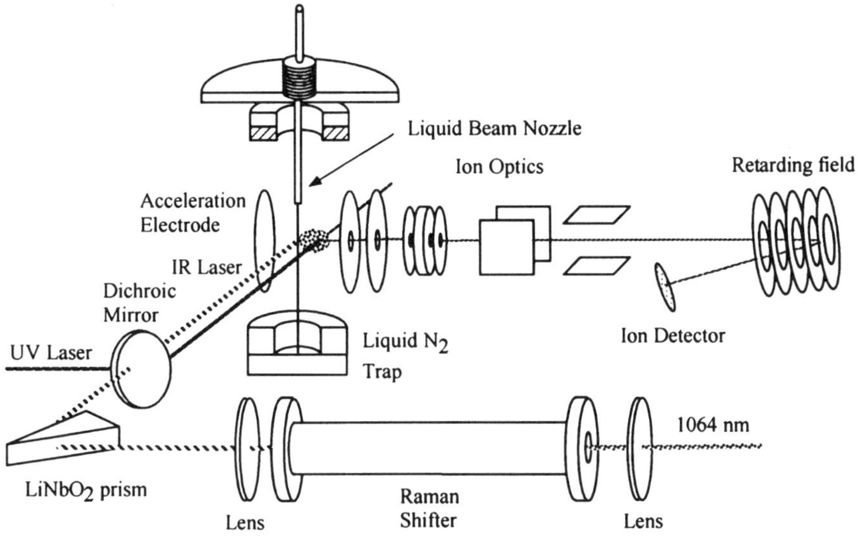

Fig. 3. Schematic diagram of a liquid beam and a reflectron time-of-flight (TOF) mass spectrometer. The liquid beam is crossed with an IR laser at 1 mm downstream from the nozzle. Neutral species isolated in the gas phase are ionized by a UV laser and mass analyzed by a TOF mass spectrometer.

tightly into a spot with 100 μm in diameter, and is illuminated outside the liquid beam. Neutral species, bare solute molecules and clusters composed of the solute and solvent molecules, are produced in the gas phase by the IR-laser irradiation, and then are ionized through resonant two-photon excitation by UV-laser irradiation. A digital delay pulse generator varies the timing of the UV-laser irradiation delayed with respect to the start of the IR-laser irradiation. Ions thus produced are accelerated by a pulsed electric field in the first acceleration region of the time-of-flight (TOF) mass spectrometer, in the direction perpendicular to both the liquid and the laser beams. The ions are then steered and focused by a set of vertical and horizontal deflectors and an einzel lens. After traveling in a 1 m field free region, the ions are reversed by the reflectron tilted by 2° off the beam axis and are detected by an electron multiplier. Signals from the multiplier are amplified by a homemade 100 MHz preamplifier and processed by a digital oscilloscope. The mass resolution, $m/\Delta m$, is found to be more than 85 at $m = 200$ in a typical experimental condition.

3. Formation Mechanism of Gas Phase Clusters from Liquid Beam

3.1. *Mass Spectrum*

Let us explain the formation mechanism of gas phase clusters from the liquid beam of a resorcinol, $C_6H_4(OH)_2$, solution in water under irradiation of a pulsed IR laser at 2.9 μm.[23–25] Resorcinol is employed as a prototype molecule for IR laser isolation of molecules from a liquid solution, because resorcinol has a low vapor pressure at a room temperature and hence a density of resorcinol molecules above the liquid beam surface is negligibly small. In addition, resorcinol is easily ionized by a pulsed UV laser at 266 nm in the gas phase, since it has an electronic transition in the wavelength range including 266 nm. Figure 4(a) shows TOF mass spectra of ions produced from the liquid beam of a 0.2 M resorcinol solution in water: the spectrum is obtained by irradiation of the IR laser onto the liquid beam before the UV laser illuminating a spot 0.4 mm outside the liquid beam. A series of ion peaks ($m/z = 110, 128, 146, 164, \ldots$) appears in the spectrum and is assigned to solvated resorcinol cluster ions, $C_6H_4(OH)_2{}^+$ $(H_2O)_n$ ($n = 1, 2, 3, \ldots$). No ions are observed, when either the IR-laser illuminates the liquid beam [trace (b)] or the UV laser illuminates outside the liquid beam [trace (c)]. These findings assure that neutral clusters composed of resorcinol and water molecules are isolated in the gas phase by the IR-laser, and then are ionized by the UV-laser. Namely, a neutral cluster, $C_6H_4(OH)_2\,(H_2O)_m$, is isolated through selective vibrational excitation of solvent H_2O molecules by the IR-laser irradiation, and is ionized to $C_6H_4(OH)_2{}^+\,(H_2O)_n$ by the UV laser irradiation. Several solvent H_2O molecules may evaporate after the ionization to release an excess energy generated in the course of photo-ionization. In summary, $C_6H_4(OH)_2{}^+$ $(H_2O)_n$ ($n \geqq 1$) is produced as:

$$C_6H_4(OH)_2(s) + h\nu_{IR} \rightarrow C_6H_4(OH)_2(H_2O)_m(g) \quad (2)$$

$$C_6H_4(OH)_2(H_2O)_m(g) + h\nu_{UV} \rightarrow C_6H_4(OH)_2{}^+(H_2O)_n(g) + (m-n)H_2O \quad (3)$$

where (s) and (g) represent species in the solution and in the gas phase, respectively. On the other hand, the precursor of $C_6H_4(OH)_2{}^+$ can be a bare resorcinol molecule, $C_6H_4(OH)_2$, and/or a cluster with a small number of

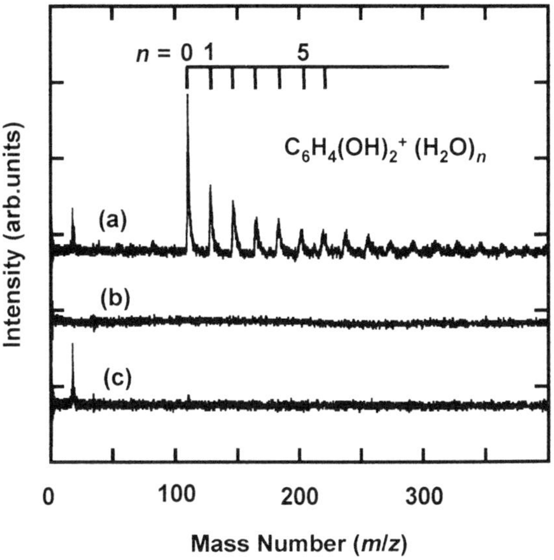

Fig. 4. Mass spectra of ions produced by irradiation of IR and UV laser beams on a 0.2 M resorcinol solution in water. In trace (a), the IR laser is focussed onto the liquid beam, and the UV laser illuminates 0.5 mm away from the liquid beam. Traces (b) and (c) show the spectra obtained by irradiation of either the IR or the UV laser.

water molecules, $C_6H_4(OH)_2\ (H_2O)_k$. Relevant experiments suggest that the precursor of $C_6H_4(OH)_2^+$ is a bare resorcinol molecule.

3.2. *Spatial Distribution*

The neutral clusters, $C_6H_4(OH)_2\ (H_2O)_m$, are found to be produced in the gas phase by irradiation of the pulsed IR-laser onto the liquid beam. The clusters continue to be released from the liquid beam surface for a period of $\sim 10\ \mu s$ after the pulsed IR-laser irradiation. The dynamics of the cluster release is further examined by observing the neutral species at a certain time after the IR-laser irradiation at a certain distance from the liquid beam. In reality, the neutral species are ionized by the pulsed UV-laser and are mass analyzed by the mass spectrometer. The spatial distribution of the neutral clusters are observed by changing the position of the ionization laser with respect to the liquid beam and the delay time of the UV-laser irradiation from the IR-laser irradiation.

Figure 5 shows mass spectra at several different delay times. Note that the spectrum is obtained by irradiation of the UV laser illuminating a spot 0.4 mm away from the liquid beam. These mass spectra demonstrate

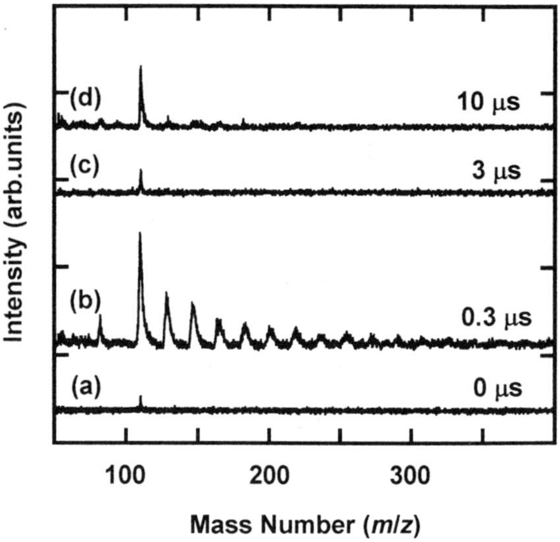

Fig. 5. Mass spectra measured at different delay times from IR-laser to UV-laser irradiation. The delay times are 0, 0.3 μs, 3 μs and 10 μs for traces (a), (b), (c) and (d), respectively. The observed cluster ions are assignable to $C_6H_4(OH)_2{}^+$ $(H_2O)_n$ $(1 \leqq n \leqq 10)$.

that both the intensity and the size-distribution of $C_6H_4(OH)_2{}^+$ $(H_2O)_n$ change significantly with the delay time. Figure 6 shows the intensity of $C_6H_4(OH)_2{}^+$ and the total intensity of all the cluster ions, $C_6H_4(OH)_2{}^+$ $(H_2O)_n$ $(n \geqq 1)$, as a function of the delay time (0–30 μs). In the delay times shorter than 3 μs (early-time domain), both the bare resorcinol ion, $C_6H_4(OH)_2{}^+$ $(H_2O)_n$, and the cluster ions, $C_6H_4(OH)_2{}^+$ $(H_2O)_n$, are observed, but in the delay times longer than 3 μs (late-time domain) only the bare resorcinol ion is observed. As the delay time increases, the total intensity of all the cluster ions observed reaches the maximum at 0.3 μs and decreases rapidly, while $C_6H_4(OH)_2{}^+$ has two maxima at 0.3 μs and 10 μs. These findings indicate the presence of two different ejection processes, which possess short and long characteristic times; an early-time domain for the production of solvated clusters, $C_6H_4(OH)_2$ $(H_2O)_m(0 \leqq m \leqq 10)$, and a late-time domain for the production of a bare resorcinol molecule, $C_6H_4(OH)_2$.

The neutral species spread out around the excitation region of the surface, with maintaining cylindrical symmetry with respect to the axis of the liquid beam. Hence, the radial distribution measured at a small stereo angle

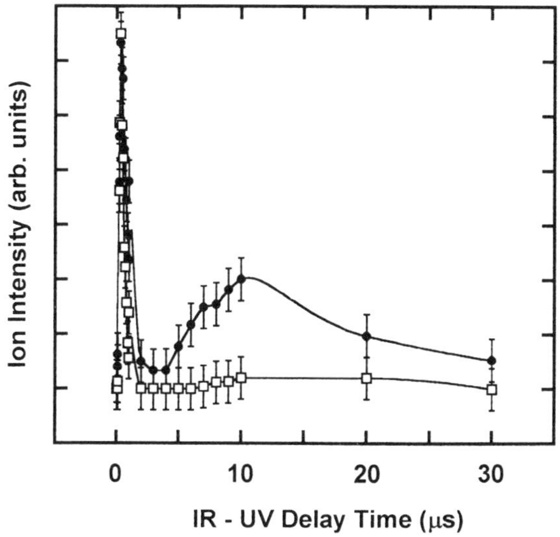

Fig. 6. The intensity of $C_6H_4(OH)_2^+$ and the total intensity of $C_6H_4(OH)_2^+(H_2O)_n (n \geq 1)$ measured as a function of the IR-UV delay time (0–30 μs). Solid and open circles represent the intensity of $C_6H_4(OH)_2^+$ and the total intensity of $C_6H_4(OH)_2^+(H_2O)_n (n \geqq 1)$, respectively.

represents the spatial distribution of the species spread toward all the directions. In practice, radial distribution is obtained by multiplying the ion intensity by the distance between the liquid beam and the ionization region, d. Figure 7 shows the radial distribution of $C_6H_4(OH)_2^+$ and $C_6H_4(OH)_2^+(H_2O)_n$ $(1 \leqq n \leqq 8)$ at different elapsed times after the IR-laser irradiation. An ion bunch consisting of $C_6H_4(OH)_2^+$ and $C_6H_4(OH)_2^+ (H_2O)_n$ $(1 \leqq n \leqq 8)$ starts to appear at the close vicinity of the liquid beam surface at the elapsed time of 50 ns (the early-time domain). The peak of the bunch shifts outward with the elapsed time; the bunch of $C_6H_4(OH)_2^+$ moves with essentially the same velocity as that of $C_6H_4(OH)_2^+ (H_2O)_n$. After the rapidly moving bunch has passed through, the other bunch consisting of $C_6H_4(OH)_2^+$ appears again at 1 μs (the late-time domain) and travels away after a certain elapsed time.

In Fig. 8, the peak positions, d_{peak}, of the rapidly and slowly moving bunches are plotted as a function of the IR-UV delay time, t_{delay}. The peak positions shift almost linearly as the delay time increases, because:

$$d_{\text{peak}} = \nu(t_{\text{delay}} - t_{\text{residence}}) \tag{4}$$

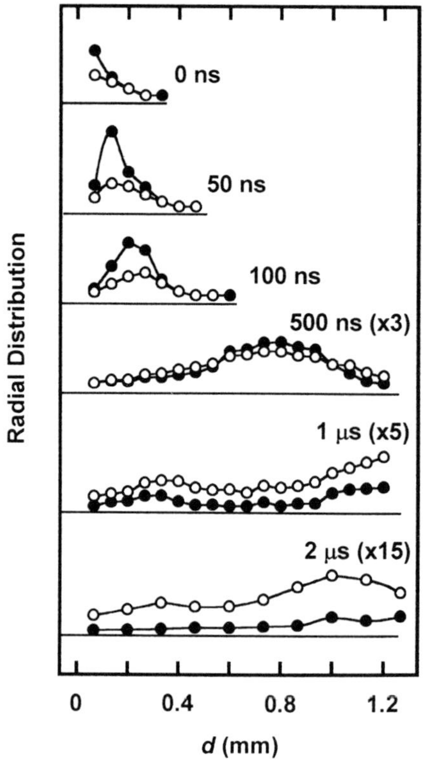

Fig. 7. Radial distributions of $C_6H_4(OH)_2$ (open circle) and $\Sigma C_6H_4(OH)_2\ (H_2O)_n$ ($1 \leqq n \leqq 8$), solid circle) isolated in the gas phase at various elapsed times after the IR-laser irradiation.

where ν is the velocity of a given ion bunch and $t_{\text{residence}}$ is the residence time of a molecule in the liquid beam after the IR-laser irradiation. The velocity and the residence time for the early-time domain are estimated to be 1300 ± 100 m s^{-1} and 0.01 ± 0.01 μs from the slopes, respectively, and those for the late-time domain are 400 ± 300 m s^{-1} and 0.4 ± 0.3 μs from the interceptions of the plots, respectively. These findings indicate that $C_6H_4(OH)_2(H_2O)_m$ in the early-time domain is liberated from the liquid surface right after the IR-laser irradiation with a relatively high velocity, while in the late-time domain, $C_6H_4(OH)_2$ is ejected continuously during ~ 10 μs at a nearly thermal velocity.

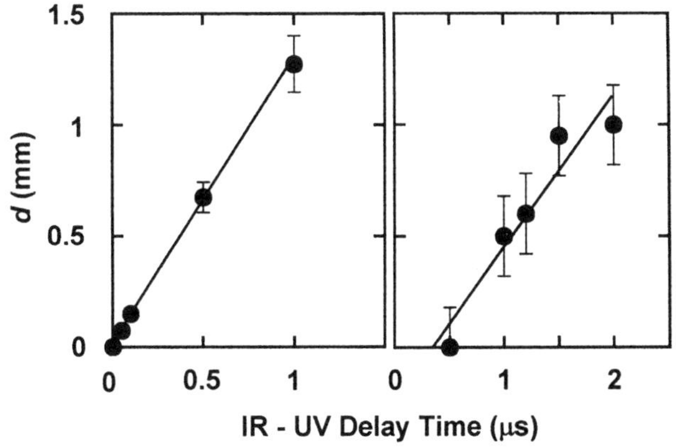

Fig. 8. Dependence of the IR-UV delay time on the peak position of the radial distribution for $\Sigma C_6H_4(OH)_2{}^+(H_2O)_n (1 \leqq n \leqq 8)$ in the early-time domain [panel (a)] and in the late-time domain [panel (b)]. Both of the peak positions shift almost linearly with increasing the delay time. The slopes of the lines shown in panels (a) and (b) give the velocities of 1300 ± 100 m s^{-1} and 400 ± 300 m s^{-1}, respectively.

3.3. *Ejection Mechanism*

As mentioned in the previous section, in the early-time domain $C_6H_4(OH)_2 (H_2O)_m (m \geqq 0)$ are ejected with a super thermal velocity into the gas phase almost concurrently with the IR-laser irradiation, while in the late-time domain $C_6H_4(OH)_2$ is ejected gradually with a thermal velocity at several ms to several tens of ms after the IR-laser irradiation.

In the early-time domain, the peaks of the radial distributions of $C_6H_4(OH)_2$ $(H_2O)_m (m \geqq 0)$ are observed to move with nearly equal velocity, irrespective of the size, m. In other words, $C_6H_4(OH)_2(H_2O)_m (m \geqq 0)$ are repelled from the liquid beam surface with a uniform velocity regardless of the size. This phenomenon is interpreted as that the clusters are repelled outwards at the liquid surface by a compression wave travelling toward the surface from its inside, which is generated by vibrational excitation of solvent molecules by the IR-laser irradiation. Note that the clusters maintain this uniform velocity, even if the clusters dissociate on moving away from the liquid beam, because the daughter clusters produced by the dissociation hold the same velocity as the parent clusters in the direction perpendicular to the liquid beam axis. The nascent neutral clusters ejected in the early-time domain must be subjected to successive unimolecular dissociation until the internal energy of the clusters is reduced below their dissociation energy.

The neutral clusters present in the early-time domain are considered to be internally cold, as a sizable amount of the solvent molecules remains in the clusters even after 1 μs has passed since their ejection from the liquid beam.

In the late-time domain, on the other hand, only $C_6H_4(OH)_2{}^+$ is observed with a thermal velocity of 400 ± 300 m s^{-1}. This finding indicates that a bare $C_6H_4(OH)_2$ is ejected from the liquid beam. Namely, the species ejected in the late-time domain seems to be internally hot, because only the bare solute molecule is observed shortly after the ejection. They are ejected with a thermal velocity by the IR-laser heating of the liquid beam.

In summary, the process occurring in the early-time domain is regarded to be non-equilibrium. When the IR laser illuminates a surface of the liquid beam, $C_6H_4(OH)_2(H_2O)_m (m \geqq 0)$ is released from it by a compression wave caused by vibrational excitation of solvent molecules before the establishment of thermal equilibrium. As a result, a solute molecule embedded in a cold solvent cluster is isolated in the gas phase at a super thermal velocity of 1300 m s^{-1}. In the late-time domain, only a bare solute molecule having a thermal velocity of 400 m s^{-1} is present because solute molecules are isolated after the evaporation of solvent by the IR-laser heating.

3.4. *Macroscopic Structure of Liquid Beam*

The liquid beam behaves like a running liquid filament and exhibited a clear optical diffraction pattern by laser illumination perpendicular to the liquid beam. The diffraction pattern and the intensity profile (see Fig. 9) are explained in terms of the Fraunhofer diffraction; the diffraction pattern is generated by the interference of light obstructed by the liquid beam. The intensity profile is given by:

$$I = I_0(\sin 2\beta)/\beta^2 \tag{5}$$

$$\beta = (\pi b \sin\theta)/\lambda \tag{6}$$

where I_0 is the intensity at the center, b is the diameter of the liquid beam, θ is the diffraction angle and λ is the wavelength of the incident laser light. We observed a change in the macroscopic profile of the liquid beam by irradiating the IR-laser before the UV laser illumination; the IR-and UV-lasers are introduced coaxially onto the liquid beam, and the UV laser illuminates the liquid beam after a certain delay time from the IR-laser. Figures 9(a) and (b) show the optical diffraction patterns by the liquid beam measured at the elapsed times of 0.2 and 5 μs after the IR-laser irradiation, respectively. The diffraction pattern at 0.2 μs coincides well with that observed

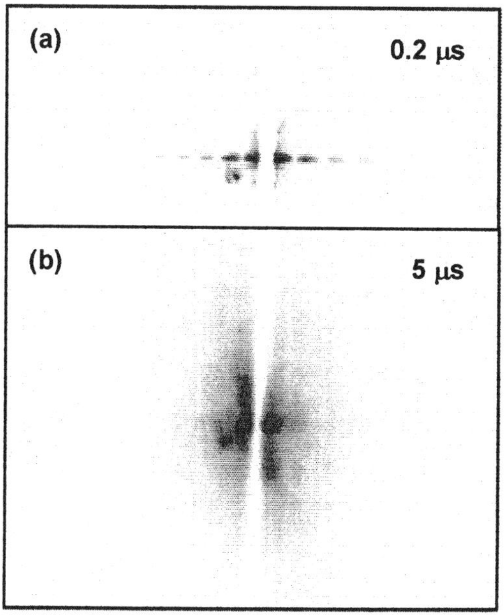

Fig. 9. Optical diffraction patterns by the liquid beam monitored at the elapsed times of 0.2 μs [panel (a)] and 5 μs [panel (b)] after the IR-laser irradiation, respectively. The diffraction pattern changes drastically at $\sim$ 5 μs in the late-time domain, when the solute molecules are partly released into the vacuum.

before the IR-laser irradiation. Namely, the macroscopic liquid beam profile does not change at the elapsed time of 0.2 μs (early-time domain) when cold neutral species have already been ejected from the liquid beam surface. On the contrary, the diffuse pattern around the diffraction center appears on the screen in the range of 5 μs (late-time domain) when hot solute molecules are partly released. These findings indicate that the surface roughness of the liquid beam overweighs the wavelength of the illumination laser.

4. Conclusion

We have reviewed several methodologies of isolation and detection of molecules and clusters from liquid media into the gas phase, with a detailed description on the method based on the liquid beam technique. The methodologies of the isolation and the detection are particularly powerful and versatile for the studies of isolated biological molecules, which provide useful information on the understanding of the functions of the biological molecules *in vivo*. In addition to the analytical importance, these are applicable to investigate the fundamental properties of liquids and liquid surfaces which are still in a premature stage.

References

1. A. Torres-Filho and J. G. Pruett, *J. Chem. Phys.* **72**, 6736 (1980).
2. C. B. Anfinsen, *Science* **181**, 223 (1973).
3. R. L. Baldwin, *Ann. Rev. Biochem.* **44**, 453 (1975).
4. T. Ebata, A. Fujii and N. Mikami, *Int. Rev. Phys. Chem.* **17**, 331 (1998).
5. A. Oikawa, H. Abe, N. Mikami and M. Ito, *J. Phys. Chem.* **87**, 5083 (1983).
6. Th. Reiners, W. Orlik, Ch. Ellert, M. Schmidt and H. Haberland, *Chem. Phys. Lett.* **215**, 357 (1993).
7. Th. Reiners, C. Ellert, M. Schmidt and H. Haberland, *Phys. Rev. Lett.* **74**, 1558 (1995).
8. C. Stoermer, J. Friedrich and M. M. Kappes, *Int. J. Mass Spectrom.* **206**, 63 (2001).
9. J. Grotemeyer and E. W. Schlag, *Acc. Chem. Res.* **22**, 399 (1989).
10. M. Yamashita and J. B. Fenn, *J. Phys. Chem.* **88**, 4451 (1984).
11. C. M. Whitehouse, R. N. Dreyer, M. Yamashita and J. B. Fenn, *Anal. Chem.* **57**, 675 (1985).
12. M. Yamashita and J. B. Fenn, *J. Phys. Chem.* **88**, 4671 (1984).
13. J. B. Fenn, M. Mann, C. K. Meng, S. F. Wong and C. M. Whitehouse, *Science* **246**, 64 (1989).
14. M. F. Jarrold, *Ann. Rev. Phys. Chem.* **51**, 179 (2000).
15. M. Barber, R. S. Bordoli, R. S. Sedgwick and A. N. Tyler, *J. Chem. Soc. Chem. Commun.* **1981**, 325 (1981).
16. M. A. Posthumus, P. G. Kistemaker, H. L. C. Meuzelaar and M. C. ten Neuver de Brauw, *Anal. Chem.* **50**, 985 (1978).
17. N. Nishi and K. Yamamoto, *J. Am. Chem. Soc.* **109**, 7353 (1987).
18. N. Nishi, *Z. Phys.* **D15**, 239 (1990).
19. K. Yamamoto and N. Nishi, *J. Am. Chem. Soc.* **112**, 549 (1990).
20. N. Nishi, K. Koga, C. Ohshima and K. Yamamoto, *J. Am. Chem. Soc.* **110**, 5246 (1988).
21. T. Kondow and F. Mafuné, *Ann. Rev. Phys. Chem.* **51**, 731 (2000).
22. M. Faubel, S. Schlemmer and J. P. Toennies, *Z. Phys.* **D10**, 269 (1988).
23. N. Horimoto, J. Kohno, F. Mafuné and T. Kondow, *J. Phys. Chem.* **A103**, 9569 (1999).
24. N. Horimoto, J. Kohno, F. Mafuné and T. Kondow, *Chem. Phys. Lett.* **318**, 536 (2000).
25. J. Kohno, F. Mafuné and T. Kondow, *J. Phys. Chem.* **A105**, 8939 (2001).

INDEX